Computational Physics

Mark Newman
University of Michigan

Computational Physics
Mark Newman

First edition 2012
Revised and expanded 2013

ISBN 978–148014551–1

Cover illustration: Computer simulation of a fluid undergoing turbulent convection when heated from below. Calculations and figure by Hans Johnston and Charles Doering. Reproduced with permission.

CONTENTS

PREFACE

COMPUTERS have become an indispensable part of the physicist's life. Physicists, like many people, use computers to read and write and communicate. But computers are also central to the work of calculating and understanding that allows one to make scientific progress in physics. Physicists use computers for solving equations, doing integrals, calculating functions, inverting matrices, and simulating physical processes of all kinds. Performing computer calculations requires skill and knowledge to get reliable, accurate answers and avoid the many pitfalls that await the unwary. This book explains the fundamentals of computational physics and describes in simple terms a wide range of techniques that every physicist should know, such as finite difference methods, numerical quadrature, and the fast Fourier transform. The book is suitable for a one-semester course in computational physics at the undergraduate level, or for advanced students or researchers who want to learn for themselves the foundational elements of this important field.

This book uses the Python programming language, an elegant, modern computer language that is easy to learn yet powerful enough for substantial physics calculations. The book assumes no prior knowledge of Python, nor indeed of computer programming of any kind. The book begins with three chapters on Python for the beginner programmer that will tell you everything you need to know to start writing programs. If, on the other hand, you are already a knowledgeable Python programmer then you can safely skip these chapters.

The remaining chapters of the book tackle the main techniques and applications of computational physics, one after another, including integration and differentiation, solution of linear and nonlinear equations, solution of ordinary and partial differential equations, stochastic processes, and Monte Carlo methods. All of the techniques introduced are illustrated with physical examples, accompanied by working Python programs. Each chapter includes a selection of exercises that the reader can use to test their comprehension of the material.

Most of the example programs in the book are also available on-line for you to download and run on your own computer if you wish. The programs, along with some data sets for use in the exercises, are packaged together in the form of a single "zip" file (of size about nine megabytes) which can be found at `http://www.umich.edu/~mejn/cpresources.zip` and you are encouraged to help yourself to a copy of this file before starting the book. Throughout the book where you see mention of programs or data in the "on-line resources," it is an indication that the items in question are included in this file. Additional resources for instructors, students, and readers of the book can also be found on the accompanying web site at `http://www.umich.edu/~mejn/cp`.

Many people have helped me with the making of this book. I would particularly like to thank the students of Physics 411: Computational Physics, the course I teach at the University of Michigan. Their enthusiastic interest and many excellent comments and questions have helped me understand computational physics and its teaching far better than I would otherwise. Special thanks also go to my colleagues Gus Evrard, Brad Orr, and Len Sander, who encouraged me to develop the course and persuaded me that Python was the right choice for teaching computational physics, to Bruce Sherwood and James Wells, who gave me useful feedback on the material as well as pointing out several mistakes, and to the many friends, colleagues, and readers who suggested improvements or corrections, including David Adams, Steve Baylor, James Binney, Chris Butenhoff, Juan Cabanela, Robert Deegan, Charlie Doering, Dave Feldman, Jonathan Keohane, Nick Kern, Nick Maher, Travis Martin, Elie Raphael, Wenbo Shen, David Strubbe, Elio Vescovo, David Walden, Mark Wilde, and Bob Ziff. Needless to say, responsibility for any remaining errors in the book rests entirely with myself, and I welcome communications from readers who find anything that needs fixing.

Finally, I would like to thank my wife Carrie for her continual support and enthusiasm. Her encouragement made writing this book far easier and more enjoyable than it would otherwise have been.

Mark Newman
Ann Arbor, Michigan
June 18, 2013

CHAPTER 1

INTRODUCTION

THE FIRST working digital computers, in the modern sense of the word, appeared in the 1940s, and almost from the moment they were switched on they were used to solve physics problems. The first computer in the United States, the ENIAC, a thirty-ton room-filling monster containing 17 000 vacuum tubes and over a thousand relays, was completed in 1946 and during its nearly ten-year lifetime performed an extraordinary range of calculations in nuclear physics, statistical physics, condensed matter, and other areas. The ENIAC was succeeded by other, faster machines, and later by the invention of the transistor and the microprocessor, which led eventually to the cheap computing power that has transformed almost every aspect of life today. Physics has particularly benefited from the widespread availability of computers, which have played a role in virtually every major physics discovery of recent years.

This book is about computational physics, the calculation of the answers to physics problems using computers. The physics we learn in school and in college focuses primarily on fundamental theories, illustrated with examples whose solution is almost always possible using nothing more than a pen, a sheet of paper, and a little perseverance. It is important to realize, however, that this is not how physics is done in the real world. It is remarkably rare these days that any new physics question can be answered by analytic methods alone. Many questions can be answered by experiment of course, but in other cases we use theoretical calculations, because experiments are impossible, or because we want to test experimental results against our understanding of basic theory, or sometimes because a theoretical calculation can give us physical insights that an experiment cannot. Whatever the reason, there are few calculations in today's theoretical physics that are so simple as to be solvable by the unaided human mind using only the tools of mathematical analysis. Far more commonly we use a computer to do all or a part of the calculation—in the jargon of computational physics, the calculation is done "numerically."

1

This is not to say that mathematical insight isn't important: many of the most elegant physics calculations, and many that we will see in this book, combine analytic steps and numerical ones. The computer is best regarded as a tool that allows us to get past the intractable parts of a calculation and make our way to the answer we want: the computer can perform a difficult integral, solve a nonlinear differential equation, or invert a 1000×1000 matrix, problems whose solution by hand would be daunting or in many cases impossible, and by doing so it has improved enormously our understanding of physical phenomena from the smallest of particles to the structure of the entire universe.

Most numerical calculations in physics fall into one of several general categories, based on the mathematical operations that their solution requires. Examples of common operations include the calculation of integrals and derivatives, linear algebra tasks such as matrix inversion or the calculation of eigenvalues, and the solution of differential equations, including both ordinary and partial differential equations. If we know how to perform each of the basic operation types then we can solve most problems we are likely to encounter as physicists.

Our approach in this book builds on this insight. We will study individually the computational techniques used to perform various operations and then we will apply them to the solution of a wide range of physics problems. There is often a real art to performing the calculations correctly. There are many ways to, say, evaluate an integral on a computer, but some of them will only give rather approximate answers while others are highly accurate. Some run faster or slower, or are appropriate for certain types of integrals but not others. Our goal in this book will be to develop a feeling for this art and learn how computational physics is done by the experts.

SOME EXAMPLES

With the rapid growth in computing power over the last few decades, we have reached a point where numerical calculations of extraordinary scale and accuracy have become possible in a wide range of fields within physics. Take a look at Figure 1.1, for instance, which shows results from an astrophysical calculation of the structure of the universe at an enormous scale, performed by physicist Volker Springel and his collaborators. In this calculation the researchers created a computer model of the evolution of the cosmos from its earliest times to the present day in a volume of space more than two billion

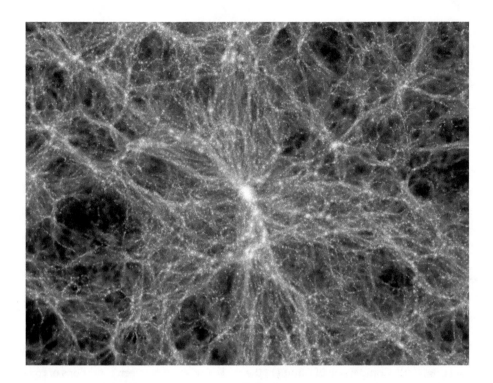

Figure 1.1: A visualization of the density of dark matter in a simulation of the evolution of the universe on scales well above the size of a galaxy—the space represented in this figure is over two billion light years on a side. Calculations by Springel, V., White, S. D. M., Jenkins, A., Frenk, C. S., Yoshida, N., Gao, L., Navarro, J., Thacker, R., Croton, D., Helly, J., Peacock, J. A., Cole, S., Thomas, P., Couchman, H., Evrard, A., Colberg, J., and Pearce, F., Simulating the joint evolution of quasars, galaxies and their large-scale distribution, *Nature* **435**, 629–636 (2005). Figure courtesy of Volker Springel. Reproduced with permission.

light-years along each side. The picture shows the extraordinary filamentary structure of the mysterious "dark matter" that is now believed to account for most of the matter in the universe. The calculation involved keeping track of billions of particles representing the dark matter distribution and occupied a supercomputer for more than a month.

Figure 1.2 shows another example, taken from a completely different area of physics, fluid dynamics. In this calculation, performed by Hans Johnston and Charles Doering, the computer followed the motion of a fluid undergoing so-called Rayleigh–Bénard convection—turbulent motion generated when a

Figure 1.2: A map of temperature variation in a computer simulation of a fluid undergoing turbulent convection when heated from below. Simulations by Johnston, H. and Doering, C. R., Comparison of turbulent thermal convection between conditions of constant temperature and constant flux, *Phys. Rev. Lett.* **102**, 064501 (2009). Figure courtesy of Hans Johnston. Reproduced with permission.

steep temperature gradient is applied to a layer of fluid, such as oil in a frying pan. In the picture the fluid is being heated at the bottom and one can see convection rolls rising from the heated surface into the cooler liquid above. The light and dark shades in the picture represent the variation of temperature in the system.

Figure 1.3 is so detailed and realistic it looks almost like a photograph, but again it's a computer-generated visualization, this time in the field of statistical physics. The system studied is diffusion-limited aggregation, the growth of clusters in a condensed-matter system when diffusing atoms or molecules stick together to create a jagged and random-looking clump. The picture, created by David Adams, shows a close-up of a portion of one of the clusters. The entire cluster contains about 12 000 atoms in this case, although clusters with a million atoms or more have been created in larger calculations.

Figure 1.3: A computer-generated visualization of a cluster of atoms created by the growth process known as diffusion-limited aggregation. Calculations and figure by David Adams. Reproduced with permission.

THE PYTHON PROGRAMMING LANGUAGE

To instruct a computer on the calculation we want it to perform we write a computer program. Programs can be written in any of a large number of programming languages—technical languages that allow us to specify the operations to be performed with mathematical precision. Different languages are suitable for different purposes, but there are a number that are appropriate for physics calculations. In this book we will learn and use the programming language called *Python*. Python is a powerful modern programming language invented in the early nineties, which finds wide use in many different fields, including physics. It is an excellent choice for computational physics for a number of reasons. First, it has a simple style that is quick to learn and easy to read, write, and understand. You may have programmed a computer before in some other language, but even if you haven't you can start learning Python from scratch and write a working physics program in just a few minutes. We will do exactly that in Chapter 2.

Second, Python is one of the most powerful of computer languages, with many features that make physics calculations easier and faster. In addition to ordinary numbers, it can handle complex numbers, vectors, matrices, tensors, sets, and many other scientific concepts effortlessly. It has built-in facilities for evaluating scientific and mathematical functions and constants and for performing calculus, matrix operations, Fourier transforms, and other staples of the physics vocabulary. It also has facilities for visualizing results, making graphs, drawing diagrams and models, and creating animations, which will prove enormously useful for picturing the physics of the systems we study.

And third, Python is free. Its creators have generously donated the Python language and the computer programs that make it work to the world at large and you may find that your computer already comes with them installed. If not, you can download them freely from the Internet and install them yourself. It will take you only a few minutes. Instructions are given in Appendix A at the end of the book.

As a demonstration of Python's straightforward nature, I give below a short example program that performs an atomic physics calculation. In 1888 Johannes Rydberg published his famous formula for the wavelengths λ of the emission lines of the hydrogen atom:

$$\frac{1}{\lambda} = R\left(\frac{1}{m^2} - \frac{1}{n^2}\right),$$

where R is the Rydberg constant $R = 1.097 \times 10^{-2}\,\text{nm}^{-1}$ and m and n are positive integers. For a given value of m, the wavelengths λ given by this formula for all $n > m$ form a series, the first three such series, for $m = 1, 2,$ and $3,$ being known as the Lyman, Balmer, and Paschen series, after their respective discoverers. Here is a Python program that prints out the wavelengths of the first five lines in each of these three series:

```
R = 1.097e-2
for m in [1,2,3]:
    print("Series for m =",m)
    for k in [1,2,3,4,5]:
        n = m + k
        invlambda = R*(1/m**2-1/n**2)
        print("   ",1/invlambda,"nm")
```

If we run this program it produces the following output:

```
Series for m = 1
   121.543603768 nm
   102.552415679 nm
   97.2348830143 nm
   94.9559404436 nm
   93.7622086209 nm
Series for m = 2
   656.335460346 nm
   486.174415071 nm
   434.084299171 nm
   410.209662716 nm
   397.042438975 nm
Series for m = 3
   1875.24417242 nm
   1281.90519599 nm
   1093.89243391 nm
   1005.01367366 nm
   954.669760504 nm
```

Apart from being calculated to more decimal places than is really justified, these are a pretty good approximation to the actual wavelengths of the emission lines of hydrogen.

Though we haven't yet started learning the Python language, you don't need to be an expert on it to understand roughly what the program does. In its first line it defines the value of the constant R which will be used later. The second line tells the program that it is to run through each of the values $m = 1, 2, 3$ in turn. It prints out each value then runs through values of another variable $k = 1, 2, 3, 4, 5$ and for each one calculates $n = m + k$. So, for instance, when $m = 1$ we will have $n = 2, 3, 4, 5, 6$ in turn. And for each of these values it then evaluates the Rydberg formula and prints out the result. We will study the details of the Python language carefully in the next chapter, but this program already shows that Python is a straightforward language, easy to read and comprehend, and yet one that is capable of doing real physics calculations with just a few lines of programming.

The Python language is being continually updated and improved by its creators. The latest version of the language (at the time of writing) is version 3, which is the version used in this book. The previous version, version 2, is also still available and in wide use since many programs were written using it. (Version 1 is considered obsolete and is no longer used.) The differences between versions 2 and 3 are quite modest and if you already have the software for version 2 on your computer, or you've used version 2 in the past and want to stick with it, then you can still use this book, but take a look at Appendix B first, which explains how to get everything working smoothly with version 2.

Python is not everyone's choice of programming language for every task. There are many other programming languages available and there are tasks for which another choice may be appropriate. If you get serious about computational physics you are, without doubt, going to need to write programs in another language someday. Common choices for scientific work (other than Python) include C, C++, Fortran, and Java. But if you learn Python first, you should find switching to another language easy. All of the languages above use basically the same concepts, and differ only in relatively small details of how you give specific commands to the computer. If you learn the fundamental ideas of computational physics laid out in this book using the Python language, then you will know most of what you need to write first-class physics programs in any programming language.

CHAPTER 2

PYTHON PROGRAMMING FOR PHYSICISTS

OUR FIRST item of business is to learn how to write computer programs in the Python programming language.

Python is easy to learn, simple to use, and enormously powerful. It has facilities and features for performing tasks of many kinds. You can do art or engineering in Python, surf the web or calculate your taxes, write words or write music, make a movie or make the next billion-dollar Internet start-up.[1] We will not attempt to learn about all of Python's features, however, but restrict ourselves to those that are most useful for doing physics calculations. We will learn about the core structure of the language first, how to put together the instructions that make up a program, but we will also learn about some of the powerful features that can make the life of a computational physicist easier, such as features for doing calculations with vectors and matrices, and features for making graphs and computer graphics. Some other features of Python that are more specialized, but still occasionally useful for physicists, will not be covered here. Luckily there is excellent documentation available on-line, so if there's something you want to do and it's not explained in this book, I encourage you to see what you can find. A good place to start when looking for information about Python is the official Python website at www.python.org.

2.1 GETTING STARTED

A Python program consists of a list of instructions, resembling a mixture of English words and mathematics and collectively referred to as *code*. We'll see exactly what form the instructions take in a moment, but first we need to know how and where to enter them into the computer.

---

[1]Some of these also require that you have a good idea.

When you are programming in Python—developing a program, as the jargon goes—you typically work in a *development environment*, which is a window or windows on your computer screen that show the program you are working on and allow you to enter or edit lines of code. There are several different development environments available for use with Python, but the most commonly used is the one called IDLE.[2] If you have Python installed on your computer then you probably have IDLE installed as well. (If not, it is available as a free download from the web.[3]) How you start IDLE depends on what kind of computer you have, but most commonly you click on an icon on the desktop or under the start menu on a PC, or in the dock or the applications folder on a Mac. If you wish, you can now start IDLE running on your computer and follow along with the developments in this chapter step by step.

The first thing that happens when you start IDLE is that a window appears on the computer screen. This is the *Python shell window*. It will have some text in it, looking something like this:

```
Python 3.2 (default, Sep 29 2012)
Type "help" for more information.
>>>
```

This tells you what version of Python you are running (your version may be different from the one above), along with some other information, followed by the symbol ">>>", which is a prompt: it tells you that the computer is ready for you to type something in. When you see this prompt you can type any command in the Python language at the keyboard and the computer will carry out that command immediately. This can be a useful way to quickly try individual Python commands when you're not sure how something works, but it's not the main way that we will use Python commands. Normally, we want to type in an entire Python program at once, consisting of many commands one after another, then run the whole program together. To do this, go to the top of the window, where you will see a set of menu headings. Click on the "File" menu and select "New Window". This will create a second window on

---

[2] IDLE stands for "Integrated Development Environment" (sort of). The name is also a joke, the Python language itself being named, allegedly, after the influential British comedy troupe *Monty Python*, one of whose members was the comedian Eric Idle.

[3] The standard versions of Python for PC and Mac computers come with IDLE. For Linux users, IDLE does not usually come installed automatically, so you may have to install it yourself. The most widely used brands of Linux, including Ubuntu and Fedora, have freely available versions of IDLE that can be installed using their built-in software installer programs.

the screen, this one completely empty. This is an *editor window*. It behaves differently from the Python shell window. You type a complete program into this window, usually consisting of many lines. You can edit it, add things, delete things, cut, paste, and so forth, in a manner similar to the way one works with a word processor. The menus at the top of the window provide a range of word-processor style features, such as cut and paste, and when you are finished writing your program you can save your work just as you would with a word processor document. Then you can run your complete program, the whole thing, by clicking on the "Run" menu at the top of the editor window and selecting "Run Module" (or you can press the F5 function key, which is quicker). This is the main way in which we will use Python and IDLE in this book.

To get the hang of how it works, try the following quick exercise. Open up an editor window if you didn't already (by selecting "New Window" from the "File" menu) and type the following (useless) two-line program into the window, just as it appears here:

```
x = 1
print(x)
```

(If it's not obvious what this does, it will be soon.) Now save your program by selecting "Save" from the "File" menu at the top of the editor window and typing in a name.[4] The names of all Python programs must end with ".py", so a suitable name might be "example.py" or something like that. (If you do not give your program a name ending in ".py" then the computer will not know that it is a Python program and will not handle it properly when you try to load it again—you will probably find that such a program will not even run at all, so the ".py" is important.)

Once you have saved your program, run it by selecting "Run module" from the "Run" menu. When you do this the program will start running, and any output it produces—anything it says or does or prints out—will appear in the Python shell window (the other window, the one that appeared first). In this

---

[4]Note that you can have several windows open at once, including the Python shell window and one or more editor windows, and that each window has its own "File" menu with its own "Save" item. When you click on one of these to save, IDLE saves the contents of the corresponding window and that window only. Thus if you want to save a program you must be careful to use the "File" menu for the window containing the program, rather than for any other window. If you click on the menu for the shell window, for instance, IDLE will save the contents of the shell window, not your program, which is probably not what you wanted.

case you should see something like this in the Python shell window:

```
1
>>>
```

The only result of this small program is that the computer prints out the number "1" on the screen. (It's the value of the variable x in the program—see Section 2.2.1 below.) The number is followed by a prompt ">>>" again, which tells you that the computer is done running your program and is ready to do something else.

This same procedure is the one you'll use for running all your programs and you'll get used to it soon. It's a good idea to save your programs, as here, when they're finished and ready to run. If you forget to do it, IDLE will ask you if you want to save before it runs your program.

IDLE is by no means the only development environment for Python. If you are comfortable with computers and enjoy trying things out, there are a wide range of others available on the Internet, mostly for free, with names like Py-Dev, Eric, BlackAdder, Komodo, Wing, and more. Feel free to experiment and see what works for you, or you can just stick with IDLE. IDLE can do everything we'll need for the material in this book. But nothing in the book will depend on what development environment you use. As far as the programming and the physics go, they are all equivalent.

2.2   BASIC PROGRAMMING

A program is a list of instructions, or *statements*, which under normal circumstances the computer carries out, or *executes*, in the order they appear in the program. Individual statements do things like performing arithmetic, asking for input from the user of the program, or printing out results. The following sections introduce the various types of statements in the Python language one by one.

2.2.1   VARIABLES AND ASSIGNMENTS

Quantities of interest in a program—which in physics usually means numbers, or sets of numbers like vectors or matrices—are represented by *variables*, which play roughly the same role as they do in ordinary algebra. Our first example of a program statement in Python is this:

```
x = 1
```

This is an *assignment statement*. It tells the computer that there is a variable called x and we are assigning it the value 1. You can think of the variable as a box that stores a value for you, so that you can come back and retrieve that value at any later time, or change it to a different value. We will use variables extensively in our computer programs to represent physical quantities like positions, velocities, forces, fields, voltages, probabilities, and wavefunctions.

In normal algebra variable names are usually just a single letter like x, but in Python (and in most other programming languages) they don't have to be— they can be two, three, or more letters, or entire words if you want. Variable names in Python can be as long as you like and can contain both letters and numbers, as well as the underscore symbol "_", but they cannot start with a number, or contain any other symbols, or spaces. Thus x and Physics_101 are fine names for variables, but 4Score&7Years is not (because it starts with a number, and also because it contains a &). Upper- and lower-case letters are distinct from one another, meaning that x and X are two different variables which can have different values.[5]

Many of the programs you will write will contain large numbers of variables representing the values of different things and keeping them straight in your head can be a challenge. It is a very good idea—one that is guaranteed to save you time and effort in the long run—to give your variables meaningful names that describe what they represent. If you have a variable that represents the energy of a system, for instance, you might call it energy. If you have a variable that represents the velocity of an object you could call it velocity. For more complex concepts, you can make use of the underscore symbol "_" to create variable names with more than one word, like maximum_energy or angular_velocity. Of course, there will be times when single-letter variable names are appropriate. If you need variables to represent the x and y positions of an object, for instance, then by all means call them x and y. And there's no reason why you can't call your velocity variable simply v if that seems natural to you. But whatever you do, choose names that help you remember what the variables represent.

---

[5]Also variables cannot have names that are "reserved words" in Python. Reserved words are the words used to assemble programming statements and include "for", "if", and "while". (We will see the special uses of each of these words in Python programming later in the chapter.)

2.2.2 VARIABLE TYPES

Variables come in several types. Variables of different types store different kinds of quantities. The main types we will use for our physics calculations are the following:

- **Integer**: Integer variables can take integer values and integer values only, such as 1, 0, or −286784. Both positive and negative values are allowed, but not fractional values like 1.5.

- **Float**: A floating-point variable, or "float" for short, can take real, or floating-point, values such as 3.14159, -6.63×10^{-34}, or 1.0. Notice that a floating-point variable can take an integer value like 1.0 (which after all is also a real number), by contrast with integer variables which cannot take noninteger values.

- **Complex**: A complex variable can take a complex value, such as $1 + 2j$ or $-3.5 - 0.4j$. Notice that in Python the unit imaginary number is called j, not i. (Despite this, we will use i in some of the mathematical formulas we derive in this book, since it is the common notation among physicists. Just remember that when you translate your formulas into computer programs you must use j instead.)

You might be asking yourself what these different types mean. What does it mean that a variable has a particular type? Why do we need different types? Couldn't all values, including integers and real numbers, be represented with complex variables, so that we only need one type of variable? In principle they could, but there are great advantages to having the different types. For instance, the values of the variables in a program are stored by the computer in its memory, and it takes twice as much memory to store a complex number as it does a float, because the computer has to store both the real and imaginary parts. Even if the imaginary part is zero (so that the number is actually real), the computer still takes up memory space storing that zero. This may not seem like a big issue given the huge amounts of memory computers have these days, but in many physics programs we need to store enormous numbers of variables—millions or billions of them—in which case memory space can become a limiting factor in writing the program.

Moreover, calculations with complex numbers take longer to complete, because the computer has to calculate both the real and imaginary parts. Again, even if the imaginary part is zero, the computer still has to do the calculation, so it takes longer either way. Many of our physics programs will involve millions or billions of operations. Big physics calculations can take days or weeks to run, so the speed of individual mathematical operations can have a big effect. Of course, if we need to work with complex numbers then we will have to use complex variables, but if our numbers are real, then it is better to use a floating-point variable.

Similar considerations apply to floating-point variables and integers. If the numbers we are working with are genuinely noninteger real numbers, then we should use floating-point variables to represent them. But if we know that the numbers are integers then using integer variables is usually faster and takes up less memory space.

Moreover, integer variables are in some cases actually more accurate than floating-point variables. As we will see in Section 4.2, floating-point calculations on computers are not infinitely accurate. Just as on a hand-held calculator, computer calculations are only accurate to a certain number of significant figures (typically about 16 on modern computers). That means that the value 1 assigned to a floating-point variable may actually be stored on the computer as 0.9999999999999999. In many cases the difference will not matter much, but what happens, for instance, if something special is supposed to take place in your program if, and only if, the number is less than 1? In that case, the difference between 1 and 0.9999999999999999 could be crucially important. Numerous bugs and problems in computer programs have arisen because of exactly this kind of issue. Luckily there is a simple way to avoid it. If the quantity you're dealing with is genuinely an integer, then you should store it in an integer variable. That way you know that 1 means 1. Integer variables are not accurate to just 16 significant figures: they are perfectly accurate. They represent the exact integer you assign to them, nothing more and nothing less. If you say "x = 1", then indeed x is equal to 1.

This is an important lesson, and one that is often missed when people first start programming computers: if you have an integer quantity, use an integer variable. In quantum mechanics most quantum numbers are integers. The number of atoms in a gas is an integer. So is the number of planets in the solar system or the number of stars in the galaxy. Coordinates on lattices in solid-state physics are often integers. Dates are integers. The population of the world is an integer. If you were representing any of these quantities in a program it would in most cases be an excellent idea to use an integer variable. More generally, whenever you create a variable to represent a quantity in one of your programs, think about what type of value that quantity will take and choose the type of your variable to match it.

And how do you tell the computer what type you want a variable to be?

The name of the variable is no help. A variable called x could be an integer or it could be a complex variable.

The type of a variable is set by the value that we give it. Thus for instance if we say "x = 1" then x will be an integer variable, because we have given it an integer value. If we say "x = 1.5" on the other hand then it will be a float. If we say "x = 1+2j" it will be complex.[6] Very large floating-point or complex values can be specified using scientific notation, in the form "x = 1.2e34" (which means 1.2×10^{34}) or "x = 1e-12 + 2.3e45j" (which means $10^{-12} + 2.3 \times 10^{45}$j).

The type of a variable can change as a Python program runs. For example, suppose we have the following two lines one after the other in our program:

```
x = 1
x = 1.5
```

If we run this program then after the first line is executed by the computer x will be an integer variable with value 1. But immediately after that the computer will execute the second line and x will become a float with value 1.5. Its type has changed from integer to float.[7]

However, although you *can* change the types of variables in this way, it doesn't mean you should. It is considered poor programming to use the same variable as two different types in a single program, because it makes the program significantly more difficult to follow and increases the chance that you may make a mistake in your programming. If x is an integer in some parts of the program and a float in others then it becomes difficult to remember which it is and confusion can ensue. A good programmer, therefore, will use a given variable to store only one type of quantity in a given program. If you need a variable to store another type, use a different variable with a different name.

Thus, in a well written program, the type of a variable will be set the first time it is given a value and will remain the same for the rest of the program. This doesn't quite tell us the whole story, however, because as we've said a

---

[6]Notice that when specifying complex values we say 1+2j, not 1+2*j. The latter means "one plus two times the variable *j*", not the complex number $1 + 2i$.

[7]If you have previously programmed in one of the so-called static-typed languages, such as C, C++, Fortran, or Java, then you'll be used to creating variables with a declaration such as "int i" which means "I'm going to be using an integer variable called i." In such languages the types of variables are fixed once they are declared and cannot change. There is no equivalent declaration in Python. Variables in Python are created when you first use them, with types which are deduced from the values they are given and which may change when they are given new values.

floating-point variable can also take an integer value. There will be times when we wish to give a variable an integer value, like 1, but nonetheless have that variable be a float. There's no contradiction in this, but how do we tell the computer that this is what we want? If we simply say "x = 1" then, as we have seen, x will be an integer variable.

There are two simple ways to do what we want here. The first is to specify a value that has an explicit decimal point in it, as in "x = 1.0". The decimal point is a signal to the computer that this is a floating-point value (even though, mathematically speaking, 1 is of course an integer) and the computer knows in this situation to make the variable x a float. Thus "x = 1.0" specifies a floating-point variable called x with the value 1.

A slightly more complicated way to achieve the same thing is to write "x = float(1)", which tells the computer to take the value 1 and convert it into a floating-point value before assigning it to the variable x. This also achieves the goal of making x a float.

A similar issue can arise with complex variables. There will be times when we want to create a variable of complex type, but we want to give it a purely real value. If we just say "x = 1.5" then x will be a real, floating-point variable, which is not what we want. So instead we say "x = 1.5 + 0j", which tells the computer that we intend x to be complex. Alternatively, we can write "x = complex(1.5)", which achieves the same thing.

There is one further type of variable, the *string*, which is often used in Python programs but which comes up only rarely in physics programming, which is why we have not mentioned it so far. A string variable stores text in the form of strings of letters, punctuation, symbols, digits, and so forth. To indicate a string value one uses quotation marks, like this:

```
x = "This is a string"
```

This statement would create a variable x of string type with the value "This is a string". Any character can appear in a string, including numerical digits. Thus one is allowed to say, for example, x = "1.234", which creates a string variable x with the value "1.234". It's crucial to understand that this is not the same as a floating-point variable with the value 1.234. A floating-point variable contains a number, the computer knows it's a number, and, as we will shortly see, one can do arithmetic with that number, or use it as the starting point for some mathematical calculation. A string variable with the value "1.234" does not represent a number. The value "1.234" is, as far as the computer is concerned, just a string of symbols in a row. The symbols happen to be digits (and a decimal point) in this case, but they could just as easily be letters or spaces or punctuation. If you try to do arithmetic with a string variable, even one that appears to contain a number, the computer will most likely either complain or give you something entirely unexpected. We will not have much need for string variables in this book and they will as a result appear only rather rarely. One place they do appear, however, is in the following section on output and input.

In all of the statements we have seen so far you are free to put spaces between parts of the statement. Thus "x=1" and "x = 1" do the exact same thing; the spaces have no effect. They can, however, do much to improve the readability of a program. When we start writing more complicated statements in the following sections, we will find it very helpful to add some spaces here and there. There are a few places where one cannot add extra spaces, the most important being at the beginning of a line, before the start of the statement. As we will see in Section 2.3.1, inserting extra spaces at the beginning of a line does have an effect on the way the program works. Thus, unless you know what you are doing, you should avoid putting spaces at the beginning of lines.

You can also include blank lines between statements in a program, at any point and as many as you like. This can be useful for separating logically distinct parts of a program from one another, again making the program easier to understand. We will use this trick many times in the programs in this book to improve their readability.

2.2.3 OUTPUT AND INPUT STATEMENTS

We have so far seen one example of a program statement, the assignment statement, which takes the form "x = 1" or something similar. The next types of statements we will examine are statements for output and input of data in Python programs. We have already seen one example of the basic output statement, or "print" statement. In Section 2.1 we gave this very short example program:

```
x = 1
print(x)
```

The first line of this program we understand: it creates an integer variable called x and gives it the value 1. The second statement tells the computer to "print" the value of x on the screen of the computer. Note that it is the *value* of the variable x that is printed, not the letter "x". The value of the variable in this case is 1, so this short program will result in the computer printing a "1" on the screen, as we saw on page 12.

The print statement always prints the current value of the variable at the moment the statement is executed. Thus consider this program:

```
x = 1
print(x)
x = 2
print(x)
```

First the variable x is set to 1 and its value is printed out, resulting in a 1 on the screen as before. Then the value of x is changed to 2 and the value is printed again, which produces a 2 on the screen. Overall we get this:

```
1
2
```

Thus the two print statements, though they look identical, produce different results in this case. Note also that each print statement starts its printing on a new line. The print statement can be used to print out more than one thing on a line. Consider this program:

```
x = 1
y = 2
print(x,y)
```

which produces this result:

```
1 2
```

Note that the two variables in the print statement are separated by a comma. When their values are printed out, however, they are printed with a space between them (not a comma).

We can also print out words, like this:

```
x = 1
y = 2
print("The value of x is",x,"and the value of y is",y)
```

which produces this on the screen:

```
The value of x is 1 and the value of y is 2
```

Adding a few words to your program like this can make its output much easier to read and understand. You can also have print statements that print out only words if you like, as in print("The results are as follows") or print("End of program").

The print statement can also print out the values of floating-point and complex variables. For instance, we can write

```
x = 1.5
z = 2+3j
print(x,z)
```

and we get

```
1.5 (2+3j)
```

In general, a print statement can include any string of quantities separated by commas, or text in quotation marks, and the computer will simply print out the appropriate things in order, with spaces in between.[8] Occasionally you may want to print things with something other than spaces in between, in which case you can write something like the following:

```
print(x,z,sep="...")
```

which would print

```
1.5...(2+3j)
```

The code sep="..." tells the computer to use whatever appears between the quotation marks as a separator between values—three dots in this case, but you could use any letters, numbers, or symbols you like. You can also have no separator between values at all by writing print(x,z,sep="") with nothing between the quotation marks, which in the present case would give

```
1.5(2+3j)
```

---

[8]The print statement is one of the things that differs between Python version 3 and earlier versions. In earlier versions there were no parentheses around the items to be printed—you would just write "print x". If you are using an earlier version of Python with this book then you will have to remember to omit the parentheses from your print statements. Alternatively, if you are using version 2.6 or later (but not version 3) then you can make the print statement behave as it does in version 3 by including the statement from __future__ import print_function at the start of your program. (Note that there are two underscore symbols before the word "future" and two after it.) See Appendix B for further discussion of the differences between Python versions.

Input statements are only a little more complicated. The basic form of an input statement in Python is like this:

```
x = input("Enter the value of x: ")
```

When the computer executes this statement it does two things. First, the statement acts something like a print statement and prints out the quantity, if any, inside the parentheses.[9] So in this case the computer would print the words "Enter the value of x: ". If there is nothing inside the parentheses, as in "x = input()", then the computer prints nothing, but the parentheses are still required nonetheless.

Next the computer will stop and wait. It is waiting for the user to type a value on the keyboard. It will wait patiently until the user types something and then the value that the user types is assigned to the variable x. However, there is a catch: the value entered is always interpreted as a *string* value, even if you type in a number.[10] (We encountered strings previously in Section 2.2.2.) Thus consider this simple two-line program:

```
x = input("Enter the value of x: ")
print("The value of x is",x)
```

This does nothing more than collect a value from the user then print it out again. If we run this program it might look something like the following:

```
Enter the value of x: 1.5
The value of x is 1.5
```

This looks reasonable. But we could also do the following:

---

[9]It doesn't act exactly like a print statement however, since it can only print a single quantity, such as a string of text in quotes (as here) or a variable, where the print statement can print many quantities in a row.

[10]Input statements are another thing that changed between versions 2 and 3 of Python. In version 2 and earlier the value generated by an input statement would have the same type as whatever the user entered. If the user entered an integer, the input statement would give an integer value. If the user entered a float it would give a float, and so forth. However, this was considered confusing, because it meant that if you then assigned that value to a variable (as in the program above) there would be no way to know in advance what the type of the variable would be—the type would depend on what the user entered at the keyboard. So in version 3 of Python the behavior was changed to its present form in which the input is always interpreted as a string. If you are using a version of Python earlier than version 3 and you want to reproduce the behavior of version 3 then you can write "x = raw_input()". The function raw_input in earlier versions is the equivalent of input in version 3.

```
Enter the value of x: Hello
The value of x is Hello
```

As you can see "value" is interpreted rather loosely. As far as the computer is concerned, anything you type in is a string, so it doesn't care whether you enter digits, letters, a complete word, or several words. Anything is fine.

For physics calculations, however, we usually want to enter numbers, and have them interpreted correctly as numbers, not strings. Luckily it is straightforward to convert a string into a number. The following will do it:

```
temp = input("Enter the value of x: ")
x = float(temp)
print("The value of x is",x)
```

This is slightly more complicated. It receives a string input from the user and assigns it to the temporary variable `temp`, which will be a string-type variable. Then the statement "`x = float(temp)`" converts the string value to a floating-point value, which is then assigned to the variable x, and this is the value that is printed out. One can also convert string input values into integers or complex numbers with statements of the form "`x = int(temp)`" or "`x = complex(temp)`".

In fact, one doesn't have to use a temporary variable. The code above can be expressed more succinctly like this:

```
x = float(input("Enter the value of x: "))
print("The value of x is",x)
```

which takes the string value given by `input`, converts it to a float, and assigns it directly to the variable x. We will use this trick many times in this book.

In order for this program to work, the value the user types must be one that makes sense as a floating-point value, otherwise the computer will complain. Thus, for instance, the following is fine:

```
Enter the value of x: 1.5
The value of x is 1.5
```

But if I enter the wrong thing, I get this:

```
Enter the value of x: Hello
ValueError: invalid literal for float(): Hello
```

This is our first example of an *error message*. The computer, in rather opaque technical jargon, is complaining that we have given it an incorrect value.

It's normal to make a few mistakes when writing or using computer programs, and you will soon become accustomed to the occasional error message (if you are not already). Working out what these messages mean is one of the tricks of the business—they are often not entirely transparent.

2.2.4 ARITHMETIC

So far our programs have done very little, certainly nothing that would be much use for physics. But we can make them much more useful by adding some arithmetic into the mix.

In most places where you can use a single variable in Python you can also use a mathematical expression, like "x+y". Thus you can write "print(x)" but you can also write "print(x+y)" and the computer will calculate the sum of x and y for you and print out the result. The basic mathematical operations—addition, subtraction, etc.—are written as follows:

x+y	addition
x-y	subtraction
x*y	multiplication
x/y	division
x**y	raising x to the power of y

Notice that we use the asterisk symbol "*" for multiplication and the slash symbol "/" for division, because there is no $\times$ or $\div$ symbol on a standard computer keyboard.

Two more obscure, but still useful operations, are integer division and the modulo operation:

x//y the integer part of x divided by y, meaning x is divided by y and the result is rounded down to the nearest integer. For instance, 14//3 gives 4 and -14//3 gives -5.

x%y modulo, which means the remainder after x is divided by y. For instance, 14%3 gives 2, because 14 divided by 3 gives 4-remainder-2. This also works for nonintegers: 1.5%0.4 gives 0.3, because 1.5 is 3×0.4, remainder 0.3. (There is, however, no modulo operation for complex numbers.) The modulo operation is particularly useful for telling when one number is divisible by another—the value of n%m will be zero if n is divisible by m. Thus, for instance, n%2 is zero if n is even (and one if n is odd).

There are a few other mathematical operations available in Python as well, but they're more obscure and rarely used.[11]

An important rule about arithmetic in Python is that the type of result a calculation gives depends on the types of the variables that go into it. Consider, for example, this statement

```
x = a + b
```

If a and b are variables of the same type—integer, float, complex—then when they are added together the result will also have the same type and this will be the type of variable x. So if a is 1.5 and b is 2.4 the end result will be that x is a floating-point variable with value 3.9. Note when adding floats like this that even if the end result of the calculation is a whole number, the variable x would still be floating point: if a is 1.5 and b is 2.5, then the result of adding them together is 4, but x will still be a floating-point variable with value 4.0 because the variables a and b that went into it are floating point.

If a and b are of different types, then the end result has the more general of the two types that went into it. This means that if you add a float and an integer, for example, the end result will be a float. If you add a float and a complex variable, the end result will be complex.

The same rules apply to subtraction, multiplication, integer division, and the modulo operation: the end result is the same type as the starting values, or the more general type if there are two different starting types. The division operation, however—ordinary non-integer division denoted by "/"—is slightly different: it follows basically the same rules except that it never gives an integer result. Only floating-point or complex values result from division. This is necessary because you can divide one integer by another and get a noninteger result (like $3 \div 2 = 1.5$ for example), so it wouldn't make sense to have integer starting values always give an integer final result.[12] Thus if you divide any

---

[11]Such as:

    x|y   bitwise (binary) OR of two integers
    x&y   bitwise (binary) AND of two integers
    x^y   bitwise (binary) XOR of two integers
  x>>y shift the bits of integer x rightwards y places
  x<<y shift the bits of integer x leftwards y places

[12]This is another respect in which version 3 of Python differs from earlier versions. In version 2 and earlier all operations gave results of the same type that went into them, including division. This, however, caused a lot of confusion for exactly the reason given here: if you divided 3 by 2, for

combination of integers or floats by one another you will always get a floating-point value. If you start with one or more complex numbers then you will get a complex value at the end.

You can combine several mathematical operations together to make a more complicated expression, like x+2*y-z/3. When you do this the operations obey rules similar to those of normal algebra. Multiplications and divisions are performed before additions and subtractions. If there are several multiplications or divisions in a row they are carried out in order from left to right. Powers are calculated before anything else. Thus

x+2*y	is equivalent to	$x + 2y$
x-y/2	is equivalent to	$x - \frac{1}{2}y$
3*x**2	is equivalent to	$3x^2$
x/2*y	is equivalent to	$\frac{1}{2}xy$

You can also use parentheses () in your algebraic expressions, just as you would in normal algebra, to mark things that should be evaluated as a unit, as in 2*(x+y). And you can add spaces between the parts of a mathematical expression to make it easier to read; the spaces don't affect the value of the expression. So "x=2*(a+b)" and "x = 2 * (a + b)" do the same thing. Thus the following are allowed statements in Python

```
x = a + b/c
x = (a + b)/c
x = a + 2*b - 0.5*(1.618**c + 2/7)
```

On the other hand, the following will *not* work:

```
2*x = y
```

You might expect that this would result in the value of x being set to half the value of y, but it's not so. In fact, if you write this line in a program the computer will simply stop when it gets to that line and print a typically cryptic error message—"SyntaxError: can't assign to operator"—because it

---

instance, the result had to be an integer, so the computer rounded it down from 1.5 to 1. Because of the difficulties this caused, the language was changed in version 3 to give the current more sensible behavior. You can still get the old behavior of dividing then rounding down using the integer divide operation //. Thus 3//2 gives 1 in all versions of Python. If you are using Python version 2 (technically, version 2.1 or later) and want the newer behavior of the divide operation, you can achieve it by including the statement "from __future__ import division" at the start of your program. The differences between Python versions are discussed in more detail in Appendix B.

doesn't know what to do. The problem is that Python does not know how to solve equations for you by rearranging them. It only knows about the simplest forms of equations, such as "x = y/2". If an equation needs to be rearranged to give the value of x then you have to do the rearranging for yourself. Python will do basic sums for you, but its knowledge of math is very limited.

To be more precise, statements like "x = a + b/c" in Python are not technically equations at all, in the mathematical sense. They are assignments. When it sees a statement like this, what your computer actually does is very simple-minded. It first examines the right-hand side of the equals sign and evaluates whatever expression it finds there, using the current values of any variables involved. When it is finished working out the value of the whole expression, and only then, it takes that value and assigns it to the variable on the left of the equals sign. In practice, this means that assignment statements in Python sometimes behave like ordinary equations, but sometimes they don't. A simple statement like "x = 1" does exactly what you would think, but what about this statement:

```
x = x + 1
```

This does not make sense, under any circumstances, as a mathematical equation. There is no way that x can ever be equal to $x + 1$—it would imply that $0 = 1$. But this statement makes perfect sense in Python. Suppose the value of x is currently 1. When the statement above is executed by the computer it first evaluates the expression on the right-hand side, which is x + 1 and therefore has the value $1 + 1 = 2$. Then, when it has calculated this value it assigns it to the variable on the left-hand side, which just happens in this case to be the same variable x. So x now gets a new value 2. In fact, no matter what value of x we start with, this statement will always end up giving x a new value that is 1 greater. So this statement has the simple (but often very useful) effect of increasing the value of x by one.

Thus consider the following lines:

```
x = 0
print(x)
x = x**2 - 2
print(x)
```

What will happen when the computer executes these lines? The first two are straightforward enough: the variable x gets the value 0 and then the 0 gets printed out. But then what? The third line says "x = x**2 - 2" which in normal mathematical notation would be $x = x^2 - 2$, which is a quadratic equation with solutions $x = 2$ and $x = -1$. However, the computer will not set x equal to either of these values. Instead it will evaluate the right-hand side of the equals sign and get $x^2 - 2 = 0^2 - 2 = -2$ and then set x to this new value. Then the last line of the program will print out "-2".

Thus the computer does not necessarily do what one might think it would, based on one's experience with normal mathematics. The computer will not solve equations for x or any other variable. It won't do your algebra for you—it's not that smart.

Another set of useful tricks are the Python *modifiers*, which allow you to make changes to a variable as follows:

x += 1	add 1 to x (i.e., make x bigger by 1)
x -= 4	subtract 4 from x
x *= -2.6	multiply x by -2.6
x /= 5*y	divide x by 5 times y
x //= 3.4	divide x by 3.4 and round down to an integer

As we have seen, you can achieve the same result as these modifiers with statements like "x = x + 1", but the modifiers are more succinct. Some people also prefer them precisely because "x = x + 1" looks like bad algebra and can be confusing.

Finally in this section, a nice feature of Python, not available in most other computer languages, is the ability to assign the values of two variables with a single statement. For instance, we can write

```
x,y = 1,2.5
```

which is equivalent to the two statements

```
x = 1
y = 2.5
```

One can assign three or more variables in the same way, listing them and their assigned values with commas in between.

A more sophisticated example is

```
x,y = 2*z+1,(x+y)/3
```

An important point to appreciate is that, like all other assignment statements, this one calculates the whole of the right-hand side of the equation before assigning values to the variables on the left. Thus in this example the computer will calculate both of the values `2*z+1` and `(x+y)/3` from the current x, y, and z, before assigning those calculated values to x and y.

One purpose for which this type of multiple assignment is commonly used is to interchange the values of two variables. If we want to swap the values of x and y we can write:

```
x,y = y,x
```

and the two will be exchanged. (In most other computer languages such swaps are more complicated, requiring the use of an additional temporary variable.)

EXAMPLE 2.1: A BALL DROPPED FROM A TOWER

Let us use what we have learned to solve a first physics problem. This is a very simple problem, one we could easily do for ourselves on paper, but don't worry—we will move onto more complex problems shortly.

The problem is as follows. A ball is dropped from a tower of height h. It has initial velocity zero and accelerates downwards under gravity. The challenge is to write a program that asks the user to enter the height in meters of the tower and a time interval t in seconds, then prints on the screen the height of the ball above the ground at time t after it is dropped, ignoring air resistance.

The steps involved are the following. First, we will use `input` statements to get the values of h and t from the user. Second, we will calculate how far the ball falls in the given time, using the standard kinematic formula $s = \frac{1}{2}gt^2$, where $g = 9.81\,\text{ms}^{-2}$ is the acceleration due to gravity. Third, we print the height above the ground at time t, which is equal to the total height of the tower minus this value, or $h - s$.

Here's what the program looks like, all four lines of it:[13]

File: dropped.py

```
h = float(input("Enter the height of the tower: "))
t = float(input("Enter the time interval: "))
s = 9.81*t**2/2
print("The height of the ball is",h-s,"meters")
```

---

[13]Many of the example programs in this book are also available on-line for you to download and run on your own computer if you wish. The programs, along with various other useful resources, are packaged together in a single "zip" file (of size about nine megabytes) which can be downloaded from http://www.umich.edu/~mejn/cpresources.zip. Throughout the book, a name printed in the margin next to a program, such as "dropped.py" above, indicates that the complete program can be found, under that name, in this file. Any mention of programs or data in the "on-line resources" also refers to the same file.

Let us use this program to calculate the height of a ball dropped from a 100 m high tower after 1 second and after 5 seconds. Running the program twice in succession we find the following:

```
Enter the height of the tower: 100
Enter the time interval: 1
The height of the ball is 95.095 meters

Enter the height of the tower: 100
Enter the time interval: 5
The height of the ball is -22.625 meters
```

Notice that the result is negative in the second case, which means that the ball would have fallen to below ground level if that were possible, though in practice the ball would hit the ground first. Thus a negative value indicates that the ball hits the ground before time t.

Before we leave this example, here's a suggestion for a possible improvement to the program above. At present we perform the calculation of the distance traveled with the single line "s = 9.81*t**2/2", which includes the constant 9.81 representing the acceleration due to gravity. When we do physics calculations on paper, however, we normally don't write out the values of constants in full like this. Normally we would write $s = \frac{1}{2}gt^2$, with the understanding that g represents the acceleration. We do this primarily because it's easier to read and understand. A single symbol g is easier to read than a row of digits, and moreover the use of the standard letter g reminds us that the quantity we are talking about is the gravitational acceleration, rather than some other constant that happens to have value 9.81. Especially in the case of constants that have many digits, such as $\pi = 3.14159265\ldots$, the use of symbols rather than digits in algebra makes life a lot easier.

The same is also true of computer programs. You can make your programs substantially easier to read and understand by using symbols for constants instead of writing the values out in full. This is easy to do—just create a variable to represent the constant, like this:

```
g = 9.81
s = g*t**2/2
```

You only have to create the variable g once in your program (usually somewhere near the beginning) and then you can use it as many times as you like thereafter. Doing this also has the advantage of decreasing the chances that you'll make a typographical error in the value of a constant. If you have to type out many digits every time you need a particular constant, odds are you are going to make a mistake at some point. If you have a variable representing the constant then you know the value will be right every time you use it, just so long as you typed it correctly when you first created the variable.[14]

Using variables to represent constants in this way is one example of a programming trick that improves your programs even though it doesn't change the way they actually work. Instead it improves readability and reliability, which can be almost as important as writing a correct program. We will see other examples of such tricks later.

Exercise 2.1: Another ball dropped from a tower

A ball is again dropped from a tower of height h with initial velocity zero. Write a program that asks the user to enter the height in meters of the tower and then calculates and prints the time the ball takes until it hits the ground, ignoring air resistance. Use your program to calculate the time for a ball dropped from a 100 m high tower.

Exercise 2.2: Altitude of a satellite

A satellite is to be launched into a circular orbit around the Earth so that it orbits the planet once every T seconds.

a) Show that the altitude h above the Earth's surface that the satellite must have is

$$h = \left(\frac{GMT^2}{4\pi^2} \right)^{1/3} - R,$$

where $G = 6.67 \times 10^{-11}\, \mathrm{m^3\, kg^{-1}\, s^{-2}}$ is Newton's gravitational constant, $M = 5.97 \times 10^{24}$ kg is the mass of the Earth, and $R = 6371$ km is its radius.

b) Write a program that asks the user to enter the desired value of T and then calculates and prints out the correct altitude in meters.

c) Use your program to calculate the altitudes of satellites that orbit the Earth once a day (so-called "geosynchronous" orbit), once every 90 minutes, and once every 45 minutes. What do you conclude from the last of these calculations?

d) Technically a geosynchronous satellite is one that orbits the Earth once per *sidereal day*, which is 23.93 hours, not 24 hours. Why is this? And how much difference will it make to the altitude of the satellite?

---

[14]In some computer languages, such as C, there are separate entities called "variables" and "constants," a constant being like a variable except that its value can be set only once in a program and is fixed thereafter. There is no such thing in Python, however; there are only variables.

2.2.5 FUNCTIONS, PACKAGES, AND MODULES

There are many operations one might want to perform in a program that are more complicated than simple arithmetic, such as multiplying matrices, calculating a logarithm, or making a graph. Python comes with facilities for doing each of these and many other common tasks easily and quickly. These facilities are divided into *packages*—collections of related useful things—and each package has a name by which you can refer to it. For instance, all the standard mathematical functions, such as logarithm and square root, are contained in a package called math. Before you can use any of these functions you have to tell the computer that you want to. For example, to tell the computer you want to use the log function, you would add the following line to your program:

```
from math import log
```

This tells the computer to "import" the logarithm function from the math package, which means that it copies the code defining the function from where it is stored (usually on the hard disk of your computer) into the computer's memory, ready for use by your program. You need to import each function you use only once per program: once the function has been imported it continues to be available until the program ends. You must import the function before the first time you use it in a calculation and it is good practice to put the "from" statement at the very start of the program, which guarantees that it occurs before the first use of the function and also makes it easy to find when you are working on your code. As we write more complicated programs, there will often be situations where we need to import many different functions into a single program with many different from statements, and keeping those statements together in a tidy block at the start of the code will make things much easier.

Once you have imported the log function you can use it in a calculation like this:

```
x = log(2.5)
```

which will calculate the (natural) logarithm of 2.5 and set the variable x equal to the result. Note that the argument of the logarithm, the number 2.5 in this case, goes in parentheses. If you miss out the parentheses the computer will complain. (Also if you use the log function without first importing it from the math package the computer will complain.)

The math package contains a good selection of the most commonly used mathematical functions:

`log`	natural logarithm
`log10`	log base 10
`exp`	exponential
`sin, cos, tan`	sine, cosine, tangent (argument in radians)
`asin, acos, atan`	arcsine, arccosine, arctangent (in radians)
`sinh, cosh, tanh`	hyperbolic sine, cosine, tangent
`sqrt`	positive square root

Note that the trigonometric functions work with angles specified in radians, not degrees. And the exponential and square root functions may seem re-dundant, since one can calculate both exponentials and square roots by taking powers. For instance, x**0.5 would give the square root of x. Because of the way the computer calculates powers and roots, however, using the functions above is usually quicker and more accurate.

The math package also contains a number of less common functions, such as the Gaussian error function and the gamma function, as well as two objects that are not functions at all but constants, namely e and π, which are denoted e and pi. This program, for instance, calculates the value of π^2:

```
from math import pi
print(pi**2)
```

which prints 9.86960440109 (which is roughly the right answer). Note that there are no parentheses after the "pi" when we use it in the print statement, because it is not a function. It's just a variable called pi with value 3.14159...

The functions in the math package do not work with complex numbers and the computer will give an error message if you try. But there is another pack-age called cmath that contains versions of most of the same functions that do work with complex numbers, plus a few additional functions that are specific to complex arithmetic.

In some cases you may find you want to use more than one function from the same package in a program. You can import two different functions—say the log and exponential functions—with two from statements, like this:

```
from math import log
from math import exp
```

but a more succinct way to do it is to use a single line like this:

```
from math import log,exp
```

You can import a list as long as you like from a single package in this way:

```
from math import log,exp,sin,cos,sqrt,pi,e
```

You can also import *all* of the functions in a package with a statement of the form

```
from math import *
```

The * here means "everything".[15] In most cases, however, I advise against using this import-everything form because it can give rise to some unexpected behaviors (for instance, if, unbeknownst to you, a package contains a function with the same name as one of your variables, causing a clash between the two). It's usually better to explicitly import only those functions you actually need to use.[16]

Finally, some large packages are for convenience split into smaller sub-packages, called *modules*. A module within a larger package is referred to as packagename.modulename. As we will see shortly, for example, there are a large number of useful mathematical facilities available in the package called numpy, including facilities for linear algebra and Fourier transforms, each in their own module within the larger package. Thus the linear algebra module is called numpy.linalg and the Fourier transform module is called numpy.fft (for "fast Fourier transform"). We can import a function from a module thus:

```
from numpy.linalg import inv
```

This would import the inv function, which calculates the inverse of a matrix.

Smaller packages, like the math package, have no submodules, in which case one could, arguably, say that the entire package is also a module, and in

---

[15]There is also another way to import the entire contents of a package in Python, with a statement of the form "import math". If you use this form, however, then when you subsequently use one of the imported functions you have to write, for example, x = math.log(2.5), instead of just x = log(2.5). Since the former is more complicated and annoying, it gets used rather rarely. Moreover the existence of the two types of import, and particularly their simultaneous use in the same program, can be quite confusing, so we will use only the "from" form in this book.

[16]A particular problem is when an imported package contains a function with the same name as a previously existing function. In such a case the newly imported one will be selected in favor of the previous one, which may not always be what you want. For instance, the packages math and cmath contain many functions with the same names, such as sqrt. But the sqrt function in cmath works with complex numbers and the one in math does not. If one did "from cmath import *" followed by "from math import *", one would end up with the version of sqrt that works only with real numbers. And if one then attempted to calculate the square root of a complex number, one would get an error message.

such cases the words package and module are often used interchangeably.

EXAMPLE 2.2: CONVERTING POLAR COORDINATES

Suppose the position of a point in two-dimensional space is given to us in polar coordinates r, θ and we want to convert it to Cartesian coordinates x, y. How would we write a program to do this? The appropriate steps are:

1. Get the user to enter the values of r and θ.

2. Convert those values to Cartesian coordinates using the standard formulas:

$$x = r \cos \theta, \qquad y = r \sin \theta. \tag{2.1}$$

3. Print out the results.

Since the formulas (2.1) involve the mathematical functions sin and cos we are going to have to import those functions from the math package at the start of the program. Also, the sine and cosine functions in Python (and in most other computer languages) take arguments in radians. If we want to be able to enter the angle θ in degrees then we are going to have to convert from degrees to radians, which means multiplying by π and dividing by 180.

Thus our program might look something like this:

File: polar.py

```
from math import sin,cos,pi

r = float(input("Enter r: "))
d = float(input("Enter theta in degrees: "))

theta = d*pi/180
x = r*cos(theta)
y = r*sin(theta)

print("x =",x," y =",y)
```

Take a moment to read through this complete program and make sure you understand what each line is doing. If we run the program, it will do something like the following:

```
Enter r: 2
Enter theta in degrees: 60
x = 1.0   y = 1.73205080757
```

(Try it for yourself if you like.)

2.2.6 BUILT-IN FUNCTIONS

There are a small number of functions in Python, called *built-in functions*, which don't come from any package. These functions are always available to you in every program; you do not have to import them. We have in fact seen several examples of built-in functions already. For instance, we saw the `float` function, which takes a number and converts it to floating point (if it's not floating point already):

```
x = float(1)
```

There are similar functions `int` and `complex` that convert to integers and complex numbers. Another example of a built-in function, one we haven't seen previously, is the `abs` function, which returns the absolute value of a number, or the modulus in the case of a complex number. Thus, `abs(-2)` returns the integer value 2 and `abs(3+4j)` returns the floating-point value 5.0.

Earlier we also used the built-in functions `input` and `print`, which are not mathematical functions in the usual sense of taking a number as argument and performing a calculation on it, but as far as the computer is concerned they are still functions. Consider, for instance, the statement

```
x = input("Enter the value of x: ")
```

Here the `input` function takes as argument the string "`Enter the value of x: `", prints it out, waits for the user to type something in response, then sets x equal to that something.

The print function is slightly different. When we say

```
print(x)
```

print is a function, but it is not here generating a value the way the log or input functions do. It does something with its argument x, namely printing it out on the screen, but it does not generate a value. This differs from the functions we are used to in mathematics, but it's allowed in Python. Sometimes you just want a function to do something but it doesn't need to generate a value.

Exercise 2.3: Write a program to perform the inverse operation to that of Example 2.2. That is, ask the user for the Cartesian coordinates x, y of a point in two-dimensional space, and calculate and print the corresponding polar coordinates, with the angle θ given in degrees.

Exercise 2.4: A spaceship travels from Earth in a straight line at relativistic speed v to another planet x light years away. Write a program to ask the user for the value of x and the speed v as a fraction of the speed of light c, then print out the time in years that the spaceship takes to reach its destination (a) in the rest frame of an observer on Earth and (b) as perceived by a passenger on board the ship. Use your program to calculate the answers for a planet 10 light years away with $v = 0.99c$.

Exercise 2.5: Quantum potential step

A well-known quantum mechanics problem involves a particle of mass m that encounters a one-dimensional potential step, like this:

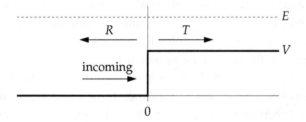

The particle with initial kinetic energy E and wavevector $k_1 = \sqrt{2mE}/\hbar$ enters from the left and encounters a sudden jump in potential energy of height V at position $x = 0$. By solving the Schrödinger equation, one can show that when $E > V$ the particle may either (a) pass the step, in which case it has a lower kinetic energy of $E - V$ on the other side and a correspondingly smaller wavevector of $k_2 = \sqrt{2m(E - V)}/\hbar$, or (b) it may be reflected, keeping all of its kinetic energy and an unchanged wavevector but moving in the opposite direction. The probabilities T and R for transmission and reflection are given by

$$T = \frac{4k_1 k_2}{(k_1 + k_2)^2}, \qquad R = \left(\frac{k_1 - k_2}{k_1 + k_2}\right)^2.$$

Suppose we have a particle with mass equal to the electron mass $m = 9.11 \times 10^{-31}$ kg and energy 10 eV encountering a potential step of height 9 eV. Write a Python program to compute and print out the transmission and reflection probabilities using the formulas above.

Exercise 2.6: Planetary orbits

The orbit in space of one body around another, such as a planet around the Sun, need not be circular. In general it takes the form of an ellipse, with the body sometimes closer in and sometimes further out. If you are given the distance ℓ_1 of closest approach that a planet makes to the Sun, also called its *perihelion*, and its linear velocity v_1 at perihelion, then any other property of the orbit can be calculated from these two as follows.

a) Kepler's second law tells us that the distance ℓ_2 and velocity v_2 of the planet at its most distant point, or *aphelion*, satisfy $\ell_2 v_2 = \ell_1 v_1$. At the same time the total energy, kinetic plus gravitational, of a planet with velocity v and distance r from the Sun is given by

$$E = \tfrac{1}{2}mv^2 - G\frac{mM}{r},$$

where m is the planet's mass, $M = 1.9891 \times 10^{30}$ kg is the mass of the Sun, and $G = 6.6738 \times 10^{-11}$ m^3 kg^{-1} s^{-2} is Newton's gravitational constant. Given that energy must be conserved, show that v_2 is the smaller root of the quadratic equation

$$v_2^2 - \frac{2GM}{v_1 \ell_1}v_2 - \left[v_1^2 - \frac{2GM}{\ell_1}\right] = 0.$$

Once we have v_2 we can calculate ℓ_2 using the relation $\ell_2 = \ell_1 v_1 / v_2$.

b) Given the values of v_1, ℓ_1, and ℓ_2, other parameters of the orbit are given by simple formulas that can be derived from Kepler's laws and the fact that the orbit is an ellipse:

$$\text{Semi-major axis:} \quad a = \tfrac{1}{2}(\ell_1 + \ell_2),$$
$$\text{Semi-minor axis:} \quad b = \sqrt{\ell_1 \ell_2},$$
$$\text{Orbital period:} \quad T = \frac{2\pi ab}{\ell_1 v_1},$$
$$\text{Orbital eccentricity:} \quad e = \frac{\ell_2 - \ell_1}{\ell_2 + \ell_1}.$$

Write a program that asks the user to enter the distance to the Sun and velocity at perihelion, then calculates and prints the quantities ℓ_2, v_2, T, and e.

c) Test your program by having it calculate the properties of the orbits of the Earth (for which $\ell_1 = 1.4710 \times 10^{11}$ m and $v_1 = 3.0287 \times 10^4$ m s^{-1}) and Halley's comet ($\ell_1 = 8.7830 \times 10^{10}$ m and $v_1 = 5.4529 \times 10^4$ m s^{-1}). Among other things, you should find that the orbital period of the Earth is one year and that of Halley's comet is about 76 years.

2.2.7 COMMENT STATEMENTS

This is a good time to mention another important feature of Python (and every other computer language), namely *comments*. In Python any program line that starts with a hash mark "#" is ignored completely by the computer. You can type anything you like on the line following a hash mark and it will have no effect:

```
# Hello!  Hi there!  This line does nothing at all.
```

Such lines are called comments. Comments make no difference whatsoever to the way a program runs, but they can be very useful nonetheless. You can use comment lines to leave reminders for yourself in your programs, saying what particular parts of the program do, what quantities are represented by which variables, changes that you mean to make later to the program, things you're not sure about, and so forth. Here, for instance, is a version of the polar coordinates program from Example 2.2, with comments added to explain what's happening:

File: polar.py

```
from math import sin,cos,pi

# Ask the user for the values of the radius and angle
r = float(input("Enter r: "))
d = float(input("Enter theta in degrees: "))

# Convert the angle to radians
theta = d*pi/180

# Calculate the equivalent Cartesian coordinates
x = r*cos(theta)
y = r*sin(theta)

# Print out the results
print("x =",x," y =",y)
```

This version of the program will perform identically to the original version on page 34, but it is easier to understand how it works.

Comments may seem unnecessary for short programs like this one, but when you get on to creating larger programs that perform complex physics calculations you will find them very useful for reminding yourself of how things work. When you're writing a program you may think you remember how everything works and there is no need to add comments, but when you return to the same program again a week later after spending the intervening time on something else you'll find it's a different story—you can't remember how anything works or why you did things this way or that, and you will be very glad if you scattered a few helpful pointers in comment lines around the program.

Comments become even more important if someone else other than you needs to understand a program you have written, for instance if you're working as part of a team that is developing a large program together. Understanding how other people's programs work can be tough at the best of times, and you will make your collaborators' lives a lot easier if you include some explanatory comments as you go along.

Comments don't have to start at the beginning of a line. Python ignores any portion of a line that follows a hash mark, whether the hash mark is at the beginning or not. Thus you can write things like this:

```
theta = d*pi/180      # Convert the angle to radians
```

and the computer will perform the calculation $\theta = d\pi/180$ at the beginning of the line but completely ignore the hash mark and the text at the end. This is a useful trick when you intend that a comment should refer to a specific single line of code only.

2.3 CONTROLLING PROGRAMS WITH "IF" AND "WHILE"

The programs we have seen so far are all very linear. They march from one statement to the next, from beginning to end of the program, then they stop. An important feature of computers is their ability to break this linear flow, to jump around the program, execute some lines but not others, or make decisions about what to do next based on given criteria. In this section we will see how this is done in the Python language.

2.3.1 THE IF STATEMENT

It will happen often in our computer programs that we want to do something only if a certain condition is met: only if $n = 0$ perhaps, or $x > \frac{1}{2}$. We can do this using an *if statement*. Consider the following example:

```
x = int(input("Enter a whole number no greater than ten: "))
if x>10:
    print("You entered a number greater than ten.")
    print("Let me fix that for you.")
    x = 10
print("Your number is",x)
```

If I run this program and type in "5", I get:

```
Enter a whole number no greater than ten: 5
Your number is 5
```

But if I break the rules and enter 11, I get:

```
Enter a whole number no greater than ten: 11
You entered a number greater than ten.
Let me fix that for you.
Your number is 10
```

This behavior is achieved using an if statement—the second line in the program above—which tests the value of the variable x to see if it is greater than ten. Note the structure of the if statement: there is the "if" part itself, which consists of the word "if" followed by the condition you are applying. In this case the condition is that $x > 10$. The condition is followed by a colon, and following that are one or more lines that tell the computer what to do if the condition is satisfied. In our program there are three of these lines, the first two printing out a message and the third fixing the value of x. Note that these three lines are *indented*—they start with a few spaces so that the text is shifted over a bit from the left-hand edge. This is how we tell the program which instructions are "part of the if." The indented instructions will be executed only if the condition in the if statement is met, i.e., only if $x > 10$ in this case. Whether or not the condition is met, the computer then moves on to the next line of the program, which prints the value of x.

(In Section 1 we saw that you are free to add spaces between the parts of a Python statement to make it more readable, as in "x = 1", and that such spaces will have no effect on the operation of the program. Here we see an exception to that rule: spaces at the beginning of lines do have an effect with an if statement. For this reason one should be careful about putting spaces at the beginning of lines—they should be added only when they are needed, as here, and not otherwise.)

A question that people sometimes ask is, "How many spaces should I put at the start of a line when I am indenting it?" The answer is that you can use any number you like. Python considers any number of spaces, from one upward, to constitute an indentation. However, it has over the years become standard practice among Python programmers to use four spaces for an indentation, and this is the number used in all the programs in this book. In fact, most Python development environments, including IDLE, automatically insert the spaces for you when they see an if statement, and they typically insert four.

There are various different types of conditions one can use in an if statement. Here are some examples:

```
if x==1:          Check if x = 1. Note the double equals sign.
if x>1:           Check if x > 1
if x>=1:          Check if x ≥ 1
if x<1:           Check if x < 1
if x<=1:          Check if x ≤ 1
if x!=1:          Check if x ≠ 1
```

Note particularly the double equals sign in the first example. It is one of the most common programming errors that people make in Python to use a single equals sign in an if statement instead of a double one. If you do this, you'll get an error message when you try to run your program.

You can also combine two conditions in a single if statement, like this:

```
if x>10 or x<1:
    print("Your number is either too big or too small.")
```

You can use "and" in a similar way:

```
if x<=10 and x>=1:
    print("Your number is just right.")
```

You can combine more than two criteria on a line as well—as many as you like.

Two useful further elaborations of the if statement are else and elif:

```
if x>10:
    print("Your number is greater than ten.")
else:
    print("Your number is fine.  Nothing to see here.")
```

This prints different messages depending on whether x is greater than 10 or not. Note that the else line, like the original if, is not indented and has a colon at the end. It is followed by one or more indented lines, the indentation indicating that the lines are "inside" the else clause.

An even more elaborate example is the following:

```
if x>10:
    print("Your number is greater than ten.")
elif x>9:
    print("Your number is OK, but you're cutting it close.")
else:
    print("Your number is fine.  Move along.")
```

The statement `elif` means "else if"—if the first criterion is not met it tells the computer to try a different one. Notice that we can use both `elif` and `else`, as here—if neither of the conditions specified in the `if` and `elif` clauses is satisfied then the computer moves on to the `else` clause. You can also have more than one `elif`, indeed you can have as many as you like, each one testing a different condition if the previous one was not satisfied.

2.3.2 THE WHILE STATEMENT

A useful variation on the if statement is the *while statement*. It looks and behaves similarly to the if statement:

```
x = int(input("Enter a whole number no greater than ten: "))
while x>10:
    print("This is greater than ten.  Please try again.")
    x = int(input("Enter a whole number no greater than ten: "))
print("Your number is",x)
```

As with the if statement, the while statement checks if the condition given is met (in this case if $x > 10$). If it is, it executes the indented block of code immediately following; if not, it skips the block. However (and this is the important difference), if the condition is met and the block is executed, the program then loops back from the end of the block to the beginning and checks the condition again. If the condition is still true, then the indented lines will be executed again. And it will go on looping around like this, repeatedly checking the condition and executing the indented code, until the condition is finally false. (And if it is never false, then the loop goes on forever.[17]) Thus, if I were to run the snippet of code above, I would get something like this:

```
Enter a whole number no greater than ten: 11
This is greater than ten.  Please try again.
Enter a whole number no greater than ten: 57
This is greater than ten.  Please try again.
Enter a whole number no greater than ten: 100
This is greater than ten.  Please try again.
Enter a whole number no greater than ten: 5
Your number is 5
```

---

[17]If you accidentally create a program with a loop that goes on for ever then you'll need to know how to stop the program: just closing the window where the program is running does the trick.

The computer keeps on going around the loop, asking for a number until it gets what it wants. This construct—sometimes also called a *while loop*—is commonly used in this way to ensure that some condition is met in a program or to keep on performing an operation until a certain point or situation is reached.

As with the if statement, we can specify two or more criteria in a single while statement using "and" or "or". The while statement can also be followed by an else statement, which is executed once (and once only) if and when the condition in the while statement fails. (This type of else statement is primarily used in combination with the break statement described in the next section.) There is no equivalent of `elif` for a while loop, but there are two other useful statements that modify its behavior, `break` and `continue`.

2.3.3 BREAK AND CONTINUE

Two useful refinements of the while statement are the break and continue statements. The break statement allows us to break out of a loop even if the condition in the while statement is not met. For instance,

```
while x>10:
    print("This is greater than ten.  Please try again.")
    x = int(input("Enter a whole number no greater than ten: "))
    if x==111:
        break
```

This loop will continue looping until you enter a number not greater than 10, *except* if you enter the number 111, in which case it will give up and proceed with the rest of the program.

If the while loop is followed by an else statement, the else statement is *not* executed after a break. This allows you to create a program that does different things if the while loop finishes normally (and executes the else statement) or via a `break` (in which case the else statement is skipped).

This example also illustrates another new concept: it contains an if statement *inside* a while loop. This is allowed in Python and used often. In the programming jargon we say the if statement is *nested* inside the while loop. While loops nested inside if statements are also allowed, or ifs within ifs, or whiles within whiles. And it doesn't have to stop at just two levels. Any number of statements within statements within statements is allowed. When we get onto some of the more complicated calculations in this book we will see examples nested four or five levels deep. In the example above, note how the break statement is doubly indented from the left margin—it is indented by an extra four spaces, for a total of eight, to indicate that it is part of a statement-within-a-statement.[18]

A variant on the idea of the break statement is the continue statement. Saying continue anywhere in a loop will make the program skip the rest of the indented code in the while loop, but instead of getting on with the rest of the program, it then goes back to the beginning of the loop, checks the condition in the while statement again, and goes around the loop again if the condition is met. (The continue statement turns out to be used rather rarely in practice. The break statement, on the other hand, gets used often and is definitely worth knowing about.)

EXAMPLE 2.3: EVEN AND ODD NUMBERS

Suppose we want to write a program that takes as input a single integer and prints out the word "even" if the number is even, and "odd" if the number is odd. We can do this by making use of the fact that n modulo 2 is zero if (and only if) n is even. Recalling that n modulo 2 is written as n%2 in Python, here's how the program would go:

```python
n = int(input("Enter an integer: "))
if n%2==0:
    print("even")
else:
    print("odd")
```

Now suppose we want a program that asks for two integers, one even and one odd—in either order—and keeps on asking until it gets what it wants. We could do this by checking all of the various combinations of even and odd, but a simpler approach is to notice that if we have one even and one odd number then their sum is odd; otherwise it's even. Thus our program might look like this:

File: evenodd.py

```python
print("Enter two integers, one even, one odd.")
m = int(input("Enter the first integer: "))
n = int(input("Enter the second integer: "))
while (m+n)%2==0:
```

---

[18] We will come across some examples in this book where we have a loop nested inside another loop and then a break statement inside the inner loop. In that case the break statement breaks out of the inner loop only, and not the outer one. (If this doesn't make sense to you, don't worry—it will become clear later when we look at examples with more than one loop.)

```
    print("One must be even and the other odd.")
    m = int(input("Enter the first integer: "))
    n = int(input("Enter the second integer: "))
print("The numbers you chose are",m,"and",n)
```

Note how the while loop checks to see if $m + n$ is *even*. If it is, then the numbers you entered must be wrong—either both are even or both are odd—so the program asks for another pair, and it will keep on doing this until it gets what it wants.

As before, take a moment to look over this program and make sure you understand what each line does and how the program works.

EXAMPLE 2.4: THE FIBONACCI NUMBERS

The Fibonacci numbers are the sequence of integers in which each is the sum of the previous two, with the first two numbers being 1, 1. Thus the first few members of the sequence are 1, 1, 2, 3, 5, 8, 13, 21. Suppose we want to calculate the Fibonacci numbers up to 1000. This would be quite a laborious task for a human, but it is straightforward for a computer program. All the program needs to do is keep a record of the most recent two numbers in the sequence, add them together to calculate the next number, then keep on repeating for as long as the numbers are less than 1000. Here's a program to do it:

```
f1 = 1
f2 = 1
while f1<=1000:
    print(f1)
    fnext = f1 + f2
    f1 = f2
    f2 = fnext
```

Observe how the program works. At all times the variables f1 and f2 store the two most recent elements of the sequence. If f1 is less than 1000, we print it out, then calculate the next element of the sequence by summing f1 and f2 and store the result in the variable fnext. Then we update the values of f1 and f2 and go around the loop again. The process continues until the value of f1 exceeds 1000, then stops.

This program works fine, but here's a neater way to solve the same problem using the "multiple assignment" feature of Python discussed in Section 2.2.4:

File: `fibonacci.py`

```
f1,f2 = 1,1
while f1<=1000:
    print(f1)
    f1,f2 = f2,f1+f2
```

If we run this program, we get the following:

```
1
1
2
3
5
8
13
21
34
55
89
144
233
377
610
987
```

Indeed, the computer will happily print out the Fibonacci numbers up to a billion or more in just a second or two. Try it if you like.

Exercise 2.7: Catalan numbers

The Catalan numbers C_n are a sequence of integers 1, 1, 2, 5, 14, 42, 132...that play an important role in quantum mechanics and the theory of disordered systems. (They were central to Eugene Wigner's proof of the so-called semicircle law.) They are given by

$$C_0 = 1, \qquad C_{n+1} = \frac{4n+2}{n+2} C_n.$$

Write a program that prints in increasing order all Catalan numbers less than or equal to one billion.

2.4 LISTS AND ARRAYS

We have seen how to work with integer, real, and complex quantities in Python and how to use variables to store those quantities. All the variables we have seen so far, however, represent only a single value, a single integer, real, or complex number. But in physics it is common for a variable to represent several numbers at once. We might use a vector **r**, for instance, to represent the position of a point in three-dimensional space, meaning that the single symbol **r** actually corresponds to three real numbers (x, y, z). Similarly, a matrix, again usually denoted by just a single symbol, can represent an entire grid of numbers, $m \times n$ of them, where m and n could be as large as we like. There are also many cases where we have a set of numbers that we would like to treat as a single entity even if they do not form a vector or matrix. We might, for instance, do an experiment in the lab and make a hundred measurements of some quantity. Rather than give a different name to each one—a, b, c, and so forth—it makes sense to denote them as say a_1, a_2, a_3, and then to consider them collectively as a set $A = \{a_i\}$, a single entity made up of a hundred numbers.

Situations like these are so common that Python provides standard features, called *containers*, for storing collections of numbers. There are several kinds of containers. In this section we look at two of them, lists and arrays.[19]

2.4.1 LISTS

The most basic type of container in Python is the *list*. A list, as the name suggests, is a list of quantities, one after another. In all the examples in this book the quantities will be numbers of some kind—integers, floats, and so forth—although any type of quantity that Python knows about is allowed in a list, such as strings for example.[20]

The quantities in a list, which are called its *elements*, do not have to be all of the same type. You can have an integer, followed by a float, followed by a complex number if you want. In most of the cases we'll deal with, however, the elements will be all of the same type—all integers, say, or all floats—because this is what physics calculations usually demand. Thus, for instance, in the example described above where we make a hundred measurements of a given quantity in the lab and we want to represent them on the computer, we could use a list one hundred elements long, and all the elements would presumably

---

[19]There are several others as well, the main ones being *tuples*, *dicts*, and *sets*. These, however, find only occasional use in physics calculations, and we will not use them in this book.

[20]If you have programmed in another computer language, then you may be familiar with "arrays," which are similar to lists but not exactly the same. Python has both lists and arrays and both have their uses in physics calculations. We study arrays in Section 2.4.2.

be of the same type (probably floats) because they all represent measurements of the same thing.

A list in Python is written like this: [3, 0, 0, -7, 24]. The elements of this particular list are all integers. Note that the elements are separated by commas and the whole list is surrounded by square brackets. Another example of a list might be [1, 2.5, 3+4.6j]. This example has three elements of different types, one integer, one real, and one complex.

A variable can be set equal to a list:

```
r = [ 1, 1, 2, 3, 5, 8, 13, 21 ]
```

Previously in this chapter all variables have represented just single numbers, but here we see that a variable can also represent a list of numbers. You can print a list variable, just as you can any other variable, and the computer will print out the entire list. If we run this program:

```
r = [ 1, 1, 2, 3, 5, 8, 13, 21 ]
print(r)
```

we get this:

```
[1, 1, 2, 3, 5, 8, 13, 21]
```

The quantities that make up the elements of a list can be specified using other variables, like this:

```
x = 1.0
y = 1.5
z = -2.2

r = [ x, y, z ]
```

This will create a three-element list with the value [1.0, 1.5, -2.2]. It is important to bear in mind, in this case, what happens when Python encounters an assignment statement like "r = [x, y, z]". Remember that in such situations Python first evaluates the expression on the right-hand side, which gives [1.0, 1.5, -2.2] in this case, then assigns that value to the variable on the left. Thus the end result is that r is equal to [1.0, 1.5, -2.2]. It is a common error to think of r as being equal to [x, y, z] so that if, say, the value of x is changed later in the program the value of r will change as well. But this is incorrect. The value of r will get set to [1.0, 1.5, -2.2] and will not change later if x is changed. If you want to change the value of

r you have to explicitly assign a new value to it, with another statement like
"r = [x, y, z]".

The elements of lists can also be calculated from entire mathematical expressions, like this:

```
r = [ 2*x, x+y, z/sqrt(x**2+y**2) ]
```

The computer will evaluate all the expressions on the right-hand side then create a list from the values it calculated.

Once we have created a list we probably want to do some calculations with the elements it contains. The individual elements in a list r are denoted r[0], r[1], r[2], and so forth. That is they are numbered in order, from beginning to end of the list, the numbers go in square brackets after the variable name, and crucially the numbers start from *zero*, not one. This may seem a little odd—it's not the way we usually do things in physics or in everyday life—and it takes a little getting used to. However, it turns out, as we'll see, to be more convenient in a lot of situations than starting from one.

The individual elements, such as r[0], behave like single variables and you can use them in the same way you would use a single variable. Thus, here is a short program that calculates and prints out the length of a vector in three dimensions:

```
from math import sqrt
r = [ 1.0, 1.5, -2.2 ]
length = sqrt( r[0]**2 + r[1]**2 + r[2]**2 )
print(length)
```

The first line imports the square root function from the math package, which we need for the calculation. The second line creates the vector, in the form of a three-element list. The third line is the one that does the actual calculation. It takes each of the three elements of the vector, which are denoted r[0], r[1], and r[2], squares them, and adds them together. Then it takes the square root of the result, which by Pythagoras' theorem gives us the length of the vector. The final line prints out the length. If we run this program it prints

```
2.84429253067
```

which is the correct answer (to twelve significant figures).

We can change the values of individual elements of a list at any time, like this:

```
r = [ 1.0, 1.5, -2.2 ]
r[1] = 3.5
print(r)
```

The first line will create a list with three elements. The second then changes the value of element 1, which is the middle of the three elements, since they are numbered starting from zero. So if we run the program it prints out this:

```
[1.0, 3.5, -2.2]
```

A powerful and useful feature of Python is its ability to perform operations on entire lists at once. For instance, it commonly happens that we want to know the sum of the values in a list. Python contains a built-in function called sum that can calculate such sums in a single line, thus:

```
r = [ 1.0, 1.5, -2.2 ]
total = sum(r)
print(total)
```

The first line here creates a three-element list and the second calculates the sum of its elements. The final line prints out the result, and if we run the program we get this:

```
0.3
```

Some other useful built-in functions are max and min, which give the largest and smallest values in a list respectively, and len, which calculates the number of elements in a list. Applied to the list r above, for instance, max(r) would give 1.5 and min(r) would give −2.2, while len(r) would give 3. Thus, for example, one can calculate the mean of the values in a list like this:

```
r = [ 1.0, 1.5, -2.2 ]
mean = sum(r)/len(r)
print(mean)
```

The second line here sums the elements in the list and then divides by the number of elements to give the mean value. In this case, the calculation would give a mean of 0.1.

A special, and especially useful, function for lists is the function map, which is a kind of meta-function—it allows you apply ordinary functions, like log or sqrt, to all the elements of a list at once. Thus map(log,r) takes the natural logarithm of each element of a list r in turn. More precisely, map creates a specialized object in the computer memory, called an *iterator*, that contains the logs, one after another in order. Normally we will want to convert this iterator into a new list, which we can do with the built-in function `list`.[21] Thus, consider this snippet of code:

```
from math import log
r = [ 1.0, 1.5, 2.2 ]
logr = list(map(log,r))
print(logr)
```

This will create a list `logr` containing the logs of the three numbers 1, 1.5, and 2.2, and print it out thus:

```
[0.0, 0.4054651081, 0.7884573603]
```

Another feature of lists in Python, one that we will use often, is the ability to add elements to an already existing list. Suppose we have a list called `r` and we want to add a new element to the end of the list with value, say, 6.1. We can do this with the statement

```
r.append(6.1)
```

This slightly odd-looking statement is a little different in form from the ones we've seen previously.[22] It consists of the name of our list, which is `r`, followed by a dot (i.e., a period), then "`append(6.1)`". Its effect is to add a new element to the end of the list and set that element equal to the value given, which is 6.1 in this case. The value can also be specified using a variable or a mathematical expression, thus:

---

[21] The difference between an iterator and a list is that the values in an iterator are not stored in the computer's memory the way the values in a list are. Instead, the computer calculates them on the fly when they are needed, which saves memory. Thus, in this case, the computer only calculates the logs of the elements of `r` when you convert the iterator to a list. In versions of Python prior to version 3, the `map` function produced a list, not an iterator, so if you are using an earlier version of the language you do not need to convert to a list using the `list` function. You can just say "`logr = map(log,r)`". For further discussion of this point, and of iterators in general, see Appendix B.

[22] This is an example of Python's object-oriented programming features. The function `append` is technically a "method" that belongs to the list "object" `r`. The function doesn't exist as an entity in its own right, only as a subpart of the list object. We will not dig into Python's object-oriented features in this book, since they are of relatively little use for the type of physics programming we will be doing. For software developers engaged in large-scale commercial or group programming projects, however, they can be invaluable.

```
r = [ 1.0, 1.5, -2.2 ]
x = 0.8
r.append(2*x+1)
print(r)
```

If we run this program we get

```
[1.0, 1.5, -2.2, 2.6]
```

Note how the computer has calculated the value of 2*x+1 to be 2.6, then added that value to the end of the list.

A particularly useful trick that we will employ frequently in this book is the following. We create an *empty* list, a list with no elements in it at all, then add elements to it one by one as we learn of or calculate their values. A list created in this way can grow as large as we like (within limitations set by the amount of memory the computer has to store the list).

To create an empty list we say

```
r = []
```

This creates a list called r with no elements. Even though it has no elements in it, the list still exists. It's like an empty set in mathematics—it exists as an object, but it doesn't contain anything (yet). Now we can add elements thus:

```
r.append(1.0)
r.append(1.5)
r.append(-2.2)
print(r)
```

which produces

```
[1.0, 1.5, -2.2]
```

We will, for instance, use this technique to make graphs in Section 3.1. Note that you must create the empty list first before adding elements. You cannot add elements to a list until it has been created—the computer will give an error message if you try.

We can also remove a value from the end of a list by saying r.pop():

```
r = [ 1.0, 1.5, -2.2, 2.6 ]
r.pop()
print(r)
```

which gives

```
[1.0, 1.5, -2.2]
```

And we can remove a value from anywhere in a list by saying `r.pop(n)`, where n is the number of the element you want to remove.[23] Bear in mind that the elements are numbered from zero, so if you want to remove the first item from a list you would say `r.pop(0)`.

2.4.2 ARRAYS

As we have seen, a list in Python is an ordered set of values, such as a set of integers or a set of floats. There is another object in Python that is somewhat similar, an *array*. An array is also an ordered set of values, but there are some important differences between lists and arrays:

1. The number of elements in an array is fixed. You cannot add elements to an array once it is created, or remove them.

2. The elements of an array must all be of the same type, such as all floats or all integers. You cannot mix elements of different types in the same array and you cannot change the type of the elements once an array is created.

Lists, as we have seen, have neither of these restrictions and, on the face of it, these seem like significant drawbacks of the array. Why would we ever use an array if lists are more flexible? The answer is that arrays have several significant advantages over lists as well:

3. Arrays can be two-dimensional, like matrices in algebra. That is, rather than just a one-dimensional row of elements, we can have a grid of them. Indeed, arrays can in principle have any number of dimensions, including three or more, although we won't use dimensions above two in this book. Lists, by contrast, are always just one-dimensional.

4. Arrays behave roughly like vectors or matrices: you can do arithmetic with them, such as adding them together, and you will get the result you expect. This is not true with lists. If you try to do arithmetic with a list

---

[23]However, removing an element from the middle (or the beginning) of a list is a slow operation because the computer then has to move all the elements above that down one place to fill the gap. For a long list this can take a long time and slow down your program, so you should avoid doing it if possible. (On the other hand, if it doesn't matter to you what order the elements of a list appear in, then you can effectively remove any element rapidly by first setting it equal to the last element in the list, then removing the last element.)

you will either get an error message, or you will not get the result you expect.

5. Arrays work faster than lists. Especially if you have a very large array with many elements then calculations may be significantly faster using an array.

In physics it often happens that we are working with a fixed number of elements all of the same type, as when we are working with matrices or vectors, for instance. In that case, arrays are clearly the tool of choice: the fact that we cannot add or remove elements is immaterial if we never need to do such a thing, and the superior speed of arrays and their flexibility in other respects can make a significant difference to our programs. We will use arrays extensively in this book—far more than we will use lists.

Before you use an array you need to create it, meaning you need to tell the computer how many elements it will have and of what type. Python provides functions that allow you do this in several different ways. These functions are all found in the package numpy.[24]

In the simplest case, we can create a one-dimensional array with n elements, all of which are initially equal to zero, using the function zeros from the numpy package. The function takes two arguments. The first is the number of elements the array is to have and the second is the type of the elements, such as int, float, or complex. For instance, to create a new array with four floating-point elements we would do the following:

```
from numpy import zeros
a = zeros(4,float)
print(a)
```

In this example the new array is denoted a. When we run the program the array is printed out as follows:

```
[ 0.  0.  0.  0.]
```

Note that arrays are printed out slightly differently from lists—there are no commas between the elements, only spaces.

We could similarly create an array of ten integers with the statement "a = zeros(10,int)" or an array of a hundred complex numbers with the statement "a = zeros(100,complex)". The size of the arrays you can create is lim-

---

[24]The word numpy is short for "numerical Python."

2.4 | LISTS AND ARRAYS

ited only by the computer memory available to hold them. Modern computers can hold arrays with hundreds of millions or even billions of elements.

To create a two-dimensional floating-point array with m rows and n columns, you say "zeros([m,n],float)", so

```
a = zeros([3,4],float)
print(a)
```

produces

```
[[ 0.  0.  0.  0.]
 [ 0.  0.  0.  0.]
 [ 0.  0.  0.  0.]]
```

Note that the first argument of zeros in this case is itself a *list* (that's why it is enclosed in brackets [...]), whose elements give the size of the array along each dimension. We could create a three-dimensional array by giving a three-element list (and so on for higher dimensions).

There is also a similar function in numpy called ones that creates an array with all elements equal to one. The form of the function is exactly the same as for the function zeros. Only the values in the array are different.

On the other hand, if we are going to change the values in an array immediately after we create it, then it doesn't make sense to have the computer set all of them to zero (or one)—setting them to zero takes some time, time that is wasted if you don't need the zeros. In that case you can use a different function, empty, again from the package numpy, to create an empty array:

```
from numpy import empty
a = empty(4,float)
```

This creates an array of four "empty" floating-point elements. In practice the elements aren't actually empty. Instead they contain whatever numbers happened to be littered around the computer's memory at the time the array is created. The computer just leaves those values as they are and doesn't waste any time changing them. You can also create empty integer or complex arrays by saying int or complex instead of float.

A different way to create an array is to take a list and convert it into an array, which you can do with the function array from the package numpy. For instance we can say:

```
r = [ 1.0, 1.5, -2.2 ]
a = array(r,float)
```

which will create an array of three floating-point elements, with values 1.0, 1.5, and −2.2. If the elements of the list (or some of them) are not already floats, they will be converted to floats. You can also create integer or complex arrays in the same fashion, and the list elements will be converted to the appropriate type if necessary.[25,26]

The two lines above can conveniently be combined into one, like this:

```
a = array([1.0,1.5,-2.2],float)
```

This is a quick and neat way to create a new array with predetermined values in its elements. We will use this trick frequently.

We can also create two-dimensional arrays with specified initial values. To do this we again use the array function, but now the argument we give it must be a *list of lists*, which gives the elements of the array row by row. For example, we can write

```
a = array([[1,2,3],[4,5,6]],int)
print(a)
```

This creates a two-dimensional array of integers and prints it out:

```
[[ 1  2  3]
 [ 4  5  6]]
```

The list of lists must have the same number of elements for each row of the array (three in this case) or the computer will complain.

We can refer to individual elements of an array in a manner similar to the way we refer to the elements of a list. For a one-dimensional array a we write a[0], a[1], and so forth. Note, as with lists, that the numbering of the elements starts at zero, not one. We can also set individual elements equal to new values thus:

---

[25]Two caveats apply here. (1) If you create an integer array from a list that has any floating-point elements, the fractional part, if any (i.e., the part after the decimal point), of the floating-point elements will be thrown away. (2) If you try to create a floating-point or integer array from a list containing complex values you will get an error message. This is not allowed.

[26]Though it's not something we will often need to do, you can also convert an array into a list using the built-in function list, thus:

```
r = list(a)
```

Note that you do not specify a type for the list, because lists don't have types. The types of the elements in the list will just be the same as the types of the elements in the array.

```
a[2] = 4
```

Note, however, that, since the elements of an array are of a particular type (which cannot be changed after the array is created), any value you specify will be converted to that type. If you give an integer value for a floating-point array element, it will be converted to floating-point. If you give a floating-point value for an integer array, the fractional part of the value will be deleted. (And if you try to assign a complex value to an integer or floating-point array you will get an error message—this is not allowed.)

For two-dimensional arrays we use two indices, separated by commas, to denote the individual elements, as in `a[2,4]`, with counting again starting at zero, for both indices. Thus, for example

```
from numpy import zeros
a = zeros([2,2],int)
a[0,1] = 1
a[1,0] = -1
print(a)
```

would produce the output

```
[[ 0  1]
 [-1  0]]
```

(You should check this example and make sure you understand why it does what it does.)

Note that when Python prints a two-dimensional array it observes the convention of standard matrix arithmetic that the first index of a two-dimensional array denotes the row of the array element and the second denotes the column.

2.4.3   READING AN ARRAY FROM A FILE

Another, somewhat different, way to create an array is to read a set of values from a computer file, which we can do with the function `loadtxt` from the package numpy. Suppose we have a text file that contains the following string of numbers, on consecutive lines:

```
1.0
1.5
-2.2
2.6
```

and suppose that this file is called `values.txt` on the computer. Then we can do the following:

```
from numpy import loadtxt
a = loadtxt("values.txt",float)
print(a)
```

When we run this program, we get the following printed on the screen:

```
[ 1.0  1.5 -2.2  2.6]
```

As you can see, the computer has read the numbers in the file and put them in a float-point array of the appropriate length. (For this to work the file `values.txt` has to be in the same folder or directory on the computer as your Python program is saved in.[27])

We can use the same trick to read a two-dimensional grid of values and put them in a two-dimensional array. If the file `values.txt` contained the following:

```
1 2 3 4
3 4 5 6
5 6 7 8
```

then the exact same program above would create a two-dimensional 3×4 array of floats with the appropriate values in it.

The `loadtxt` function is a very useful one for physics calculations. It happens often that we have a file or files containing numbers we need for a calculation. They might be data from an experiment, for example, or numbers calculated by another computer program. We can use the `loadtxt` function to transfer those numbers into an array so that we can perform further calculations on them.

2.4.4 ARITHMETIC WITH ARRAYS

As with lists, the individual elements of an array behave like ordinary variables, and we can do arithmetic with them in the usual way. We can write things like

```
a[0] = a[1] + 1
```

---

[27] You can also give a full path name for the file, specifying explicitly the folder as well as the file name, in which case the file can be in any folder.

or

```
x = a[2]**2 - 2*a[3]/y
```

But we can also do arithmetic with entire arrays at once, a powerful feature that can be enormously useful in physics calculations. In general, when doing arithmetic with entire arrays, the rule is that whatever arithmetic operation you specify is done independently to each element of the array or arrays involved. Consider this short program:

```
from numpy import array
a = array([1,2,3,4],int)
b = 2*a
print(b)
```

When we run this program it prints

```
[2 4 6 8]
```

As you can see, when we multiply the array a by 2 the computer simply multiplies each individual element by 2. A similar thing happens if you divide. Notice that when we run this program, the computer creates a new array b holding the results of our multiplication. This is another way to create arrays, different from the methods we mentioned before. We do not have to create the array b explicitly, using for instance the empty function. When we perform a calculation with arrays, Python will automatically create a new array for us to hold the results.

If you add or subtract two arrays, the computer will add or subtract each element separately, so that

```
a = array([1,2,3,4],int)
b = array([2,4,6,8],int)
print(a+b)
```

results in

```
[3 6 9 12]
```

(For this to work, the arrays must have the same size. If they do not, the computer will complain.)

All of these operations give the same result as the equivalent mathematical operations on vectors in normal algebra, which makes arrays well suited to representing vectors in physics calculations.[28] If we represent a vector using an array then arithmetic operations such as multiplying or dividing by a scalar quantity or adding or subtracting vectors can be written just as they would in normal mathematics. You can also add a scalar quantity to an array (or subtract one), which the computer interprets to mean it should add that quantity to every element. So

```
a = array([1,2,3,4],int)
print(a+1)
```

results in

```
[2 3 4 5]
```

However, if we multiply two arrays together the outcome is perhaps not exactly what you would expect—you do not get the vector (dot) product of the two. If we do this:

```
a = array([1,2,3,4],int)
b = array([2,4,6,8],int)
print(a*b)
```

we get

```
[2 8 18 32]
```

What has the computer done here? It has multiplied the two arrays together element by corresponding element. The first elements of the two arrays are multiplied together, then the second elements, and so on. This is logical in a sense—it is the exact equivalent of what happens when you add. Each element of the first array is multiplied by the corresponding element of the second. (Division works similarly.) Occasionally this may be what you want the computer to do, but more often in physics calculations we want the true vector dot product of our arrays. This can be calculated using the function dot from the package numpy:

```
from numpy import array,dot
a = array([1,2,3,4],int)
b = array([2,4,6,8],int)
print(dot(a,b))
```

[28]The same operations, by contrast, do not work with lists, so lists are less good for storing vector values.

The function dot takes two arrays as arguments and calculates their dot product, which would be 60 in this case.

All of the operations above also work with two-dimensional arrays, which makes such arrays convenient for storing matrices. Multiplying and dividing by scalars as well as addition and subtraction of two-dimensional arrays all work as in standard matrix algebra. Multiplication will multiply element by element, which is usually not what you want, but the dot function calculates the standard matrix product. Consider, for example, this matrix calculation:

$$\begin{pmatrix} 1 & 3 \\ 2 & 4 \end{pmatrix} \begin{pmatrix} 4 & -2 \\ -3 & 1 \end{pmatrix} + 2 \begin{pmatrix} 1 & 2 \\ 2 & 1 \end{pmatrix} = \begin{pmatrix} -3 & 5 \\ 0 & 2 \end{pmatrix}$$

In Python we would do this as follows:

```
a = array([[1,3],[2,4]],int)
b = array([[4,-2],[-3,1]],int)
c = array([[1,2],[2,1]],int)
print(dot(a,b)+2*c)
```

You can also multiply matrices and vectors together. If v is a one-dimensional array then dot(a,v) treats it as a column vector and multiplies it on the left by the matrix a, while dot(v,a) treats it as a row vector and multiplies on the right by a. Python is intelligent enough to know the difference between row and column vectors, and between left- and right-multiplication.

Functions can be applied to arrays in much the same way as to lists. The built-in functions sum, min, max, and len described in Section 2.4.1 can be applied to one-dimensional arrays to calculate sums of elements, minimum and maximum values, and the number of elements. The map function also works, applying any ordinary function to all elements of a one-dimensional array and producing an iterator (see Section 2.4.1), which can then be converted into a list with the list function, or into an array using the list function then the array function:

```
b = array(list(map(sqrt,a)),float)
```

This will create an iterator whose elements are the square roots of the elements of a, then convert the iterator into an array b containing those values.

Applying functions to arrays with two or more dimensions produces more erratic results. For instance, the len function applied to a two-dimensional array returns the number of rows in the array and the functions max and min produce only error messages. However, the numpy package contains functions that perform similar duties and work more predictably with two-dimensional arrays, such as functions `min` and `max` that find minimum and maximum values. In place of the `len` function, there are two different features, called `size` and `shape`. Consider this example:

```
a = array([[1,2,3],[4,5,6]],int)
print(a.size)
print(a.shape)
```

which produces

```
6
(2, 3)
```

That is, `a.size` tells you the total number of elements in all rows and columns of the array a (which is roughly the equivalent of the `len` function for lists and one-dimensional arrays), and `a.shape` returns a list giving the dimensions of the array along each axis. (Technically it is a "tuple" not a list, but for our purposes it is roughly the same thing. You can say n = a.shape, and then n[0] is the number of rows of a and n[1] is the number of columns.) For one-dimensional arrays there is not really any difference between `size` and `shape`. They both give the total number of elements.

There are a number of other functions in the numpy package that are useful for performing calculations with arrays. The full list can be found in the on-line documentation at www.scipy.org.

EXAMPLE 2.5: AVERAGE OF A SET OF VALUES IN A FILE

Suppose we have a set of numbers stored in a file `values.txt` and we want to calculate their mean. Even if we don't know how many numbers there are we can do the calculation quite easily:

```
from numpy import loadtxt
values = loadtxt("values.txt",float)
mean = sum(values)/len(values)
print(mean)
```

The first line imports the `loadtxt` function and the second uses it to read the values in the file and put them in an array called `values`. The third line calculates the mean as the sum of the values divided by the number of values and the fourth prints out the result.

Now suppose we want to calculate the mean-square value. To do this, we first need to calculate the squares of the individual values, which we can do by multiplying the array `values` by itself. Recall, that the product of two arrays in Python is calculated by multiplying together each pair of corresponding elements, so `values*values` is an array with elements equal to the squares of the original values. (We could also write `values**2`, which would produce the same result.) Then we can use the function `sum` again to add up the squares. Thus our program might look like this:

```
from numpy import loadtxt
values = loadtxt("values.txt",float)
mean = sum(values*values)/len(values)
print(mean)
```

On the other hand, suppose we want to calculate the *geometric* mean of our set of numbers. (We'll assume our numbers are all positive, since one cannot take the geometric mean of negative numbers.) The geometric mean of a set of n values x_i is defined to be the nth root of their product, thus:

$$\overline{x} = \left[\prod_{i=1}^{n} x_i\right]^{1/n}.$$

(2.2)

Taking natural logs of both sides we get

$$\ln\overline{x} = \ln\left[\prod_{i=1}^{n} x_i\right]^{1/n} = \frac{1}{n}\sum_{i=1}^{n}\ln x_i$$

(2.3)

or

$$\overline{x} = \exp\left(\frac{1}{n}\sum_{i=1}^{n}\ln x_i\right).$$

(2.4)

In other words, the geometric mean is the exponential of the arithmetic mean of the logarithms. We can modify our previous program for the arithmetic mean to calculate the geometric mean thus:

```
from numpy import loadtxt
from math import log,exp
values = loadtxt("values.txt",float)
logs = array(list(map(log,values)),float)
geometric = exp(sum(logs)/len(logs))
print(geometric)
```

Note how we combined the `map` function and the `log` function to calculate the logarithms and then calculated the arithmetic mean of the resulting values.

If we want to be clever, we can streamline this program by noting that the numpy package contains its own log function that will calculate the logs of all the elements of an array. Thus we can rewrite our program as

```
from numpy import loadtxt,log
from math import exp
values = loadtxt("values.txt",float)
geometric = exp(sum(log(values))/len(values))
print(geometric)
```

As well as being more elegant, this version of the program will probably also run a little faster, since the log function in numpy is designed specifically to work efficiently with arrays.

Finally in this section, here is a word of warning. Consider the following program:

```
from numpy import array
a = array([1,1],int)
b = a
a[0] = 2
print(a)
print(b)
```

Take a look at this program and work out for yourself what you think it will print. If we actually run it (and you can try this for yourself) it prints the following:

```
[2 1]
[2 1]
```

This is probably not what you were expecting. Reading the program, it looks like array a should be equal to [2,1] and b should be equal to [1,1] when the program ends, but the output of the program appears to indicate that both are equal to [2,1]. What has gone wrong?

The answer lies in the line "b = a" in the program. In Python direct assignment of arrays in this way, setting the value of one array equal to another, does not work as you might expect it to. You might imagine that "b = a" would cause Python to create a new array b holding a copy of the numbers in the array a, but this is not what happens. In fact, all that "b = a" does is it declares "b" to be a new name for the array previously called "a". That is, "a" and "b" now both refer to the same array of numbers, stored somewhere in the memory of the computer. If we change the value of an element in array a, as we do in the program above, then we also change the same element of array b, because a and b are really just the same array.[29]

This is a tricky point, one that can catch you out if you are not aware of it. You can do all sorts of arithmetic operations with arrays and they will work just fine, but this one operation, setting an array equal to another array, does not work the way you expect it to.

Why does Python work like this? At first sight it seems peculiar, annoying even, but there is a good reason for it. Arrays can be very large, with millions or even billions of elements. So if a statement like "b = a" caused the computer to create a new array b that was a complete copy of the array a, it might have to copy very many numbers in the process, potentially using a lot of time and memory space. But in many cases it's not actually necessary to make a copy of the array. Particularly if you are interested only in reading the numbers in an array, not in changing them, then it does not matter whether a and b are separate arrays that happen to contain the same values or are actually just two names for the same array—everything will work the same either way. Creating a new name for an old array is normally far faster than making a copy of the entire contents, so, in the interests of efficiency, this is what Python does.

Of course there are times when you really do want to make a new copy of an array, so Python also provides a way of doing this. To make a copy of an array a we can use the function copy from the numpy package thus:

```
from numpy import copy
b = copy(a)
```

This will create a separate new array b whose elements are an exact copy of those of array a. If we were to use this line, instead of the line "b = a", in the program above, then run the program, it would print this:

```
[2 1]
[1 1]
```

which is now the "correct" answer.

---

[29]If you have worked with the programming languages C or C++ you may find this behavior familiar, since those languages treat arrays the same way. In C, the statement "b = a", where a and b are arrays, also merely creates a new name for the array a, not a new array.

Exercise 2.8: Suppose arrays a and b are defined as follows:

```
from numpy import array
a = array([1,2,3,4],int)
b = array([2,4,6,8],int)
```

What will the computer print upon executing the following lines? (Try to work out the answer before trying it on the computer.)

    a) `print(b/a+1)`
    b) `print(b/(a+1))`
    c) `print(1/a)`

2.4.5 SLICING

Here's another useful trick, called *slicing*, which works with both arrays and lists. Suppose we have a list r. Then r[m:n] is another list composed of a subset of the elements of r, starting with element m and going up to *but not including* element n. Here's an example:

```
r = [ 1, 3, 5, 7, 9, 11, 13, 15 ]
s = r[2:5]
print(s)
```

which produces

```
[5, 7, 9]
```

Observe what happens here. The variable s is a new list, which is a sublist of r consisting of elements 2, 3, and 4 of r, but not element 5. Since the numbering of elements starts at zero, not one, element 2 is actually the third element of the list, which is the 5, and elements 3 and 4 are the 7 and 9. So s has three elements equal to 5, 7, and 9.

Slicing can be useful in many physics calculations, particularly, as we'll see, in matrix calculations, calculations on lattices, and in the solution of differential equations. There are a number of variants on the basic slicing formula above. You can write r[2:], which means all elements of the list from element 2 up to the end of the list, or r[:5], which means all elements from the start of the list up to, but not including, element 5. And r[:] with no numbers at all means all elements from the beginning to the end of the list, i.e., the entire list. This last is not very useful—if we want to refer to the whole list we can just say r. We get the same thing, for example, whether we write print(r[:]) or print(r). However, we will see a use for this form in a moment.

Slicing also works with arrays: you can specify a subpart of an array and you get another array, of the same type as the first, as in this example:

```
from numpy import array
a = array([2,4,6,8,10,12,14,16],int)
b = a[3:6]
print(b)
```

which prints

```
[ 8 10 12]
```

You can also write a[3:], or a[:6], or a[:], as with lists.

Slicing works with two-dimensional arrays as well. For instance a[2,3:6] gives you a one-dimensional array with three elements equal to a[2,3], a[2,4], and a[2,5]. And a[2:4,3:6] gives you a two-dimensional array of size 2×3 with values drawn from the appropriate subblock of a, starting at a[2,3].

Finally, a[2,:] gives you the whole of row 2 of array a, which means the third row since the numbering starts at zero. And a[:,3] gives you the whole of column 3, which is the fourth column. These forms will be particularly useful to us for doing vector and matrix arithmetic.

2.5 "FOR" LOOPS

In Section 2.3.2 we saw a way to make a program loop repeatedly around a given section of code using a while statement. It turns out, however, that while statements are used only rather rarely. There is another, much more commonly used loop construction in the Python language, the *for loop*. A for loop is a loop that runs through the elements of a list or array in turn. Consider this short example:

```
r = [ 1, 3, 5 ]
for n in r:
    print(n)
    print(2*n)
print("Finished")
```

If we run this program it prints out the following:

```
1
2
3
6
5
10
Finished
```

What's happening here is as follows. The program first creates a list called r, then the for statement sets n equal to each value in the list in turn. For each value the computer carries out the steps in the following two lines, printing out n and $2n$, then loops back around to the for statement again and sets n to the next value in the list. Note that the two print statements are indented, in a manner similar to the if and while statements we saw earlier. This is how we tell the program which instructions are "in the loop." Only the indented instructions will be executed each time round the loop. When the loop has worked its way through all the values in the list, it stops looping and moves on to the next line of the program, which in this case is a third print statement which prints the word "Finished." Thus in this example the computer will go around the loop three times, since there are three elements in the list r.

The same construction works with arrays as well—you can use a for loop to go through the elements of a (one-dimensional) array in turn. Also the statements break and continue (see Section 2.3.3) can be used with for loops the same way they're used with while loops: break ends the loop and moves to the next statement after the loop; continue abandons the current iteration of the loop and moves on to the next iteration.

The most common use of the for loop is simply to run through a given piece of code a specified number of times, such as ten, say, or a million. To achieve this, Python provides a special built-in function called range, which creates a list of a given length, usually for use with a for loop. For example range(5) gives a list [0, 1, 2, 3, 4]—that is, a list of consecutive integers, starting at zero and going up to, but not including, 5. Note that this means the list contains exactly five elements but does not include the number 5 itself.

(Technically, range produces an iterator, not a list. Iterators were discussed previously in Section 2.4.1. If you actually want to produce a list using range then you would write something of the form "list(range(5))", which creates an iterator and then converts it to a list. In practice, however, we do this very rarely, and never in this book—the main use of the range function is in for loops and you are allowed to use an iterator directly in a for loop without converting it into a list first.[30])

Thus

```
r = range(5)
for n in r:
    print("Hello again")
```

produces the following output

```
Hello again
Hello again
Hello again
Hello again
Hello again
```

The for loop gives n each of the values in r in turn, of which there are five, and for each of them it prints out the words Hello again. So the end result is that the computer prints out the same message five times. In this case we are not actually interested in the values r contains, only the fact that there are five of them—they merely provide a convenient tool that allows us to run around the same piece of code a given number of times.

A more interesting use of the range function is the following:

```
r = range(5)
for n in r:
    print(n**2)
```

Now we are making use of the actual values r contains, printing out the square of each one in turn:

```
0
1
4
9
16
```

In both these examples we used a variable r to store the results of the range function, but it's not necessary to do this. Often one takes a shortcut and just

---

[30]In versions of Python prior to version 3, the range function actually did produce a list, not an iterator. If you wanted to produce an iterator you used the separate function xrange, which no longer exists in version 3. Both list and iterator give essentially the same results, however, so the for loops in this book will work without modification with either version of Python. For further discussion of this point, and of iterators in general, see Appendix B.

writes something like

```
for n in range(5):
    print(n**2)
```

which achieves the same result with less fuss. This is probably the most common form of the for loop and we will see many loops of this form throughout this book.

There are a number of useful variants of the range function, as follows:

range(5)	gives	[0, 1, 2, 3, 4]
range(2,8)	gives	[2, 3, 4, 5, 6, 7]
range(2,20,3)	gives	[2, 5, 8, 11, 14, 17]
range(20,2,-3)	gives	[20, 17, 14, 11, 8, 5]

When there are two arguments to the function it generates integer values that run from the first up to, but not including, the second. When there are three arguments, the values run from the first up to but not including second, *in steps of* the third. Thus in the third example above the values increase in steps of 3. In the fourth example, which has a negative argument, the values *decrease* in steps of 3. Note that in each case the values returned by the function do not include the value at the end of the given range—the first value in the range is always included; the last never is.

Thus, for example, we can print out the first ten powers of two with the following lines:

```
for n in range(1,11):
    print(2**n)
```

Notice how the upper limit of the range must be given as 11. This program will print out the powers 2, 4, 8, 16, and so forth up to 1024. It stops at 2^{10}, not 2^{11}, because range always excludes the last value.

A further point to notice about the range function is that all its arguments must be integers. The function doesn't work if you give it noninteger arguments, such as floats, and you will get an error message if you try. It is particularly important to remember this when the arguments are calculated from the values of other variables. This short program, for example, will not work:

```
p = 10
q = 2
for n in range(p/q):
    print(n)
```

You might imagine these lines would print out the integers from zero to four, but if you try them you will just get an error message because, as discussed in Section 2.2.4, the division operation always returns a floating-point quantity, even if the result of the division is, mathematically speaking, an integer. Thus the quantity "p/q" in the program above is a floating-point quantity equal to 5.0 and is not allowed as an argument of the range function. We can fix this problem by using an integer division instead:

```
for n in range(p//q):
    print(n)
```

This will now work as expected. (See Section 2.2.4, page 23 for a discussion of integer division.)

Another useful function is the function arange from the numpy package, which is similar to range but generates arrays, rather than lists or iterators[31] and, more importantly, works with floating-point arguments as well as integer ones. For example arange(1,8,2) gives a one-dimensional array of integers [1,3,5,7] as we would expect, but arange(1.0,8.0,2.0) gives an array of floating-point values [1.0,3.0,5.0,7.0] and arange(2.0,2.8,0.2) gives [2.0,2.2,2.4,2.6]. As with range, arange can be used with one, two, or three arguments, and does the equivalent thing to range in each case.

Another similar function is the function linspace, also from the numpy package, which generates an array with a given number of floating-point values between given limits. For instance, linspace(2.0,2.8,5) divides the interval from 2.0 to 2.8 into 5 values, creating an array with floating-point elements [2.0,2.2,2.4,2.6,2.8]. Similarly, linspace(2.0,2.8,3) would create an array with elements [2.0,2.4,2.8]. Note that, unlike both range and arange, linspace includes the last point in the range. Also note that although linspace can take either integer or floating-point arguments, it always generates floating-point values, even when the arguments are integers.

---

[31]The function arange generates an actual array, calculating all the values and storing them in the computer's memory. This can cause problems if you generate a very large array because the computer can run out of memory, crashing your program, an issue that does not arise with the iterators generated by the range function. For instance, arange(10000000000) will produce an error message on most computers, while the equivalent expression with range will not. See Appendix B for more discussion of this point.

EXAMPLE 2.6: PERFORMING A SUM

It happens often in physics calculations that we need to evaluate a sum. If we have the values of the terms in the sum stored in a list or array then we can calculate the sum using the built-in function sum described in Section 2.4.1. In more complicated situations, however, it is often more convenient to use a for loop to calculate a sum. Suppose, for instance, we want to know the value of the sum $s = \sum_{k=1}^{100}(1/k)$. The standard way to program this is as follows:

1. First create a variable to hold the value of the sum, and initially set it equal to zero. As above we'll call the variable s, and we want it to be a floating-point variable, so we'd do "s = 0.0".

2. Now use a for loop to take the variable k through all values from 1 to 100. For each value, calculate 1/k and add it to the variable s.

3. When the for loop ends the variable s will contain the value of the complete sum.

Thus our program looks like this:

```
s = 0.0
for k in range(1,101):
    s += 1/k
print(s)
```

Note how we use range(1,101) so that the values of k start at 1 and end at 100. We also used the "+=" modifier, which adds to a variable as described in Section 2.2.4. If we run this program it prints the value of the sum thus:

```
5.1873775176
```

As another example, suppose we have a set of real values stored in a computer file called values.txt and we want to compute and print the sum of their squares. We could achieve this as follows:

```
from numpy import loadtxt
values = loadtxt("values.txt",float)
s = 0.0
for x in values:
    s += x**2
print(s)
```

Here we have used the function loadtxt from Section 2.4.3 to read the values in the file and put them in an array called values. Note also how this example does not use the range function, but simply goes through the list of values directly.

For loops and the sum function give us two different ways to compute sums of quantities. It is not uncommon for there to be more than one way to achieve a given goal in a computer program, and in particular it's often the case that one can use either a for loop or a function or similar array operation to perform the same calculation. In general for loops are more flexible, but direct array operations are often faster and can save significant amounts of time if you are dealing with large arrays. Thus both approaches have their advantages. Part of the art of good computer programming is learning which approach is best in which situation.

EXAMPLE 2.7: EMISSION LINES OF HYDROGEN

Let us revisit an example we saw in Chapter 1. On page 6 we gave a program for calculating the wavelengths of emission lines in the spectrum of the hydrogen atom, based on the Rydberg formula

$$\frac{1}{\lambda} = R\left(\frac{1}{m^2} - \frac{1}{n^2}\right). \tag{2.5}$$

Our program looked like this:

```
R = 1.097e-2
for m in [1,2,3]:
    print("Series for m =",m)
    for k in [1,2,3,4,5]:
        n = m + k
        invlambda = R*(1/m**2-1/n**2)
        print("   ",1/invlambda,"nm")
```

We can now understand exactly how this program works. It uses two nested for loops—a loop within another loop—with the code inside the inner loop doubly indented. We discussed nesting previously in Section 2.3.3. The first for loop takes the integer variable m through the values 1, 2, 3. And for each value of m the second, inner loop takes k though the values 1, 2, 3, 4, 5, adds those values to m to calculate n and then applies the Rydberg formula. The end result will be that the program prints out a wavelength for each combination of values of *m* and *n*, which is what we want.

In fact, knowing what we now know, we can write a simpler version of this program, by making use of the range function, thus:

File: rydberg.py

```
R = 1.097e-2
for m in range(1,4):
    print("Series for m =",m)
    for n in range(m+1,m+6):
        invlambda = R*(1/m**2-1/n**2)
        print("   ",1/invlambda,"nm")
```

Note how we were able to eliminate the variable k in this version by specifying a range for n that depends directly on the value of m.

Exercise 2.9: The Madelung constant

In condensed matter physics the Madelung constant gives the total electric potential felt by an atom in a solid. It depends on the charges on the other atoms nearby and their locations. Consider for instance solid sodium chloride—table salt. The sodium chloride crystal has atoms arranged on a cubic lattice, but with alternating sodium and chlorine atoms, the sodium ones having a single positive charge $+e$ and the chlorine ones a single negative charge $-e$, where e is the charge on the electron. If we label each position on the lattice by three integer coordinates (i, j, k), then the sodium atoms fall at positions where $i + j + k$ is even, and the chlorine atoms at positions where $i + j + k$ is odd.

Consider a sodium atom at the origin, $i = j = k = 0$, and let us calculate the Madelung constant. If the spacing of atoms on the lattice is a, then the distance from the origin to the atom at position (i, j, k) is

$$\sqrt{(ia)^2 + (ja)^2 + (ka)^2} = a\sqrt{i^2 + j^2 + k^2},$$

and the potential at the origin created by such an atom is

$$V(i, j, k) = \pm \frac{e}{4\pi\epsilon_0 a\sqrt{i^2 + j^2 + k^2}},$$

with ϵ_0 being the permittivity of the vacuum and the sign of the expression depending on whether $i + j + k$ is even or odd. The total potential felt by the sodium atom is then the sum of this quantity over all other atoms. Let us assume a cubic box around the sodium at the origin, with L atoms in all directions. Then

$$V_{\text{total}} = \sum_{\substack{i,j,k=-L \\ \text{not } i=j=k=0}}^{L} V(i, j, k) = \frac{e}{4\pi\epsilon_0 a} M,$$

where M is the Madelung constant, at least approximately—technically the Madelung constant is the value of M when $L \to \infty$, but one can get a good approximation just by using a large value of L.

Write a program to calculate and print the Madelung constant for sodium chloride. Use as large a value of L as you can, while still having your program run in reasonable time—say in a minute or less.

Exercise 2.10: The semi-empirical mass formula

In nuclear physics, the semi-empirical mass formula is a formula for calculating the approximate nuclear binding energy B of an atomic nucleus with atomic number Z and mass number A:

$$B = a_1 A - a_2 A^{2/3} - a_3 \frac{Z^2}{A^{1/3}} - a_4 \frac{(A - 2Z)^2}{A} + \frac{a_5}{A^{1/2}},$$

where, in units of millions of electron volts, the constants are $a_1 = 15.8$, $a_2 = 18.3$, $a_3 = 0.714$, $a_4 = 23.2$, and

$$a_5 = \begin{cases} 0 & \text{if } A \text{ is odd,} \\ 12.0 & \text{if } A \text{ and } Z \text{ are both even,} \\ -12.0 & \text{if } A \text{ is even and } Z \text{ is odd.} \end{cases}$$

a) Write a program that takes as its input the values of A and Z, and prints out the binding energy for the corresponding atom. Use your program to find the binding energy of an atom with $A = 58$ and $Z = 28$. (Hint: The correct answer is around 500 MeV.)

b) Modify your program to print out not the total binding energy B, but the binding energy per nucleon, which is B/A.

c) Now modify your program so that it takes as input just a single value of the atomic number Z and then goes through all values of A from $A = Z$ to $A = 3Z$, to find the one that has the largest binding energy per nucleon. This is the most stable nucleus with the given atomic number. Have your program print out the value of A for this most stable nucleus and the value of the binding energy per nucleon.

d) Modify your program again so that, instead of taking Z as input, it runs through all values of Z from 1 to 100 and prints out the most stable value of A for each one. At what value of Z does the maximum binding energy per nucleon occur? (The true answer, in real life, is $Z = 28$, which is nickel.)

2.6 USER-DEFINED FUNCTIONS

We saw in Section 2.2.5 how to use functions, such as `log` or `sqrt`, to do mathematics in our programs, and Python comes with a broad array of functions for performing all kinds of calculations. There are many situations in computational physics, however, where we need a specialized function to perform a particular calculation and Python allows you to define your own functions in such cases.

Suppose, for example, we are performing a calculation that requires us to calculate the factorials of integers. Recall that the factorial of n is defined as the product of all integers from 1 to n:

$$n! = \prod_{k=1}^{n} k. \tag{2.6}$$

We can calculate this in Python with a loop like this:

```
f = 1.0
for k in range(1,n+1):
    f *= k
```

When the loop finishes, the variable f will be equal to the factorial we want.

If our calculation requires us to calculate factorials many times in various different parts of the program we could write out a loop, as above, each time, but this could get tedious quickly and would increase the chances that we would make an error. A more convenient approach is to define our own function to calculate the factorial, which we do like this:

```
def factorial(n):
    f = 1.0
    for k in range(1,n+1):
        f *= k
    return f
```

Note how the lines of code that define the function are indented, in a manner similar to the if statements and for loops of previous sections. The indentation tells Python where the function ends.

Now, anywhere later in the program we can simply say something like

```
a = factorial(10)
```

or

```
b = factorial(r+2*s)
```

and the program will calculate the factorial of the appropriate number. In effect what happens when we write "a = factorial(10)"—when the function is *called*—is that the program jumps to the definition of the function (the part starting with def above), sets n = 10, and then runs through the instructions in the function. When it gets to the final line "return f" it jumps back to where it came from and the value of the factorial function is set equal to whatever quantity appeared after the word return—which is the final value of the variable f in this case. The net effect is that we calculate the factorial of 10 and set the variable a equal to the result.

An important point to note is that any variables created inside the definition of a function exist only inside that function. Such variables are called *local variables*. For instance the variables f and k in the factorial function above are local variables. This means we can use them only when we are inside the function and they disappear when we leave. Thus, for example, you could print the value of the variable k just fine if you put the print statement inside the function, but if you were to try to print the variable anywhere outside the function then you would get an error message telling you that no such variable exists.[32] Note, however, that the reverse is not true—you can use a variable inside a function that is defined outside it.

User-defined functions allow us to encapsulate complex calculations inside a single function definition and can make programs much easier to write and to read. We will see many uses for them in this book.

User-defined functions can have more than one argument. Suppose, for example, we have a point specified in cylindrical coordinates r, θ, z, and we want to know the distance d between the point and the origin. The simplest way to do the calculation is to convert r and θ to Cartesian coordinates first, then apply Pythagoras' Theorem to calculate d:

$$x = r \cos \theta, \qquad y = r \sin \theta, \qquad d = \sqrt{x^2 + y^2 + z^2}. \tag{2.7}$$

If we find ourselves having to do such a conversion many times within a program we might want to define a function to do it. Here's a suitable function in Python:

```
def distance(r,theta,z):
    x = r*cos(theta)
    y = r*sin(theta)
    d = sqrt(x**2+y**2+z**2)
    return d
```

---

[32]To make things more complicated, you can separately define a variable called k outside the function and then you are allowed to print that (or do any other operation with it), but in that case it is a *different* variable—now you have two variables called k that have separate values and which value you get depends on whether you are inside the function or not.

(This assumes that we have already imported the functions `sin`, `cos`, and `sqrt` from the `math` package.)

Note how the function takes three different arguments now. When we call the function we must now supply it with three different arguments and they must come in the same order—r, θ, z—that they come in in the definition of the function. Thus if we say

```
D = distance(2.0,0.1,-1.5)
```

the program will calculate the distance for $r = 2$, $\theta = 0.1$, and $z = -1.5$.

The values used as arguments for a function can be any type of quantity Python knows about, including integers and real and complex numbers, but also including, for instance, lists and arrays. This allows us, for example, to create functions that perform operations on vectors or matrices stored in arrays. We will see examples of such functions when we look at linear algebra methods in Chapter 6.

The value returned by a function can also be of any type, including integer, real, complex, or a list or array. Using lists or arrays allows us to return more than one value if want to, or to return a vector or matrix. For instance, we might write a function to convert from polar coordinates to Cartesian coordinates like this:

```
def cartesian(r,theta):
    x = r*cos(theta)
    y = r*sin(theta)
    position = [x,y]
    return position
```

This function takes a pair of values r, θ and returns a two-element list containing the corresponding values of x and y. In fact, we could combine the two final lines here into one and say simply

```
return [x,y]
```

Or we could return x and y in the form of a two-element array by saying

```
return array([x,y],float)
```

An alternative way to return multiple values from a function is to use the "multiple assignment" feature of Python, which we examined in Section 2.2.4. We saw there that one can write statements of the form "`x,y = a,b`" which will simultaneously set x = a and y = b. The equivalent maneuver with a user-defined function is to write

```
def f(z):
    # Some calculations here...
    return a,b
```

which will make the function return the values of a and b both. To call such a function we write something like

```
x,y = f(1)
```

and the two returned values would get assigned to the variables x and y. One can also specify three or more returned values in this fashion, and the individual values themselves can again be lists, arrays, or other objects, in addition to single numbers, which allows functions to return very complex sets of values when necessary.

User-defined functions can also return no value at all—it is permitted for functions to end without a return statement. The body of the function is marked by indenting the lines of code and the function ends when the indentation does, whether or not there is a return statement. If the function ends without a return statement then the program will jump back to wherever it came from, to the statement where it was called, without giving a value. Why would you want to do this? In fact there are many cases where this is a useful thing to do. For example, suppose you have a program that uses three-element arrays to hold vectors and you find that you frequently want to print out the values of those vectors. You could write something like

```
print("(",r[0],r[1],r[2],")")
```

every time you want to print a vector, but this is difficult to read and prone to typing errors. A better way to do it would be to define a function that prints a vector, like this:

```
def print_vector(r):
    print("(",r[0],r[1],r[2],")")
```

Then when you want to print a vector you simply say "print_vector(r)" and the computer handles the rest. Note how, when calling a function that returns no value you simply give the name of the function. One just says "print_vector(r)", and not "x = print_vector(r)" or something like that.

This is different from the functions we are used to in mathematics, which always return a value. Perhaps a better name for functions like this would be "user-defined statements" or something similar, but by convention they are still called "functions" in Python.[33]

The definition of a user-defined function—the code starting with def—can occur anywhere in a program, except that it must occur before the first time you use the function. It is good programming style to put all your function definitions (you will often have more than one) at or near the beginning of your programs. This guarantees that they come before their first use, and also makes them easier to find if you want to look them up or change them later. Once defined, your functions can be used in the same way as any other Python functions. You can use them in mathematical expressions. You can use them in print statements. You can even use them inside the definitions of other functions. You can also apply a user-defined function to all the elements of a list or array with the map function. For instance, to multiply every element of a list by 2 and subtract 1, we could do the following:

```
def f(x):
    return 2*x-1

newlist = list(map(f,oldlist))
```

This applies the function f to every element in turn of the list oldlist and makes a list of the results called newlist.

One more trick is worth mentioning, though it is more advanced and you should feel free to skip it if you're not interested. The functions you define do not have to be saved in the same file on your computer as your main program. You could, for example, place a function definition for a function called myfunction in a separate file called mydefinitions.py. You can put the definitions for many different functions in the same file if you want. Then, when you want to use a function in a program, you say

```
from mydefinitions import myfunction
```

This tells Python to look in the file mydefinitions.py to find the definition of myfunction and magically that function will now become available in your program. This is a very convenient feature if you write a function that you

---

[33]We have already seen one other example of a function with no return value, the standard print function itself.

want to use in many different programs: you need write it only once and store it in a file; then you can import it into as many other programs as you like.

As you will no doubt have realized by now, this is what is happening when we say things like "from math import sqrt" in a program. Someone wrote a function called sqrt that calculates square roots and placed it in a file so that you can import it whenever you need it. The math package in Python is nothing other than a large collection of function definitions for useful mathematical functions, gathered together in one file.[34]

EXAMPLE 2.8: PRIME FACTORS AND PRIME NUMBERS

Suppose we have an integer n and we want to know its prime factors. The prime factors can be calculated relatively easily by dividing repeatedly by all integers from 2 up to n and checking to see if the remainder is zero. Recall that the remainder after division can be calculated in Python using the modulo operation "%". Here is a function that takes the number n as argument and returns a list of its prime factors:

```
def factors(n):
    factorlist = []
    k = 2
    while k<=n:
        while n%k==0:
            factorlist.append(k)
            n //= k
        k += 1
    return factorlist
```

This is a slightly tricky piece of code—make sure you understand how it does the calculation. Note how we have used the integer division operation "//" to perform the divisions, which ensures that the result returned is another integer. (Remember that the ordinary division operation "/" always produces a float.) Note also how we change the value of the variable n (which is the argument of the function) inside the function. This is allowed: the argument variable behaves like any other variable and can be modified, although, like all variables inside functions, it is a local variable—it exists only inside the

---

[34]In fact the functions in the math package aren't written in Python—they're written in the C programming language, and one has to do some additional trickery to make these C functions work in Python, but the same basic principle still applies.

function and gets erased when the function ends.

Now if we say "print(factors(17556))", the computer prints out the list of factors "[2, 2, 3, 7, 11, 19]". On the other hand, if we specify a prime number in the argument, such as "print(factors(23))", we get back "[23]"—the only prime factor of a prime number is itself. We can use this fact to make a program that prints out the prime numbers up to any limit we choose by checking to see if they have only a single prime factor:

```
for n in range(2,10000):
    if len(factors(n))==1:
        print(n)
```

Run this program, and in a matter of seconds we have a list of the primes up to 10 000. (This is, however, not a very efficient way of calculating the primes—see Exercise 2.12 for a faster way of doing it.)

Exercise 2.11: Binomial coefficients

The binomial coefficient $\binom{n}{k}$ is an integer equal to

$$\binom{n}{k} = \frac{n!}{k!(n-k)!} = \frac{n \times (n-1) \times (n-2) \times \ldots \times (n-k+1)}{1 \times 2 \times \ldots \times k}$$

when $k \geq 1$, or $\binom{n}{0} = 1$ when $k = 0$.

a) Using this form for the binomial coefficient, write a Python user-defined function binomial(n,k) that calculates the binomial coefficient for given n and k. Make sure your function returns the answer in the form of an integer (not a float) and gives the correct value of 1 for the case where $k = 0$.

b) Using your function write a program to print out the first 20 lines of "Pascal's triangle." The nth line of Pascal's triangle contains $n + 1$ numbers, which are the coefficients $\binom{n}{0}$, $\binom{n}{1}$, and so on up to $\binom{n}{n}$. Thus the first few lines are

1 1
1 2 1
1 3 3 1
1 4 6 4 1

c) The probability that an unbiased coin, tossed n times, will come up heads k times is $\binom{n}{k}/2^n$. Write a program to calculate (a) the total probability that a coin tossed 100 times comes up heads exactly 60 times, and (b) the probability that it comes up heads 60 or more times.

Exercise 2.12: Prime numbers

The program in Example 2.8 is not a very efficient way of calculating prime numbers: it checks each number to see if it is divisible by any number less than it. We can develop a much faster program for prime numbers by making use of the following observations:

a) A number n is prime if it has no prime factors less than n. Hence we only need to check if it is divisible by other primes.

b) If a number n is non-prime, having a factor r, then $n = rs$, where s is also a factor. If $r \geq \sqrt{n}$ then $n = rs \geq \sqrt{n}s$, which implies that $s \leq \sqrt{n}$. In other words, any non-prime must have factors, and hence also prime factors, less than or equal to $\sqrt{n}$. Thus to determine if a number is prime we only have to check its prime factors up to and including $\sqrt{n}$—if there are none then the number is prime.

c) If we find even a single prime factor less than $\sqrt{n}$ then we know that the number is non-prime, and hence there is no need to check any further—we can abandon this number and move on to something else.

Write a Python program that finds all the primes up to ten thousand. Create a list to store the primes, which starts out with just the one prime number 2 in it. Then for each number n from 3 to 10 000 check whether the number is divisible by any of the primes in the list up to and including $\sqrt{n}$. As soon as you find a single prime factor you can stop checking the rest of them—you know n is not a prime. If you find no prime factors $\sqrt{n}$ or less then n is prime and you should add it to the list. You can print out the list all in one go at the end of the program, or you can print out the individual numbers as you find them.

Exercise 2.13: Recursion

A useful feature of user-defined functions is *recursion*, the ability of a function to call itself. For example, consider the following definition of the factorial $n!$ of a positive integer n:

$$n! = \begin{cases} 1 & \text{if } n = 1, \\ n \times (n-1)! & \text{if } n > 1. \end{cases}$$

This constitutes a complete definition of the factorial which allows us to calculate the value of $n!$ for any positive integer. We can employ this definition directly to create a Python function for factorials, like this:

```
def factorial(n):
    if n==1:
        return 1
    else:
        return n*factorial(n-1)
```

Note how, if n is not equal to 1, the function calls itself to calculate the factorial of $n - 1$. This is recursion. If we now say "`print(factorial(5))`" the computer will correctly print the answer 120.

a) We encountered the Catalan numbers C_n previously in Exercise 2.7 on page 46. With just a little rearrangement, the definition given there can be rewritten in the form

$$C_n = \begin{cases} 1 & \text{if } n = 0, \\ \dfrac{4n-2}{n+1} C_{n-1} & \text{if } n > 0. \end{cases}$$

Write a Python function, using recursion, that calculates C_n. Use your function to calculate and print C_{100}.

b) Euclid showed that the greatest common divisor $g(m, n)$ of two nonnegative integers m and n satisfies

$$g(m,n) = \begin{cases} m & \text{if } n = 0, \\ g(n, m \bmod n) & \text{if } n > 0. \end{cases}$$

Write a Python function $g(m,n)$ that employs recursion to calculate the greatest common divisor of m and n using this formula. Use your function to calculate and print the greatest common divisor of 108 and 192.

Comparing the calculation of the Catalan numbers in part (a) above with that of Exercise 2.7, we see that it's possible to do the calculation two ways, either directly or using recursion. In most cases, if a quantity can be calculated *without* recursion, then it will be faster to do so, and we normally recommend taking this route if possible. There are some calculations, however, that are essentially impossible (or at least much more difficult) without recursion. We will see some examples later in this book.

2.7 GOOD PROGRAMMING STYLE

When writing a program to solve a physics problem there are, usually, many ways to do it, many programs that will give you the solution you are looking for. For instance, you can use different names for your variables, use either lists or arrays for storing sets of numbers, break up the code by using user-defined functions to do some operations, and so forth. Although all of these approaches may ultimately give the same answer, not all of them are equally satisfactory. There are well written programs and poorly written ones. A well written program will, as far as possible, have a simple structure, be easy to read and understand, and, ideally, run fast. A poorly written one may be convoluted or unnecessarily long, difficult to follow, or may run slowly. Making programs easy to read is a particularly important—and often overlooked—goal. An easy-to-read program makes it easier to find problems, easier to modify the code, and easier for other people to understand how things work.

Good programming is, to some extent, a matter of experience, and you will quickly get the hang of it as you start to write programs. But here are a few general rules of thumb that may help.

1. **Include comments in your programs.** Leave comments in the code to remind yourself what particular variables mean, what calculations are being performed in different sections of the code, what arguments functions require, and so forth. It's amazing how you can come back to a program you wrote only a week ago and not remember how it works. You will thank yourself later if you include comments. And comments are even more important if you are writing programs that other people will have to read and understand. It's frustrating to be the person who has to fix or modify someone else's code if they neglected to include any comments to explain how it works.

2. **Use meaningful variable names.** Give your variables names that help you remember what they represent. The names don't have to be long. In fact, very long names are usually harder to read. But choose your names sensibly. Use E for energy and t for time. Use full words where appropriate or even pairs of words to spell out what a variable represents, like mass or angular_momentum. If you are writing a program to calculate the value of a mathematical formula, give your variables the same names as in the formula. If variables are called x and β in the formula, call them x and beta in the program.

3. **Use the right types of variables.** Use integer variables to represent quantities that actually are integers, like vector indices or quantum numbers. Use floats and complex variables for quantities that really are real or complex numbers.

4. **Import functions first.** If you are importing functions from packages, put your import statements at the start of your program. This makes them easy to find if you need to check them or add to them, and it ensures that you import functions before the first time they are used.

5. **Give your constants names.** If there are constants in your program, such as the number of atoms N in a gas or the mass m of a particle, create similarly named variables at the beginning of your program to represent these quantities, then use those variables wherever those quantities appear in your program. This makes formulas easier to read and understand and it allows you to later change the value of a constant by changing only a single line at the beginning of the program, even if the constant

appears many times throughout your calculations. Thus, for example, you might have a line "A = 58" that sets the atomic mass of an atom for a calculation at the beginning of the program, then you would use A everywhere else in the program that you need to refer to the atomic mass. If you later want to perform the same calculation for atomic mass 59, you need only change the single line at the beginning to "A = 59". Most physics programs have a section near the beginning (usually right after the import statements) that defines all the constants and parameters of the program, making them easy to find when you need to change their values.

6. **Employ user-defined functions, where appropriate.** Use user-defined functions to represent repeated operations, especially complicated operations. Functions can greatly increase the legibility of your program. Avoid overusing them, however: simple operations, ones that can be represented by just a line or two of code, are often better left in the main body of the program, because it allows you to follow the flow of the calculation when reading through your program without having to jump to a function and back. Normally you should put your function definitions at the start of your program (probably after imports and constant definitions). This ensures that each definition appears before the first use of the function it defines and that the definitions can be easily found and modified when necessary.

7. **Print out partial results and updates throughout your program.** Large computational physics calculations can take a long time—minutes, hours, or even days. You will find it helpful to include print statements in your program that print updates about where the program has got to or partial results from the calculations, so you know how the program is progressing. It's difficult to tell whether a calculation is working correctly if the computer simply sits silently, saying nothing, for hours on end. Thus, for example, if there is a for loop in your program that repeats many times, it is useful to include code like this at the beginning of the loop:

```
for n in range(1000000):
    if n%1000==0:
        print("Step",n)
```

These lines will cause the program to print out what step it has reached every time n is exactly divisible by 1000, i.e., every thousandth step. So it will print:

```
Step 0
Step 1000
Step 2000
Step 3000
```

and so forth as it goes along.

8. **Lay out your programs clearly.** You can add spaces or blank lines in most places within a Python program without changing the operation of the program and doing so can improve readability. Make use of blank lines to split code into logical blocks. Make use of spaces to divide up complicated algebraic expressions or particularly long program lines.

  You can also split long program lines into more than one line if necessary. If you place a backslash symbol "\" at the end of a line it tells the computer that the following line is a continuation of the current one, rather than a new line in its own right. Thus, for instance you can write:

```
energy = mass*(vx**2 + vy**2)/2 + mass*g*y \
         + moment_of_inertia*omega**2/2
```

and the computer will interpret this as a single formula. If a program line is very long indeed you can spread it over three or more lines on the screen with backslashes at the end of each one, except the last.[35]

9. **Don't make your programs unnecessarily complicated.** A short simple program is enormously preferable to a long involved one. If the job can be done in ten or twenty lines, then it's probably worth doing it that way—the code will be easier to understand, for you or anyone else, and if there are mistakes in the program it will be easier to work out where they lie.

Good programming, like good science, is a matter of creativity as well as technical skill. As you gain more experience with programming you will no doubt develop your own programming style and learn to write code in a way that makes sense to you and others, creating programs that achieve their scientific goals quickly and elegantly.

---

[35]Under certain circumstances, you do not need to use a backslash. If a line does not make sense on its own but it does make sense when the following line is interpreted as a continuation, then Python will automatically assume the continuation even if there is no backslash character. This, however, is a complicated rule to remember, and there are no adverse consequences to using a backslash when it's not strictly needed, so in most cases it is simpler just to use the backslash and not worry about the rules.

CHAPTER **3**

GRAPHICS AND VISUALIZATION

SO FAR we have created programs that print out words and numbers, but often we will also want our programs to produce graphics, meaning pictures of some sort. In this chapter we will see how to produce the two main types of computer graphics used in physics. First, we look at that most common of scientific visualizations, the graph: a depiction of numerical data displayed on calibrated axes. Second, we look at methods for making scientific diagrams and animations: depictions of the arrangement or motion of the parts of a physical system, which can be useful in understanding the structure or behavior of the system.

3.1 GRAPHS

A number of Python packages include features for making graphs. In this book we will use the powerful, easy-to-use, and popular package pylab.[1] The package contains features for generating graphs of many different types. We will concentrate of three types that are especially useful in physics: ordinary line graphs, scatter plots, and density (or heat) plots. We start by looking at line graphs.[2]

To create an ordinary graph in Python we use the function plot from the

---

[1] The name pylab is a reference to the scientific calculation program Matlab, whose graph-drawing features pylab is intended to mimic. The pylab package is a part of a larger package called matplotlib, some of whose features can occasionally be useful in computational physics, although we will use only the ones in pylab in this book. If you're interested in the other features of matplotlib, take a look at the on-line documentation, which can be found at matplotlib.org.

[2] The pylab package can also make contour plots, polar plots, pie charts, histograms, and more, and all of these find occasional use in physics. If you find yourself needing one of these more specialized graph types, you can find instructions for making them in the on-line documentation at matplotlib.org.

pylab package. In the simplest case, this function takes one argument, which is a list or array of the values we want to plot. The function creates a graph of the given values in the memory of the computer, but it doesn't actually display it on the screen of the computer—it's stored in the memory but not yet visible to the computer user. To display the graph we use a second function from pylab, the show function, which takes the graph in memory and draws it on the screen. Here is a complete program for plotting a small graph:

```
from pylab import plot,show
y = [ 1.0, 2.4, 1.7, 0.3, 0.6, 1.8 ]
plot(y)
show()
```

After importing the two functions from pylab, we create the list of values to be plotted, create a graph of those values with plot(y), then display that graph on the screen with show(). Note that show() has parentheses after it—it is a function that has no argument, but the parentheses still need to be there.

If we run the program above, it produces a new window on the screen with a graph in it like this:

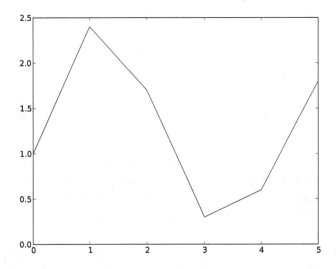

The computer has plotted the values in the list y at unit intervals along the *x*-axis (starting from zero in the standard Python style) and joined them up with straight lines.

While it's better than nothing, this is not a very useful kind of graph for physics purposes. Normally we want to specify both the x- and y-coordinates for the points in the graph. We can do this using a plot statement with two list arguments, thus:

```
from pylab import plot,show
x = [ 0.5, 1.0, 2.0, 4.0, 7.0, 10.0 ]
y = [ 1.0, 2.4, 1.7, 0.3, 0.6, 1.8 ]
plot(x,y)
show()
```

which produces a graph like this:

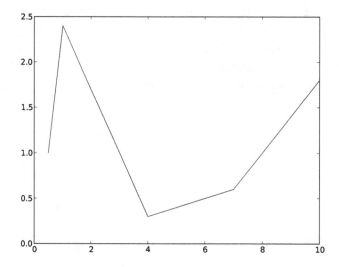

The first of the two lists now specifies the x-coordinates of each point, the second the y-coordinates. The computer plots the points at the given positions and then again joins them with straight lines. The two lists must have the same number of entries, as here. If they do not, you'll get an error message and no graph.

Why do we need two commands, plot and show, to make a graph? In the simple examples above it seems like it would be fine to combine the two into a single command that both creates a graph and shows it on the screen. However, there are more complicated situations where it is useful to have separate commands. In particular, in cases where we want to plot two or more different curves on the same graph, we can do so by using the plot function two or more times, once for each curve. Then we use the show function once to make a single graph with all the curves. We will see examples of this shortly.

Once you have displayed a graph on the screen you can do other things with it. You will notice a number of buttons along the bottom of the window in which the graph appears (not shown in the figures here, but you will see them if you run the programs on your own computer). Among other things, these buttons allow you to zoom in on portions of the graph, move your view around the graph, or save the graph on your computer in various file formats, allowing you to view it again later, print it out on a printer, or insert it as a figure in a word processor document.

Let us apply the plot and show functions to the creation of a slightly more interesting graph, a graph of the function sin x from $x = 0$ to $x = 10$. To do this we first create an array of the x values, then we take the sines of those values to get the y-coordinates of the points:

```
from pylab import plot,show
from numpy import linspace,sin

x = linspace(0,10,100)
y = sin(x)
plot(x,y)
show()
```

Notice how we used the linspace function from numpy (see Section 2.5) to generate the array of x-values, and the sin function from numpy, which is a special version of sine that works with arrays—it just takes the sine of every element in the array. (We could alternatively have used the ordinary sin function from the math package and taken the sines of each element individually using a for loop, or using map(sin,x). As is often the case, there's more than one way to do the job.)

If we run this program we get the classic sine curve graph shown in Fig. 3.1. Notice that we have not really drawn a curve at all here: our plot consists of a finite set of points—a hundred of them in this case—and the computer draws straight lines joining these points. So the end result is not actually curved; it's a set of straight-line segments. To us, however, it looks like a convincing sine wave because our eyes are not sharp enough to see the slight kinks where the segments meet. This is a useful and widely used trick for making curves in computer graphics: choose a set of points spaced close enough together that when joined with straight lines the result looks like a curve even though it really isn't.

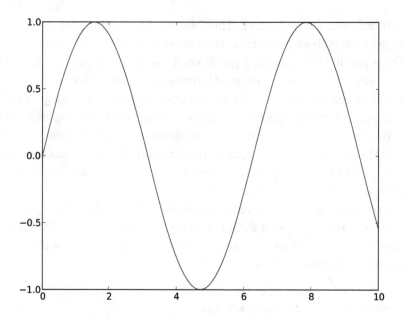

Figure 3.1: Graph of the sine function. A simple graph of the sine function produced by the program given in the text.

As another example of the use of the plot function, suppose we have some experimental data in a computer file values.txt, stored in two columns, like this:

```
0          12121.71
1          12136.44
2          12226.73
3          12221.93
4          12194.13
5          12283.85
6          12331.6
7          12309.25
...
```

We can make a graph of these data as follows:

```
from numpy import loadtxt
from pylab import plot,show

data = loadtxt("values.txt",float)
```

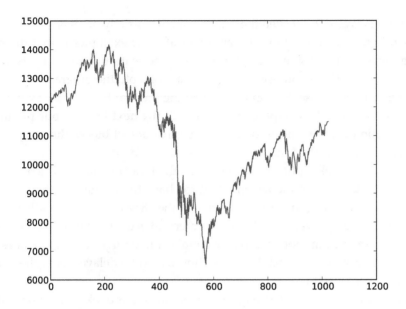

Figure 3.2: Graph of data from a file. This graph was produced by reading two columns of data from a file using the program given in the text.

```
x = data[:,0]
y = data[:,1]
plot(x,y)
show()
```

In this example we have used the `loadtxt` function from `numpy` (see Section 2.4.3) to read the values in the file and put them in an array and then we have used Python's array slicing facilities (Section 2.4.5) to extract the first and second columns of the array and put them in separate arrays x and y for plotting. The end result is a plot as shown in Fig. 3.2.

In fact, it's not necessary in this case to use the separate arrays x and y. We could shorten the program by saying instead

```
data = loadtxt("values.txt",float)
plot(data[:,0],data[:,1])
show()
```

which achieves the same result. (Arguably, however, this program is more difficult to read. As we emphasized in Section 2.7, it is a good idea to make programs easy to read where possible, so you might, in this case, want to use the extra arrays x and y even though they are not strictly necessary.)

An important point to notice about all of these examples is that the program *stops* when it displays the graph. To be precise it stops when it gets to the show function. Once you use show to display a graph, the program will go no further until you close the window containing the graph. Only once you close the window will the computer proceed with the next line of your program. The function show is said to be a *blocking* function—it blocks the progress of the program until the function is done with its job. We have seen one other example of a blocking function previously, the function input, which collects input from the user at the keyboard. It too halts the running of the program until its job is done. (The blocking action of the show function has little impact in the programs above, since the show statement is the last line of the program in each case. But in more complex examples there might be further lines after the show statement and their execution would be delayed until the graph window was closed.)

A useful trick that we will employ frequently in this book is to build the lists of x- and y-coordinates for a graph step by step as we go through a calculation. It will happen often that we do not know all of the x or y values for a graph ahead of time, but work them out one by one as part of some calculation we are doing. In that case, a good way to create a graph of the results is to start with two empty lists for the x- and y-coordinates and add points to them one by one, as we calculate the values. Going back to the sine wave example, for instance, here is an alternative way to make a graph of $\sin x$ that calculates the individual values one by one and adds them to a growing list:

```
from pylab import plot,show
from math import sin
from numpy import linspace

xpoints = []
ypoints = []
for x in linspace(0,10,100):
    xpoints.append(x)
    ypoints.append(sin(x))

plot(xpoints,ypoints)
show()
```

If you run it, this program produces a picture of a sine wave identical to the one in Fig. 3.1 on page 92. Notice how we created the two empty lists and then appended values to the end of each one, one by one, using a for loop. We will use this technique often. (See Section 2.4.1 for a discussion of the append function.)

The graphs we have seen so far are very simple, but there are many extra features we can add to them, some of which are illustrated in Fig. 3.3. For instance, in all the previous graphs the computer chose the range of x and y values for the two axes. Normally the computer makes good choices, but occasionally you might like to make different ones. In our picture of a sine wave, Fig. 3.1, for instance, you might decide that the graph would be clearer if the curve did not butt right up against the top and bottom of the frame— a little more space at top and bottom would be nice. You can override the computer's choice of x- and y-axis limits with the functions xlim and ylim. These functions take two arguments each, for the lower and upper limits of the range of the respective axes. Thus, for instance, we might modify our sine wave program as follows:

```
from pylab import plot,ylim,show
from numpy import linspace,sin

x = linspace(0,10,100)
y = sin(x)
plot(x,y)
ylim(-1.1,1.1)
show()
```

The resulting graph is shown in Fig. 3.3a and, as we can see, it now has a little extra space above and below the curve because the y-axis has been modified to run from -1.1 to $+1.1$. Note that the ylim statement has to come after the plot statement but before the show statement—the plot statement has to create the graph first before you can modify its axes.

It's good practice to label the axes of your graphs, so that you and anyone else knows what they represent. You can add labels to the x- and y-axes with the functions xlabel and ylabel, which take a string argument—a string of letters or numbers in quotation marks. Thus we could again modify our program above by importing xlabel and ylabel from pylab then writing

```
plot(x,y)
ylim(-1.1,1.1)
xlabel("x axis")
ylabel("y = sin x")
show()
```

95

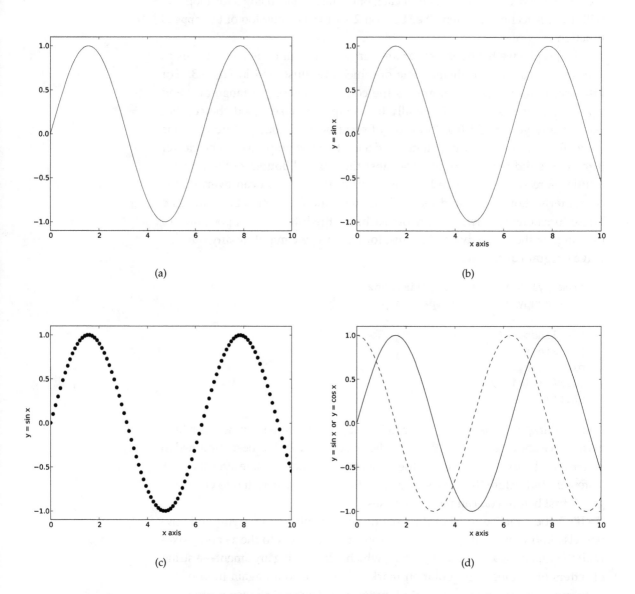

(a)

(b)

(c)

(d)

Figure 3.3: Graph styles. Several different versions of the same sine wave plot. (a) A basic graph, but with a little extra space added above and below the curve to make it clearer; (b) a graph with labeled axes; (c) a graph with the curve replaced by circular dots; (d) sine and cosine curves on the same graph.

which produces the graph shown in Fig. 3.3b.

You can also vary the style in which the computer draws the curve on the graph. To do this a third argument is added to the `plot` function, which takes the form of a (slightly cryptic) string of characters, like this:

```
plot(x,y,"g--")
```

The first letter of the string tells the computer what color to draw the curve with. Allowed letters are r, g, b, c, m, y, k, and w, for red, green, blue, cyan, magenta, yellow, black, and white, respectively. The remainder of the string says what style to use for the line. Here there are many options, but the ones we'll use most often are "-" for a solid line (like the ones we've seen so far), "--" for a dashed line, "o" to mark points with a circle (but not connect them with lines), and "s" to mark points with a square. Thus, for example, this modification:

```
plot(x,y,"ko")
ylim(-1.1,1.1)
xlabel("x axis")
ylabel("y = sin x")
show()
```

tells the computer to plot our sine wave as a set of black circular points. The result is shown in Fig. 3.3c.

Finally, we will often need to plot more than one curve or set of points on the same graph. This can be achieved by using the `plot` function repeatedly. For instance, here is a complete program that plots both the sine function and the cosine function on the same graph, one as a solid curve, the other as a dashed curve:

```
from pylab import plot,ylim,xlabel,ylabel,show
from numpy import linspace,sin,cos

x = linspace(0,10,100)
y1 = sin(x)
y2 = cos(x)
plot(x,y1,"k-")
plot(x,y2,"k--")
ylim(-1.1,1.1)
xlabel("x axis")
ylabel("y = sin x   or   y = cos x")
show()
```

The result is shown in Fig. 3.3d. You could also, for example, use a variant of the same trick to make a plot that had both dots and lines for the same data—just plot the data twice on the same graph, using two plot statements, one with dots and one with lines.

There are many other variations and styles available in the pylab package. You can add legends and annotations to your graphs. You can change the color, size, or typeface used in the labels. You can change the color or style of the axes, or add a background color to the graph. These and many other possibilities are described in the on-line documentation at matplotlib.org.

Exercise 3.1: Plotting experimental data

In the on-line resources[3] you will find a file called sunspots.txt, which contains the observed number of sunspots on the Sun for each month since January 1749. The file contains two columns of numbers, the first being the month and the second being the sunspot number.

a) Write a program that reads in the data and makes a graph of sunspots as a function of time.

b) Modify your program to display only the first 1000 data points on the graph.

c) Modify your program further to calculate and plot the running average of the data, defined by

$$Y_k = \frac{1}{2r+1} \sum_{m=-r}^{r} y_{k+m}$$

where $r = 5$ in this case (and the y_k are the sunspot numbers). Have the program plot both the original data and the running average on the same graph, again over the range covered by the first 1000 data points.

Exercise 3.2: Curve plotting

Although the plot function is designed primarily for plotting standard xy graphs, it can be adapted for other kinds of plotting as well.

a) Make a plot of the so-called *deltoid* curve, which is defined parametrically by the equations

$$x = 2\cos\theta + \cos 2\theta, \qquad y = 2\sin\theta - \sin 2\theta,$$

where $0 \le \theta < 2\pi$. Take a set of values of θ between zero and 2π and calculate x and y for each from the equations above, then plot y as a function of x.

------

[3]The on-line resources for this book can be downloaded in the form of a single "zip" file from http://www.umich.edu/~mejn/cpresources.zip.

b) Taking this approach a step further, one can make a polar plot $r = f(\theta)$ for some function f by calculating r for a range of values of θ and then converting r and θ to Cartesian coordinates using the standard equations $x = r\cos\theta$, $y = r\sin\theta$. Use this method to make a plot of the Galilean spiral $r = \theta^2$ for $0 \leq \theta \leq 10\pi$.

c) Using the same method, make a polar plot of "Fey's function"

$$r = e^{\cos\theta} - 2\cos 4\theta + \sin^5\frac{\theta}{12}$$

in the range $0 \leq \theta \leq 24\pi$.

3.2 SCATTER PLOTS

In an ordinary graph, such as those of the previous section, there is one independent variable, usually placed on the horizontal axis, and one dependent variable, on the vertical axis. The graph is a visual representation of the variation of the dependent variable as a function of the independent one—voltage as a function of time, say, or temperature as a function of position. In other cases, however, we measure or calculate two dependent variables. A classic example in physics is the temperature and brightness—also called the magnitude—of stars. Typically we might measure temperature and magnitude for each star in a given set and we would like some way to visualize how the two quantities are related. A standard approach is to use a *scatter plot*, a graph in which the two quantities are placed along the axes and we make a dot on the plot for each pair of measurements, i.e., for each star in this case.

There are two different ways to make a scatter plot using the pylab package. One of them we have already seen: we can make an ordinary graph, but with dots rather than lines to represent the data points, using a statement of the form:

```
plot(x,y,"ko")
```

This will place a black dot at each point. A slight variant of the same idea is this:

```
plot(x,y,"k.")
```

which will produce smaller dots.

Alternatively, pylab provides the function scatter, which is designed specifically for making scatter plots. It works in a similar fashion to the plot function: you give it two lists or arrays,

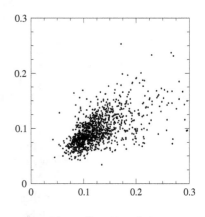

A small scatter plot.

one containing the *x*-coordinates of the points and the other containing the *y*-coordinates, and it creates the corresponding scatter plot:

```
scatter(x,y)
```

You do not have to give a third argument telling `scatter` to plot the data as dots—all scatter plots use dots automatically. As with the `plot` function, `scatter` only creates the scatter plot in the memory of the computer but does not display it on the screen. To display it you need to use the function `show`.

Suppose, for example, that we have the temperatures and magnitudes of a set of stars in a file called `stars.txt` on our computer, like this:

```
4849.4 5.97
5337.8 5.54
4576.1 7.72
4792.4 7.18
5141.7 5.92
6202.5 4.13
...
```

The first column is the temperature and the second is the magnitude. Here's a Python program to make a scatter plot of these data:

File: hrdiagram.py

```
from pylab import scatter,xlabel,ylabel,xlim,ylim,show
from numpy import loadtxt

data = loadtxt("stars.txt",float)
x = data[:,0]
y = data[:,1]

scatter(x,y)
xlabel("Temperature")
ylabel("Magnitude")
xlim(0,13000)
ylim(-5,20)
show()
```

If we run this program it produces the figure shown in Fig. 3.4.

Many of the same variants illustrated in Fig. 3.3 for the `plot` function work for the `scatter` function also. In this program we used `xlabel` and `ylabel` to label the temperature and magnitude axes, and `xlim` and `ylim` to set the ranges of the axes. You can also change the size and style of the dots and many other things. In addition, as with the `plot` function, you can use `scatter`

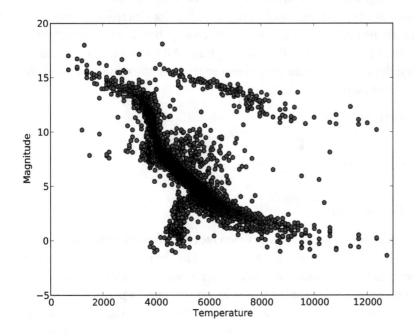

Figure 3.4: The Hertzsprung–Russell diagram. A scatter plot of the magnitude (i.e., brightness) of stars against their approximate surface temperature (which is estimated from the color of the light they emit). Each dot on the plot represents one star out of a catalog of 7860 stars that are close to our solar system.

two or more times in succession to plot two or more sets of data on the same graph, or you can use any combination of `scatter` and `plot` functions to draw scatter data and curves on the same graph. Again, see the on-line manual at `matplotlib.org` for more details.

The scatter plot of the magnitudes and temperatures in Fig. 3.4 reveals an interesting pattern in the data: a substantial majority of the points lie along a rough band running from top left to bottom right of the plot. This is the so-called *main sequence* to which most stars belong. Rarer types of stars, such as red giants and white dwarfs, stand out in the figure as dots that lie well off the main sequence. A scatter plot of stellar magnitude against temperature is called a *Hertzsprung–Russell diagram* after the astronomers who first drew it. The diagram is one of the fundamental tools of stellar astrophysics.

In fact, Fig. 3.4 is, in a sense, upside down, because the Hertzsprung–

Russell diagram is, for historical reasons,[4] normally plotted with both the magnitude and temperature axes *decreasing*, rather than increasing. One of the nice things about `pylab` is that it is easy to change this kind of thing with just a small modification of the Python program. All we need to do in this case is change the `xlim` and `ylim` statements so that the start and end points of each axis are reversed, thus:

```
xlim(13000,0)
ylim(20,-5)
```

Then the figure will be magically turned around.

3.3 DENSITY PLOTS

There are many times in physics when we need to work with two-dimensional grids of data. A condensed matter physicist might measure variations in charge or temperature or atomic deposition on a solid surface; a fluid dynamicist might measure the heights of waves in a ripple tank; a particle physicist might measure the distribution of particles incident on an imaging detector; and so on. Two-dimensional data are harder to visualize on a computer screen than the one-dimensional lists of values that go into an ordinary graph. But one tool that is helpful in many cases is the *density plot*, a two-dimensional plot where color or brightness is used to indicate data values. Figure 3.5 shows an example.

In Python density plots are produced by the function `imshow` from `pylab`. Here's the program that produced Fig. 3.5:

```
from pylab import imshow,show
from numpy import loadtxt

data = loadtxt("circular.txt",float)
imshow(data)
show()
```

The file `circular.txt` contains a simple array of values, like this:

---

[4]The magnitude of a star is defined in such a way that it actually increases as the star gets fainter, so reversing the vertical axis makes sense since it puts the brightest stars at the top. The temperature axis is commonly plotted not directly in terms of temperature but in terms of the so-called *color index*, which is a measure of the color of light a star emits, which is in turn a measure of temperature. Temperature decreases with increasing color index, which is why the standard Hertzsprung–Russell diagram has temperature decreasing along the horizontal axis.

```
0.0050 0.0233 0.0515 0.0795 0.1075 ...
0.0233 0.0516 0.0798 0.1078 0.1358 ...
0.0515 0.0798 0.1080 0.1360 0.1639 ...
0.0795 0.1078 0.1360 0.1640 0.1918 ...
0.1075 0.1358 0.1639 0.1918 0.2195 ...
 ...    ...    ...    ...    ...
```

The program reads the values in the file and puts them in the two-dimensional array data using the loadtxt function, then creates the density plot with the imshow function and displays it with show. The computer automatically adjusts the color-scale so that the picture uses the full range of available shades.

The computer also adds numbered axes along the sides of the figure, which measure the rows and columns of the array, though it is possible to change the calibration of the axes to use different units—we'll see how to do this in a moment. The image produced is a direct picture of the array, laid out in the usual fashion for matrices, row by row, starting at the top and working downwards. Thus the top left corner in Fig. 3.5 represents the value stored in the array element data[0,0], followed to the right by data[0,1], data[0,2], and so on. Immediately below those, the next row is data[1,0], data[1,1], data[1,2], and so on.

Note that the numerical labels on the axes reflect the array indices, with the origin of the figure being at the top left and the vertical axis increasing downwards. While this is natural from the point of view of matrices, it is a little odd for a graph. Most of us are accustomed to

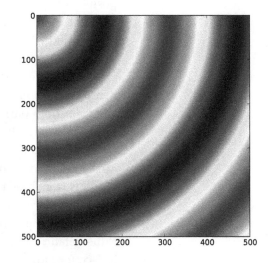

Figure 3.5: A example of a density plot

graphs whose vertical axes increase upwards. What's more, the array elements data[i,j] are written (as is the standard with matrices) with the *row* index first—i.e., the vertical index—and the column, or horizontal, index second. This is the opposite of the convention normally used with graphs where we list the coordinates of a point in x, y order—i.e., horizontal first, then vertical. There's nothing much we can do about the conventions for matrices: they are the ones that mathematicians settled upon centuries ago and it's too late to change them now. But the conflict between those conventions and the conventions used when plotting graphs can be confusing, so take this opportunity to make a mental note.

In fact, Python provides a way to deal with the first problem, of the origin in a density plot being at the top. You can include an additional argument with the imshow function thus:

```
imshow(data,origin="lower")
```

which flips the density plot top-to-bottom, putting the array element data[0,0] in the lower left corner, as is conventional, and changing the labeling of the vertical axis accordingly, so that it increases in the upward direction. The resulting plot is shown in Fig. 3.6a. We will use this trick for most of the density plots in this book. Note, however, that this does not fix our other problem: indices i and j for the element data[i,j] still correspond to vertical and horizontal positions respectively, not the reverse. That is, the index i corresponds to the y-coordinate and the index j corresponds to the x-coordinate. You need to keep this in mind when making density plots—it's easy to get the axes swapped by mistake.

The black-and-white printing in this book doesn't really do justice to the density plot in Fig. 3.6a. The original is in bright colors, ranging through the spectrum from dark blue for the lowest values to red for the highest. If you wish, you can run the program yourself to see the density plot in its full glory—both the program, which is called circular.py, and the data file circular.txt can be found in the on-line resources. Density plots with this particular choice of colors from blue to red (or similar) are sometimes called *heat maps*, because the same color scheme is often used to denote temperature, with blue being the coldest temperature and red being the hottest.[5] The heat map color scheme is the default choice for density maps in Python, but it's not always the best. In fact, for most purposes, a simple gray-scale from black to white is easier to read. Luckily, it's simple to change the color scheme. To change to gray-scale, for instance, you use the function gray, which takes no arguments:

```
from pylab import imshow,gray,show
from numpy import loadtxt

data = loadtxt("circular.txt",float)
imshow(data,origin="lower")
gray()
show()
```

---

[5]It's not completely clear why people use these colors. As every physicist knows, red light has the longest wavelength of the visible colors and corresponds to the coolest objects, while blue has the shortest wavelengths and corresponds to the hottest—the exact opposite of the traditional choices. The hottest stars, for instance, are blue and the coolest are red.

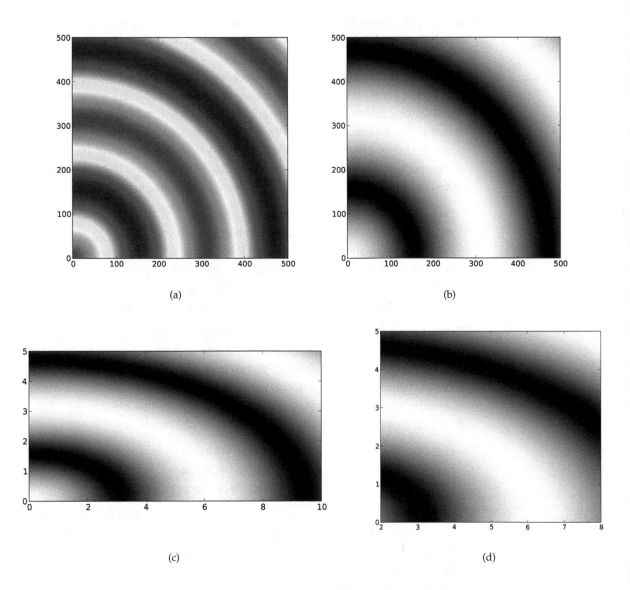

(a)

(b)

(c)

(d)

Figure 3.6: Density plots. Four different versions of the same density plot. (a) A plot using the default "heat map" color scheme, which is colorful on the computer screen but doesn't make much sense with the black-and-white printing in this book. (b) The gray color scheme, which runs from black for the lowest values to white for the highest. (c) The same plot as in panel (b) but with the calibration of the axes changed. Because the range chosen is different for the horizontal and vertical axes, the computer has altered the shape of the figure to keep distances equal in all directions. (d) The same plot as in (c) but with the horizontal range reduced so that only the middle portion of the data is shown.

Figure 3.6b shows the result. Even in black-and-white it looks somewhat different from the heat-map version in panel (a), and on the screen it looks entirely different. Try it if you like.[6]

All of the density plots in this book use the gray scale (except Figs. 3.5 and 3.6a of course). It may not be flashy, but it's informative, easy to read, and suitable for printing on monochrome printers or for publications (like many scientific books and journals) that are in black-and-white only. However, `pylab` provides many other color schemes, which you may find useful occasionally. A complete list, with illustrations, is given in the on-line documentation at `matplotlib.org`, but here are a few that might find use in physics:

`jet`	The default heat-map color scheme
`gray`	Gray-scale running from black to white
`hot`	An alternative heat map that goes black-red-yellow-white
`spectral`	A spectrum with 7 clearly defined colors, plus black and white
`bone`	An alternative gray-scale with a hint of blue
`hsv`	A rainbow scheme that starts and ends with red

Each of these has a corresponding function, `jet()`, `spectral()`, and so forth, that selects the relevant color scheme for use in future density plots. Many more color schemes are given in `pylab` and one can also define one's own schemes, although the definitions involve some slightly tricky programming. Example code is given in Appendix E and in the on-line resources to define three additional schemes that can be useful for physics:[7]

`redblue`	Runs from red to blue via black
`redwhiteblue`	Runs from red to blue via white
`inversegray`	Runs from white to black, the opposite of gray

There is also a function `colorbar()` in the `pylab` package that instructs Python to add a bar to the side of your figure showing the range of colors used in the plot along with a numerical scale indicating which values correspond to which colors, something that can be helpful when you want to make a more precise quantitative reading of a density plot.

---

[6]The function `gray` works slightly differently from other functions we have seen that modify plots, such as `xlabel` or `ylim`. Those functions modified only the current plot, whereas `gray` (and the other color scheme functions in `pylab`) changes the color scheme for all subsequent density plots. If you write a program that makes more than one plot, you only need to call `gray` once.

[7]To use these color schemes copy the file `colormaps.py` from the on-line resources into the folder containing your program and then in your program say, for example, "`from colormaps import redblue`". Then the statement "`redblue()`" will switch to the `redblue` color map.

As with graphs and scatter plots, you can modify the appearance of density plots in various ways. The functions `xlabel` and `ylabel` work as before, adding labels to the two axes. You can also change the scale marked on the axes. By default, the scale corresponds to the elements of the array holding the data, but you might want to calibrate your plot with a different scale. You can do this by adding an extra parameter to `imshow`, like this:

```
imshow(data,origin="lower",extent=[0,10,0,5])
```

which results in a modified plot as shown in Fig. 3.6c. The argument consists of "extent=" followed by a list of four values, which give, in order, the beginning and end of the horizontal scale and the beginning and end of the vertical scale. The computer will use these numbers to mark the axes, but the actual content displayed in the body of the density plot remains unchanged—the `extent` argument affects only how the plot is labeled. This trick can be very useful if you want to calibrate your plot in "real" units. If the plot is a picture of the surface of the Earth, for instance, you might want axes marked in units of latitude and longitude; if it's a picture of a surface at the atomic scale you might want axes marked in nanometers.

Note also that in Fig. 3.6c the computer has changed the shape of the plot— its *aspect ratio*—to accommodate the fact that the horizontal and vertical axes have different ranges. The `imshow` function attempts to make unit distances equal along the horizontal and vertical directions where possible. Sometimes, however, this is not what we want, in which case we can tell the computer to use a different aspect ratio. For instance, if we wanted the present figure to remain square we would say:

```
imshow(data,origin="lower",extent=[0,10,0,5],aspect=2.0)
```

This tells the computer to use unit distances twice as large along the vertical axis as along the horizontal one, which will make the plot square once more.

Note that, as here, we are free to use any or all of the `origin`, `extent`, and aspect arguments together in the same function. We don't have to use them all if we don't want to—any selection is allowed—and they can come in any order.

We can also limit our density plot to just a portion of the data, using the functions `xlim` and `ylim`, just as with graphs and scatter plots. These functions work with the scales specified by the `extent` argument, if there is one, or with the row and column indices otherwise. So, for instance, we could say `xlim(2,8)` to reduce the density plot of Fig. 3.6b to just the middle portion of the horizontal scale, from 2 to 8. Figure 3.6d shows the result. Note that, unlike the `extent` argument, `xlim` and `ylim` do change which data are displayed in the body of the density plot—the `extent` argument makes purely cosmetic changes to the labeling of the axes, but `xlim` and `ylim` actually change which data appear.

Finally, you can use the functions `plot` and `scatter` to superimpose graphs or scatter plots of data on the same axes as a density plot. You can use any combination of `imshow`, `plot`, and `scatter` in sequence, followed by `show`, to create a single graph with density data, curves, or scatter data, all on the same set of axes.

EXAMPLE 3.1: WAVE INTERFERENCE

Suppose we drop a pebble in a pond and waves radiate out from the spot where it fell. We could create a simple representation of the physics with a sine wave, spreading out in a uniform circle, to represent the height of the waves at some later time. If the center of the circle is at x_1, y_1 then the distance r_1 to the center from a point x, y is

$$r_1 = \sqrt{(x - x_1)^2 + (y - y_1)^2} \tag{3.1}$$

and the sine wave for the height is

$$\xi_1(x, y) = \xi_0 \sin k r_1, \tag{3.2}$$

where ξ_0 is the amplitude of the waves and k is the wavevector, related to the wavelength λ by $k = 2\pi/\lambda$.

Now suppose we drop another pebble in the pond, creating another circular set of waves with the same wavelength and amplitude but centered on a different point x_2, y_2:

$$\xi_2(x, y) = \xi_0 \sin k r_2 \quad \text{with} \quad r_2 = \sqrt{(x - x_2)^2 + (y - y_2)^2}. \tag{3.3}$$

Then, assuming the waves add linearly (which is a reasonable assumption for water waves, provided they are not too big), the total height of the surface at a point x, y is

$$\xi(x, y) = \xi_0 \sin k r_1 + \xi_0 \sin k r_2. \tag{3.4}$$

Suppose the wavelength of the waves is $\lambda = 5\,\text{cm}$, the amplitude is $1\,\text{cm}$, and the centers of the circles are $20\,\text{cm}$ apart. Here is a program to make an image of the height over a $1\,\text{m}$ square region of the pond. To make the image we create an array of values representing the height ζ at a grid of points and then use that array to make a density plot. In this example we use a grid of 500×500 points to cover the 1 m square, which means the grid points have a separation of $100/500 = 0.2$ cm.

```
from math import sqrt,sin,pi
from numpy import empty
from pylab import imshow,gray,show

wavelength = 5.0
k = 2*pi/wavelength
xi0 = 1.0
separation = 20.0      # Separation of centers in cm
side = 100.0           # Side of the square in cm
points = 500           # Number of grid points along each side
spacing = side/points  # Spacing of points in cm

# Calculate the positions of the centers of the circles
x1 = side/2 + separation/2
y1 = side/2
x2 = side/2 - separation/2
y2 = side/2

# Make an array to store the heights
xi = empty([points,points],float)

# Calculate the values in the array
for i in range(points):
    y = spacing*i
    for j in range(points):
        x = spacing*j
        r1 = sqrt((x-x1)**2+(y-y1)**2)
        r2 = sqrt((x-x2)**2+(y-y2)**2)
        xi[i,j] = xi0*sin(k*r1) + xi0*sin(k*r2)

# Make the plot
imshow(xi,origin="lower",extent=[0,side,0,side])
gray()
show()
```

File: ripples.py

This is the longest and most involved program we have seen so far, so it may be worth taking a moment to make sure you understand how it works. Note in particular how the height is calculated and stored in the array xi. The

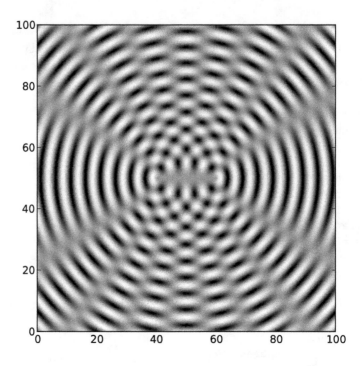

Figure 3.7: Interference pattern. This plot, produced by the program given in the text, shows the superposition of two circular sets of sine waves, creating an interference pattern with fringes that appear as the gray bars radiating out from the center of the picture.

variables i and j go through the rows and columns of the array respectively, and from these we calculate the values of the coordinates x and y. Since, as discussed earlier, the rows correspond to the vertical axis and the columns to the horizontal axis, the value of x is calculated from j and the value of y is calculated from i. Other than this subtlety, the program is a fairly straightforward translation of Eqs. (3.1–3.4).[8]

If we run the program above, it produces the picture shown in Fig. 3.7. The picture shows clearly the interference of the two sets of waves. The interference

---

[8]One other small detail is worth mentioning. We called the variable for the wavelength "wavelength". You might be tempted to call it "lambda" but if you did you would get an error message and the program would not run. The word "lambda" has a special meaning in the Python language and cannot be used as a variable name, just as words like "for" and "if" cannot be used as variable names. (See footnote 5 on page 13.) The names of other Greek letters—alpha, beta, gamma, and so on—are allowed as variable names.

fringes are visible as the gray bands radiating from the center.

Exercise 3.3: There is a file in the on-line resources called stm.txt, which contains a grid of values from scanning tunneling microscope measurements of the (111) surface of silicon. A scanning tunneling microscope (STM) is a device that measures the shape of a surface at the atomic level by tracking a sharp tip over the surface and measuring quantum tunneling current as a function of position. The end result is a grid of values that represent the height of the surface and the file stm.txt contains just such a grid of values. Write a program that reads the data contained in the file and makes a density plot of the values. Use the various options and variants you have learned about to make a picture that shows the structure of the silicon surface clearly.

3.4 3D GRAPHICS

One of the flashiest applications of computers today is the creation of 3D graphics and computer animation. In any given week millions of people flock to cinemas worldwide to watch the latest computer-animated movie from the big animation studios. 3D graphics and animation find a more humble, but very useful, application in computational physics as a tool for visualizing the behavior of physical systems. Python provides some excellent tools for this purpose, which we'll use extensively in this book.

There are a number of different packages available for graphics and animation in Python, but we will focus on the package visual, which is specifically designed with physicists in mind. This package provides a way to create simple pictures and animations with a minimum of effort, but also has enough power to handle complex situations when needed.

The visual package works by creating specified objects on the screen, such as spheres, cylinders, cones, and so forth, and then, if necessary, changing their position, orientation, or shape to make them move around. Here's a short first program using the package:

```
from visual import sphere
sphere()
```

When we run this program a window appears on the screen with a large sphere in it, like this:

The window of course is two-dimensional, but the computer stores the shape and position of the sphere in three dimensions and automatically does a perspective rendering of the sphere with a 3D look to it that aids the eye in understanding the scene.

You can choose the size and position of the sphere like this

```
sphere(radius=0.5,pos=[1.0,-0.2,0.0])
```

The radius is specified as a single number. The units are arbitrary and the computer will zoom in or out as necessary to make the sphere visible. So you can set the radius to 0.5 as here, or to 10^{-15} if you're drawing a picture of a proton. Either will work fine.

The position of the sphere is a three-dimensional vector, which you give as a list or array of three real numbers x, y, z (we used a list in this case). The x- and y-axes run to the right and upwards in the window, as normal, and the z-axis runs directly out of the screen towards you. You can also specify the position as a list or array of just two numbers, x and y, in which case Python will assume the z-coordinate to be zero. This can be useful for drawing pictures of two-dimensional systems, which have no z-coordinate.

You can also change the color of the sphere thus:

```
from visual import sphere,color
sphere(color=color.green)
```

Note how we have imported the object called `color` from the visual package, then individual colors are called things like `color.green` and `color.red`. The available colors are the same as those for drawing graphs with pylab: red, green, blue, cyan, magenta, yellow, black, and white.[9] The `color` argument can be used at the same time as the `radius` and `position` arguments, so one can control all features of the sphere at the same time.

We can also create several spheres, all in the same window on the screen, by using the `sphere` function repeatedly, putting different spheres in different places to build up an entire scene made of spheres. The following exercise gives an example.

EXAMPLE 3.2: PICTURING AN ATOMIC LATTICE

Suppose we have a solid composed of atoms arranged on a simple cubic lattice. We can visualize the arrangement of the atoms using the `visual` package by creating a picture with many spheres at positions (i, j, k) with $i, j, k = -L \ldots L$, thus:

```
from visual import sphere
L = 5
R = 0.3
for i in range(-L,L+1):
    for j in range(-L,L+1):
        for k in range(-L,L+1):
            sphere(pos=[i,j,k],radius=R)
```
File: `lattice.py`

Notice how this program has three nested for loops that run through all combinations of the values of i, j, and k. Run this program and it produces the picture shown in Fig. 3.8. Download the program and try it if you like.

After running the program, you can rotate your view of the lattice to look at it from different angles by moving the mouse while holding down either the right mouse button or the Ctrl key on the keyboard (the Command key on a Mac). You can also hold down both mouse buttons (if you have two), or the Alt key (the Option key on a Mac) and move the mouse in order to zoom in and out of the picture.

---

[9] All visible colors can be represented as mixtures of the primary colors red, green, and blue, and this is how they are stored inside the computer. A "color" in the `visual` package is actually just a list of three floating-point numbers giving the intensities of red, green, and blue (in that order) on a scale of 0 to 1 each. Thus red is [1.0, 0.0, 0.0], yellow is [1.0, 1.0, 0.0], and white is [1.0, 1.0, 1.0]. You can create your own colors if you want by writing things like midgray = [0.5, 0.5, 0.5]. Then you can use "midgray" just like any other color. (You would just say midgray, not `color.midgray`, because the color you defined is an ordinary variable, not a part of the `color` object in `visual`.)

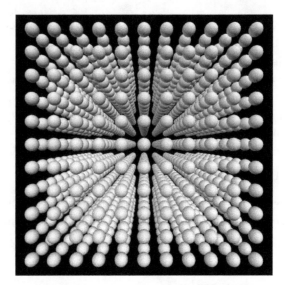

Figure 3.8: Visualization of atoms in a simple cubic lattice. A perspective rendering of atoms in a simple cubic lattice, generated using the `visual` package and the program `lattice.py` given in the text.

Exercise 3.4: Using the program from Example 3.2 above as a starting point, or starting from scratch if you prefer, do the following:

a) A sodium chloride crystal has sodium and chlorine atoms arranged on a cubic lattice but the atoms alternate between sodium and chlorine, so that each sodium is surrounded by six chlorines and each chlorine is surrounded by six sodiums. Create a visualization of the sodium chloride lattice using two different colors to represent the two types of atoms.

b) The face-centered cubic (fcc) lattice, which is the most common lattice in naturally occurring crystals, consists of a cubic lattice with atoms positioned not only at the corners of each cube but also at the center of each face. Create a visualization of an fcc lattice with a single species of atom (such as occurs in metallic nickel, for instance).

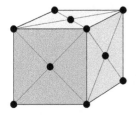

Atoms in the fcc lattice lie at the corners and center of each face of a cubic cell.

It is possible to change the properties of a sphere after it is first created, including its position, size, and color. When we do this the sphere will actually move or change on the screen. In order to refer to a particular sphere on the screen we must use a slightly different form of the sphere function to create it, like this:

```
s = sphere()
```

This form, in addition to drawing a sphere on the computer screen, creates a variable s in a manner similar to the way functions like zeros or empty create arrays (see Section 2.4.2). The new variable s is a variable of type "sphere", in the same way that other variables are of type int or float. This is a special variable type used only in the visual package to store the properties of spheres. Each sphere variable corresponds to a sphere on the screen and when we change the properties stored in the sphere variable the on-screen sphere changes accordingly. Thus, for example, we can say

```
s.radius = 0.5
```

and the radius of the corresponding sphere on the screen will change to 0.5, right before our eyes. Or we can say

```
s.color = color.blue
```

and the color will change. You can also change the position of a sphere in this way, in which case the sphere will move on the screen. We will use this trick in Section 3.5 to create animations of physical systems.

You can use variables of the sphere type in similar ways to other types of variable. A useful trick, for instance, is to create an array of spheres thus:

```
from visual import sphere
from numpy import empty
s = empty(10,sphere)
```

This creates an array, initially empty, of ten sphere-type variables that you can then fill with actual spheres thus:

```
for n in range(10):
    s[n] = sphere()
```

As each sphere is created, a corresponding sphere will appear on the screen.

This technique can be useful if you are creating a visualization or animation with many spheres and you want to be able to change the properties of any of them at will. Exercise 3.5 involves exactly such a situation, and the trick above would be a good one to use in solving that exercise.

Spheres are by no means the only shape one can draw. There is a large selection of other elements provided by the visual package, including boxes, cones, cylinders, pyramids, and arrows. Here are the functions that create each of these objects:

```
from visual import box,cone,cylinder,pyramid,arrow

box(pos=[x,y,z], axis=[a,b,c], \
    length=L, height=H, width=W, up=[q,r,s])
cone(pos=[x,y,z], axis=[a,b,c], radius=R)
cylinder(pos=[x,y,z], axis=[a,b,c], radius=R)
pyramid(pos=[x,y,z], size=[z,b,c])
arrow(pos=[x,y,z], axis=[a,b,c], \
      headwidth=H, headlength=L, shaftwidth=W)
```

For a detailed explanation of the meaning of all the parameters, take a look at the on-line documentation at www.vpython.org. In addition to the parameters above, standard ones like color can also be used to give the objects a different appearance. And each element has a corresponding variable type—box, cone, cylinder, and so forth—that is used for storing and changing the properties of elements after they are created.

Another useful feature of the visual package is the ability to change various properties of the screen window in which your objects appear. You can, for example, change the window's size and position on the screen, you can change the background color, and you can change the direction that the "camera" is looking in. All of these things you do with the function display. Here is an example:

```
from visual import display
display(x=100,y=100,width=600,height=600, \
        center=[5,0,0],forward=[0,0,-1], \
        background=color.blue,foreground=color.yellow)
```

This will produce a window 600 × 600 in size, where size is measured in pixels (the small dots that make up the picture on a computer screen). The window will be 100 pixels in from the left and top of the screen. The argument "center=[5,0,0]" sets the point in 3D space that will be in the center of the window, and "forward=[0,0,-1]" chooses the direction in which we are looking. Between the two of them these two arguments determine the position and direction of our view of the scene. The background color of the window will be blue in this case and objects appearing in the window—the "foreground"—will be yellow by default, although you can specify other colors for individual objects in the manner described above for spheres.

(Notice also how we used the backslash character "\" in the code above to indicate to the computer that a single logical line of code has been spread over more than one line in the text of the program. We discussed this use of the backslash previously in Section 2.7.)

The arguments for the display function can be in any order and you do not have to include all of them. You need include only those you want. The ones you don't include have sensible default values. For example, the default background color is black and the default foreground color is white, so if you don't specify any colors you get white objects on a black background.

As with the sphere function you can assign a variable to keep track of the display window by writing, for example,

```
d = display(background=color.blue)
```

or even just

```
d = display()
```

This allows you to change display parameters later in your program. For instance, you can change the background color to black at any time by writing "d.background = color.black". Some parameters, however, cannot be changed later, notably the size and position of the window, which are fixed when the window is created (although you can change the size and position manually by dragging the window around the screen with your mouse).

There are many other features of the visual package that are not listed here. For more details take a look at www.vpython.org.

3.5 ANIMATION

As we have seen, the visual package allows you to change the properties of an on-screen object, such as its size, color, orientation, or position. If you change the position of an object repeatedly and rapidly, you can make the object appear to be moving and you have an animation. We will use such animations in this book to help us understand the behavior of physical systems.

For example, to create a sphere and then change its position you could do the following:

```
from visual import sphere
s = sphere(pos=[0,0,0])
s.pos = [1,4,3]
```

This will create a sphere at the origin, then move it to the new position $(1, 4, 3)$.

This is not not a very useful program, however. The computer is so fast that you probably wouldn't even see the sphere in its first position at the origin before it gets moved. To slow down movements to a point where they are visible, visual provides a function called rate. Saying rate(x) tells the computer to wait until $1/x$ of a second has passed since the last time you called rate. Thus if you call rate(30) immediately before each change you make on the screen, you will ensure that changes never get made more than 30 times a second, which is very useful for making smooth animations.

EXAMPLE 3.3: A MOVING SPHERE

Here is a program to move a sphere around on the screen:

File: revolve.py

```
from visual import sphere,rate
from math import cos,sin,pi
from numpy import arange

s = sphere(pos=[1,0,0],radius=0.1)
for theta in arange(0,10*pi,0.1):
    rate(30)
    x = cos(theta)
    y = sin(theta)
    s.pos = [x,y,0]
```

Here the value of the angle variable theta increases by 0.1 radians every 30th of a second, the rate function ensuring that we go around the for loop 30 times each second. The angle is converted into Cartesian coordinates and used to update the position of the sphere. The net result, if we run the program is that a sphere appears on the screen and moves around in a circle. Download the program and try it if you like. This simple animation could be the basis, for instance, for an animation of the simultaneous motions of the planets of the solar system. Exercise 3.5 below invites you to create exactly such an animation.

Exercise 3.5: Visualization of the solar system

The innermost six planets of our solar system revolve around the Sun in roughly circular orbits that all lie approximately in the same (ecliptic) plane. Here are some basic parameters:

Object	Radius of object (km)	Radius of orbit (millions of km)	Period of orbit (days)
Mercury	2440	57.9	88.0
Venus	6052	108.2	224.7
Earth	6371	149.6	365.3
Mars	3386	227.9	687.0
Jupiter	69173	778.5	4331.6
Saturn	57316	1433.4	10759.2
Sun	695500	–	–

Using the facilities provided by the visual package, create an animation of the solar system that shows the following:

a) The Sun and planets as spheres in their appropriate positions and with sizes proportional to their actual sizes. Because the radii of the planets are tiny compared to the distances between them, represent the planets by spheres with radii c_1 times larger than their correct proportionate values, so that you can see them clearly. Find a good value for c_1 that makes the planets visible. You'll also need to find a good radius for the Sun. Choose any value that gives a clear visualization. (It doesn't work to scale the radius of the Sun by the same factor you use for the planets, because it'll come out looking much too large. So just use whatever works.) For added realism, you may also want to make your spheres different colors. For instance, Earth could be blue and the Sun could be yellow.

b) The motion of the planets as they move around the Sun (by making the spheres of the planets move). In the interests of alleviating boredom, construct your program so that time in your animation runs a factor of c_2 faster than actual time. Find a good value of c_2 that makes the motion of the orbits easily visible but not unreasonably fast. Make use of the rate function to make your animation run smoothly.

Hint: You may find it useful to store the sphere variables representing the planets in an array of the kind described on page 115.

Here's one more trick that can prove useful. As mentioned above, you can make your objects small or large and the computer will automatically zoom in or out so that they remain visible. And if you make an animation in which your objects move around the screen the computer will zoom out when objects move out of view, or zoom in as objects recede into the distance. While this is useful in many cases, it can be annoying in others. The display function provides a parameter for turning the automatic zooming off if it becomes distracting, thus:

```
display(autoscale=False)
```

More commonly, one calls the `display` function at the beginning of the program and then turns off the zooming separately later, thus:

```
d = display()
d.autoscale = False
```

One can also turn it back on with

```
d.autoscale = True
```

A common approach is to place all the objects of your animation in their initial positions on the screen first, allow the computer to zoom in or out appropriately, so that they are all visible, then turn zooming off with "d.autoscale = False" before beginning the animation proper, so that the view remains fixed as objects move around.

FURTHER EXERCISES

3.6 Deterministic chaos and the Feigenbaum plot: One of the most famous examples of the phenomenon of chaos is the *logistic map*, defined by the equation

$$x' = rx(1 - x).$$

For a given value of the constant r you take a value of x—say $x = \frac{1}{2}$—and you feed it into the right-hand side of this equation, which gives you a value of x'. Then you take that value and feed it back in on the right-hand side again, which gives you another value, and so forth. This is an *iterative map*. You keep doing the same operation over and over on your value of x, and one of three things happens:

1. The value settles down to a fixed number and stays there. This is called a *fixed point*. For instance, $x = 0$ is always a fixed point of the logistic map. (You put $x = 0$ on the right-hand side and you get $x' = 0$ on the left.)

2. It doesn't settle down to a single value, but it settles down into a periodic pattern, rotating around a set of values, such as say four values, repeating them in sequence over and over. This is called a *limit cycle*.

3. It goes crazy. It generates a seemingly random sequence of numbers that appear to have no rhyme or reason to them at all. This is *deterministic chaos*. "Chaos" because it really does look chaotic, and "deterministic" because even though the

values look random, they're not. They're clearly entirely predictable, because they are given to you by one simple equation. The behavior is *determined*, although it may not look like it.

Write a program that calculates and displays the behavior of the logistic map. Here's what you need to do. For a given value of r, start with $x = \frac{1}{2}$, and iterate the logistic map equation a thousand times. That will give it a chance to settle down to a fixed point or limit cycle if it's going to. Then run for another thousand iterations and plot the points (r, x) on a graph where the horizontal axis is r and the vertical axis is x. You can either use the `plot` function with the options `"ko"` or `"k."` to draw a graph with dots, one for each point, or you can use the `scatter` function to draw a scatter plot (which always uses dots). Repeat the whole calculation for values of r from 1 to 4 in steps of 0.01, plotting the dots for all values of r on the same figure and then finally using the function `show` once to display the complete figure.

Your program should generate a distinctive plot that looks like a tree bent over onto its side. This famous picture is called the *Feigenbaum plot*, after its discoverer Mitchell Feigenbaum, or sometimes the *figtree plot*, a play on the fact that it looks like a tree and Feigenbaum means "figtree" in German.[10]

Give answers to the following questions:

a) For a given value of r what would a fixed point look like on the Feigenbaum plot? How about a limit cycle? And what would chaos look like?

b) Based on your plot, at what value of r does the system move from orderly behavior (fixed points or limit cycles) to chaotic behavior? This point is sometimes called the "edge of chaos."

The logistic map is a very simple mathematical system, but deterministic chaos is seen in many more complex physical systems also, including especially fluid dynamics and the weather. Because of its apparently random nature, the behavior of chaotic systems is difficult to predict and strongly affected by small perturbations in outside conditions. You've probably heard of the classic exemplar of chaos in weather systems, the *butterfly effect*, which was popularized by physicist Edward Lorenz in 1972 when he gave a lecture to the American Association for the Advancement of Science entitled, "Does the flap of a butterfly's wings in Brazil set off a tornado in Texas?"[11]

---

[10]There is another approach for computing the Feigenbaum plot, which is neater and faster, making use of Python's ability to perform arithmetic with entire arrays. You could create an array r with one element containing each distinct value of r you want to investigate: [1.0, 1.01, 1.02, ...]. Then create another array x of the same size to hold the corresponding values of x, which should all be initially set to 0.5. Then an iteration of the logistic map can be performed for all values of r at once with a statement of the form x = r*x*(1-x). Because of the speed with which Python can perform calculations on arrays, this method should be significantly faster than the more basic method above.

[11]Although arguably the first person to suggest the butterfly effect was not a physicist at all, but the science fiction writer Ray Bradbury in his famous 1952 short story *A Sound of Thunder*, in which a time traveler's careless destruction of a butterfly during a tourist trip to the Jurassic era

3.7 The Mandelbrot set: The Mandelbrot set, named after its discoverer, the French mathematician Benoît Mandelbrot, is a *fractal*, an infinitely ramified mathematical object that contains structure within structure within structure, as deep as we care to look. The definition of the Mandelbrot set is in terms of complex numbers as follows.

Consider the equation

$$z' = z^2 + c,$$

where z is a complex number and c is a complex constant. For any given value of c this equation turns an input number z into an output number z'. The definition of the Mandelbrot set involves the repeated iteration of this equation: we take an initial starting value of z and feed it into the equation to get a new value z'. Then we take that value and feed it in again to get another value, and so forth. The Mandelbrot set is the set of points in the complex plane that satisfies the following definition:

> *For a given complex value of c, start with $z = 0$ and iterate repeatedly. If the magnitude $|z|$ of the resulting value is ever greater than 2, then the point in the complex plane at position c is not in the Mandelbrot set, otherwise it is in the set.*

In order to use this definition one would, in principle, have to iterate infinitely many times to prove that a point is in the Mandelbrot set, since a point is in the set only if the iteration never passes $|z| = 2$ ever. In practice, however, one usually just performs some large number of iterations, say 100, and if $|z|$ hasn't exceeded 2 by that point then we call that good enough.

Write a program to make an image of the Mandelbrot set by performing the iteration for all values of $c = x + iy$ on an $N \times N$ grid spanning the region where $-2 \le x \le 2$ and $-2 \le y \le 2$. Make a density plot in which grid points inside the Mandelbrot set are colored black and those outside are colored white. The Mandelbrot set has a very distinctive shape that looks something like a beetle with a long snout—you'll know it when you see it.

Hint: You will probably find it useful to start off with quite a coarse grid, i.e., with a small value of N—perhaps $N = 100$—so that your program runs quickly while you are testing it. Once you are sure it is working correctly, increase the value of N to produce a final high-quality image of the shape of the set.

If you are feeling enthusiastic, here is another variant of the same exercise that can produce amazing looking pictures. Instead of coloring points just black or white, color points according to the number of iterations of the equation before $|z|$ becomes greater than 2 (or the maximum number of iterations if $|z|$ never becomes greater than 2). If you use one of the more colorful color schemes Python provides for density plots, such as the "hot" or "jet" schemes, you can make some spectacular images this way. Another interesting variant is to color according to the logarithm of the number of iterations, which helps reveal some of the finer structure outside the set.

---

changes the course of history.

3.8 Least-squares fitting and the photoelectric effect: It's a common situation in physics that an experiment produces data that lies roughly on a straight line, like the dots in this figure:

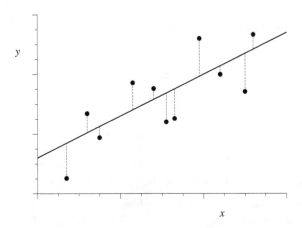

The solid line here represents the underlying straight-line form, which we usually don't know, and the points representing the measured data lie roughly along the line but don't fall exactly on it, typically because of measurement error.

The straight line can be represented in the familiar form $y = mx + c$ and a frequent question is what the appropriate values of the slope m and intercept c are that correspond to the measured data. Since the data don't fall perfectly on a straight line, there is no perfect answer to such a question, but we can find the straight line that gives the best compromise fit to the data. The standard technique for doing this is the *method of least squares*.

Suppose we make some guess about the parameters m and c for the straight line. We then calculate the vertical distances between the data points and that line, as represented by the short vertical lines in the figure, then we calculate the sum of the squares of those distances, which we denote χ^2. If we have N data points with coordinates (x_i, y_i), then χ^2 is given by

$$\chi^2 = \sum_{i=1}^{N}(mx_i + c - y_i)^2.$$

The least-squares fit of the straight line to the data is the straight line that minimizes this total squared distance from data to line. We find the minimum by differentiating with respect to both m and c and setting the derivatives to zero, which gives

$$m\sum_{i=1}^{N}x_i^2 + c\sum_{i=1}^{N}x_i - \sum_{i=1}^{N}x_i y_i = 0,$$

$$m\sum_{i=1}^{N}x_i + cN - \sum_{i=1}^{N}y_i = 0.$$

123

For convenience, let us define the following quantities:

$$E_x = \frac{1}{N}\sum_{i=1}^{N} x_i, \qquad E_y = \frac{1}{N}\sum_{i=1}^{N} y_i, \qquad E_{xx} = \frac{1}{N}\sum_{i=1}^{N} x_i^2, \qquad E_{xy} = \frac{1}{N}\sum_{i=1}^{N} x_i y_i,$$

in terms of which our equations can be written

$$mE_{xx} + cE_x = E_{xy},$$
$$mE_x + c = E_y.$$

Solving these equations simultaneously for m and c now gives

$$m = \frac{E_{xy} - E_x E_y}{E_{xx} - E_x^2}, \qquad c = \frac{E_{xx} E_y - E_x E_{xy}}{E_{xx} - E_x^2}.$$

These are the equations for the least-squares fit of a straight line to N data points. They tell you the values of m and c for the line that best fits the given data.

a) In the on-line resources you will find a file called millikan.txt. The file contains two columns of numbers, giving the x and y coordinates of a set of data points. Write a program to read these data points and make a graph with one dot or circle for each point.

b) Add code to your program, before the part that makes the graph, to calculate the quantities E_x, E_y, E_{xx}, and E_{xy} defined above, and from them calculate and print out the slope m and intercept c of the best-fit line.

c) Now write code that goes through each of the data points in turn and evaluates the quantity $mx_i + c$ using the values of m and c that you calculated. Store these values in a new array or list, and then graph this new array, as a solid line, on the same plot as the original data. You should end up with a plot of the data points plus a straight line that runs through them.

d) The data in the file millikan.txt are taken from a historic experiment by Robert Millikan that measured the *photoelectric effect*. When light of an appropriate wavelength is shone on the surface of a metal, the photons in the light can strike conduction electrons in the metal and, sometimes, eject them from the surface into the free space above. The energy of an ejected electron is equal to the energy of the photon that struck it minus a small amount ϕ called the *work function* of the surface, which represents the energy needed to remove an electron from the surface. The energy of a photon is $h\nu$, where h is Planck's constant and ν is the frequency of the light, and we can measure the energy of an ejected electron by measuring the voltage V that is just sufficient to stop the electron moving. Then the voltage, frequency, and work function are related by the equation

$$V = \frac{h}{e}\nu - \phi,$$

where e is the charge on the electron. This equation was first given by Albert Einstein in 1905.

The data in the file `millikan.txt` represent frequencies ν in hertz (first column) and voltages V in volts (second column) from photoelectric measurements of this kind. Using the equation above and the program you wrote, and given that the charge on the electron is 1.602×10^{-19} C, calculate from Millikan's experimental data a value for Planck's constant. Compare your value with the accepted value of the constant, which you can find in books or on-line. You should get a result within a couple of percent of the accepted value.

This calculation is essentially the same as the one that Millikan himself used to determine of the value of Planck's constant, although, lacking a computer, he fitted his straight line to the data by eye. In part for this work, Millikan was awarded the Nobel prize in physics in 1923.

CHAPTER 4

ACCURACY AND SPEED

W E HAVE now seen the basic elements of programming in Python: input and output, variables and arithmetic, loops and if statements. With these we can perform a wide variety of calculations. We have also seen how to visualize our results using various types of computer graphics. There are many additional features of the Python language that we haven't covered. In later chapters of the book, for example, we will introduce a number of specialized features, such as facilities for doing linear algebra and Fourier transforms. But for now we have the main components in place to start doing physics.

There is, however, one fundamental issue that we have not touched upon. Computers have limitations. They cannot store real numbers with an infinite number of decimal places. There is a limit to the largest and smallest numbers they can store. They can perform calculations quickly, but not infinitely quickly. In many cases these issues need not bother us—the computer is fast enough and accurate enough for many of the calculations we do in physics. However, there are also situations in which the computer's limitations will affect us significantly, so it will be crucial that we understand those limitations, as well as methods for mitigating them or working around them when necessary.

4.1 VARIABLES AND RANGES

We have seen examples of the use of variables in computer programs, including integer, floating-point, and complex variables, as well as lists and arrays. Python variables can hold numbers that span a wide range of values, including very large numbers, but they cannot hold numbers that are arbitrarily large. For instance, the largest value you can give a floating-point variable is about 10^{308}. (There is also a corresponding largest negative value of about -10^{308}.) This is enough for most physics calculations, but we will see occasional ex-

126

amples where we run into problems. Complex numbers are similar: both their real and imaginary parts can go up to about $\pm 10^{308}$ but not larger.[1] Large numbers can be specified in scientific notation, using an "e" to denote the exponent. For instance, 2e9 means 2×10^9 and 1.602e-19 means 1.602×10^{-19}. Note that numbers specified in scientific notation are always floats. Even if the number is, mathematically speaking, an integer (like 2e9), the computer will still treat it as a float.

If the value of a variable exceeds the largest floating-point number that can be stored on the computer we say the variable has *overflowed*. For instance, if a floating-point variable x holds a number close to the maximum allowed value of 10^{308} and then we execute a statement like "y = 10*x" it is likely that the result will be larger than the maximum and the variable y will overflow (but not the variable x, whose value is unchanged).

If this happened in the course of a calculation you might imagine that the program would stop, perhaps giving an error message, but in Python this is not what happens. Instead the computer will set the variable to the special value "inf," which means infinity. If you print such a variable with a print statement, the computer will actually print the word "inf" on the screen. In effect, every number over 10^{308} is infinity as far as the computer is concerned. Unfortunately, this is usually not what you want, and when it happens your program will probably give incorrect answers, so you need to watch out for this problem. It's rare, but it'll probably happen to you at some point.

There is also a smallest number (meaning smallest magnitude) that can be represented by a floating-point variable. In Python this number is 10^{-308} roughly.[2] If you go any smaller than this—if the calculation *underflows*—the computer will just set the number to zero. Again, this usually messes things up and gives wrong answers, so you need to be on the lookout.

What about integers? Here Python does something clever. There is no largest integer value in Python: it can represent integers to *arbitrary precision*. This means that no matter how many digits an integer has, Python stores all of them—provided you have enough memory on your computer. Be aware, however, that calculations with integers, even simple arithmetic operations, take longer with more digits, and can take a very long time if there are very

---

[1] The actual largest number is 1.79769×10^{308}, which is the decimal representation of the binary number 2^{1024}, the largest number that can be represented in the IEEE 754 double-precision floating-point format used by the Python language.

[2] Actually 2.22507×10^{-308}, which is 2^{-1022}.

many digits. Try, for example, doing print(2**1000000) in Python. The calculation can be done—it yields a number with 301 030 digits—but it's so slow that you might as well forget about using your computer for anything else for the next few minutes.[3]

Exercise 4.1: Write a program to calculate and print the factorial of a number entered by the user. If you wish you can base your program on the user-defined function for factorial given in Section 2.6, but write your program so that it calculates the factorial using *integer* variables, not floating-point ones. Use your program to calculate the factorial of 200.

Now modify your program to use floating-point variables instead and again calculate the factorial of 200. What do you find? Explain.

4.2 NUMERICAL ERROR

Floating-point numbers (unlike integers) are represented on the computer to only a certain precision. In Python, at least at the time of writing of this book, the standard level of precision is 16 significant digits. This means that numbers like π or $\sqrt{2}$, which have an infinite number of digits after the decimal point, can only be represented approximately. Thus, for instance:

True value of π:	3.1415926535897932384626 ...
Value in Python:	3.141592653589793
Difference:	0.0000000000000002384626 ...

The difference between the true value of a number and its value on the computer is called the *rounding error* on the number. It is the amount by which the computer's representation of the number is wrong.

A number does not have to be irrational like π to suffer from rounding error—any number whose true value has more than 16 significant figures will get rounded off. What's more, when one performs arithmetic with floating-point numbers, the answers are only guaranteed accurate to about 16 figures, even if the numbers that went into the calculation were expressed exactly. If

---

[3]If you do actually try this, then you might want to know how to stop your program if you get bored waiting for it to finish. The simplest thing to do is just to close the window where it's running.

you add 1.1 and 2.2 in Python, then obviously the answer should be 3.3, but the computer might give 3.299999999999999 instead.

Usually this is accurate enough, but there are times when it can cause problems. One important consequence of rounding error is that you should *never use an if statement to test the equality of two floats.* For instance, you should never, in any program, have a statement like

```
if x==3.3:
    print(x)
```

because it may well not do what you want it to do. If the value of x is supposed to be 3.3 but it's actually 3.299999999999999, then as far as the computer is concerned it's not 3.3 and the if statement will fail. In fact, it rarely occurs in physics calculations that you need to test the equality of floats, but if you do, then you should do something like this instead:

```
epsilon = 1e-12
if abs(x-3.3)<epsilon:
    print(x)
```

As we saw in Section 2.2.6, the built-in function abs calculates the absolute value of its argument, so abs(x-3.3) is the absolute difference $|x - 3.3|$. The code above tests whether this difference is less than the small number epsilon. In other words, the if statement will succeed whenever x is very close to 3.3, but the two don't have to be exactly equal. If x is 3.299999999999999 things will still work as expected. The value of epsilon has to be chosen appropriately for the situation—there's nothing special or universal about the value of 10^{-12} used above and a different value may be appropriate in another calculation.

The rounding error on a number, which we will denote ϵ, is defined to be the amount you would have to add to the value calculated by the computer to get the true value. For instance, if we do the following:

```
from math import sqrt
x = sqrt(2)
```

then we will end up not with $x = \sqrt{2}$, but rather with $x + \epsilon = \sqrt{2}$, where ϵ is the rounding error, or equivalently $x = \sqrt{2} - \epsilon$. This is the same definition of error that one uses when discussing measurement error in experiments. When we say, for instance, that the age of the universe is 13.80 ± 0.04 billion years, we mean that the measured value is 13.80 billion years, but the true value is possibly greater or less than this by an amount of order 0.04 billion years.

The error ϵ in the example above could be either positive or negative, depending on how the variable x gets rounded off. If we are lucky ϵ could be small, but we cannot count on it. In general if x is accurate to a certain number of significant digits, say 16, then the rounding error will have a typical size of $x/10^{16}$. It's usually a good assumption to consider the error to be a (uniformly distributed) random number with standard deviation $\sigma = Cx$, where $C \simeq 10^{-16}$ in this case. We will refer to the constant C as the *error constant*. When quoting the error on a calculation we typically give the value of the standard deviation σ. (We can't give the value of the error ϵ itself, since we don't know it—if we did, then we could calculate $x + \epsilon$ and recover the exact value for the quantity of interest, so there would in effect be no error in the calculation at all.)

Rounding error is important, as described above, if we are testing the equality of two floating-point numbers, but in other respects it may appear to be only a minor annoyance. An error of one part in 10^{16} does not seem very bad. But what happens if we now add, subtract, or otherwise combine several different numbers, each with its own error? In many ways the rounding error on a number behaves similarly to measurement error in a laboratory experiment, and the rules for combining errors are the same. For instance, if we add or subtract two numbers x_1 and x_2, with standard deviations σ_1 and σ_2, then standard results about combinations of random variables tell us that the *variance* σ^2 of the sum or difference is equal to the sum of the individual variances:

$$\sigma^2 = \sigma_1^2 + \sigma_2^2. \tag{4.1}$$

Hence the standard deviation of the sum or difference is

$$\sigma = \sqrt{\sigma_1^2 + \sigma_2^2}. \tag{4.2}$$

Similarly if we multiply or divide two numbers then the variance of the result x obeys

$$\frac{\sigma^2}{x^2} = \frac{\sigma_1^2}{x_1^2} + \frac{\sigma_2^2}{x_2^2}. \tag{4.3}$$

But, as discussed above, the standard deviations on x_1 and x_2 are given by $\sigma_1 = Cx_1$ and $\sigma_2 = Cx_2$, so that if, for example, we are adding or subtracting our two numbers, meaning Eq. (4.2) applies, then

$$\sigma = \sqrt{C^2 x_1^2 + C^2 x_2^2} = C\sqrt{x_1^2 + x_2^2}. \tag{4.4}$$

I leave it as an exercise to show that the corresponding result for the error on the product of two numbers $x = x_1 x_2$ or their ratio $x = x_1/x_2$ is

$$\sigma = \sqrt{2}\, Cx. \tag{4.5}$$

We can extend these results to combinations of more than two numbers. If, for instance, we are calculating the sum of N numbers $x_1 \ldots x_N$ with errors having standard deviation $\sigma_i = Cx_i$, then the variance on the final result is the sum of the variances on the individual numbers:

$$\sigma^2 = \sum_{i=1}^{N} \sigma_i^2 = \sum_{i=1}^{N} C^2 x_i^2 = C^2 N \overline{x^2}, \tag{4.6}$$

where $\overline{x^2}$ is the mean-square value of x. Thus the standard deviation on the final result is

$$\sigma = C\sqrt{N}\,\sqrt{\overline{x^2}}. \tag{4.7}$$

As we can see, this quantity increases in size as N increases—the more numbers we combine, the larger the error on the result—although the increase is a relatively slow one, proportional to the square root of N.

We can also ask about the *fractional* error on $\sum_i x_i$, i.e., the total error divided by the value of the sum. The size of the fractional error is given by

$$\frac{\sigma}{\sum_i x_i} = \frac{C\sqrt{N \overline{x^2}}}{N \overline{x}} = \frac{C}{\sqrt{N}} \frac{\sqrt{\overline{x^2}}}{\overline{x}}, \tag{4.8}$$

where $\overline{x} = N^{-1} \sum_i x_i$ is the mean value of x. In other words the fractional error in the sum actually *goes down* as we add more numbers.

At first glance this appears to be pretty good. So what's the problem? Actually, there are a couple of them. One is when the sizes of the numbers you are adding vary widely. If some are much smaller than others then the smaller ones may get lost. But the most severe problems arise when you are not adding but subtracting numbers. Suppose, for instance, that we have the following two numbers:

$$x = 1000000000000000$$
$$y = 1000000000000001.2345678901234$$

and we want to calculate the difference $y - x$. Unfortunately, the computer only represents these two numbers to 16 significant figures, which means that as far as the computer is concerned:

$$x = 100000000000000$$

$$y = 100000000000001.2$$

The first number is represented exactly in this case, but the second has been truncated. Now when we take the difference we get $y - x = 1.2$, when the true result would be 1.2345678901234. In other words, instead of 16-figure accuracy, we now only have two figures and the fractional error is several percent of the true value. This is much worse than before.

To put this in more general terms, if the difference between two numbers is very small, comparable with the error on the numbers, i.e., with the accuracy of the computer, then the fractional error can become large and you may have a problem.

EXAMPLE 4.1: THE DIFFERENCE OF TWO NUMBERS

To see an example of this in practice, consider the two numbers

$$x = 1, \qquad y = 1 + 10^{-14}\sqrt{2}. \tag{4.9}$$

Trivially we see that

$$10^{14}(y - x) = \sqrt{2}. \tag{4.10}$$

Let us perform the same calculation in Python and see what we get. Here is the program:

```python
from math import sqrt
x = 1.0
y = 1.0 + (1e-14)*sqrt(2)
print((1e14)*(y-x))
print(sqrt(2))
```

The penultimate line calculates the value in Eq. (4.10) while the last line prints out the true value of $\sqrt{2}$ (at least to the accuracy of the computer). Here's what we get when we run the program:

```
1.42108547152
1.41421356237
```

As we can see, the calculation is accurate to only the first decimal place—after that the rest is garbage.

This issue, of large errors in calculations that involve the subtraction of numbers that are nearly equal, arises with some frequency in physics calculations. We will see various examples throughout the book. It is perhaps the most common cause of significant numerical error in computations and you need to be aware of it at all times when writing programs.

Exercise 4.2: Quadratic equations

Consider a quadratic equation $ax^2 + bx + c = 0$ that has real solutions.

a) Write a program that takes as input the three numbers, a, b, and c, and prints out the two solutions using the standard formula

$$x = \frac{-b \pm \sqrt{b^2 - 4ac}}{2a}.$$

Use your program to compute the solutions of $0.001x^2 + 1000x + 0.001 = 0$.

b) There is another way to write the solutions to a quadratic equation. Multiplying top and bottom of the solution above by $-b \mp \sqrt{b^2 - 4ac}$, show that the solutions can also be written as

$$x = \frac{2c}{-b \mp \sqrt{b^2 - 4ac}}.$$

Add further lines to your program to print these values in addition to the earlier ones and again use the program to solve $0.001x^2 + 1000x + 0.001 = 0$. What do you see? How do you explain it?

c) Using what you have learned, write a new program that calculates both roots of a quadratic equation accurately in all cases.

This is a good example of how computers don't always work the way you expect them to. If you simply apply the standard formula for the quadratic equation, the computer will sometimes get the wrong answer. In practice the method you have worked out here is the correct way to solve a quadratic equation on a computer, even though it's more complicated than the standard formula. If you were writing a program that involved solving many quadratic equations this method might be a good candidate for a user-defined function: you could put the details of the solution method inside a function to save yourself the trouble of going through it step by step every time you have a new equation to solve.

Exercise 4.3: Calculating derivatives

Suppose we have a function $f(x)$ and we want to calculate its derivative at a point x. We can do that with pencil and paper if we know the mathematical form of the function, or we can do it on the computer by making use of the definition of the derivative:

$$\frac{df}{dx} = \lim_{\delta \to 0} \frac{f(x + \delta) - f(x)}{\delta}.$$

On the computer we can't actually take the limit as δ goes to zero, but we can get a reasonable approximation just by making δ small.

a) Write a program that defines a function f(x) returning the value $x(x-1)$, then calculates the derivative of the function at the point $x = 1$ using the formula above with $\delta = 10^{-2}$. Calculate the true value of the same derivative analytically and compare with the answer your program gives. The two will not agree perfectly. Why not?

b) Repeat the calculation for $\delta = 10^{-4}$, 10^{-6}, 10^{-8}, 10^{-10}, 10^{-12}, and 10^{-14}. You should see that the accuracy of the calculation initially gets better as δ gets smaller, but then gets worse again. Why is this?

We will look at numerical derivatives in more detail in Section 5.10, where we will study techniques for dealing with these issues and maximizing the accuracy of our calculations.

4.3 PROGRAM SPEED

As we have seen, computers are not infinitely accurate. And neither are they infinitely fast. Yes, they work at amazing speeds, but many physics calculations require the computer to perform millions of individual computations to get a desired overall result and collectively those computations can take a significant amount of time. Some of the example calculations described in Chapter 1 took months to complete, even though they were run on some of the most powerful computers in the world.

One thing we need to get a feel for is how fast computers really are. As a general guide, performing a million mathematical operations is no big problem for a computer—it usually takes less than a second. Adding a million numbers together, for instance, or finding a million square roots, can be done in very little time. Performing a billion operations, on the other hand, could take minutes or hours, though it's still possible provided you are patient. Performing a trillion operations, however, will basically take forever. So a fair rule of thumb is that the calculations we can perform on a computer are ones that can be done with *about a billion operations or less.*

This is only a rough guide. Not all operations are equal and it makes a difference whether we are talking about additions or multiplications of single numbers (which are easy and quick) versus, say, calculating Bessel functions or multiplying matrices (which are not). Moreover, the billion-operation rule will change over time because computers get faster. However, computers have been getting faster a lot less quickly in the last few years—progress has slowed. So we're probably stuck with a billion operations for a while.

Example 4.2: Quantum harmonic oscillator at finite temperature

The quantum simple harmonic oscillator has energy levels $E_n = \hbar\omega(n + \frac{1}{2})$, where $n = 0, 1, 2, \ldots, \infty$. As shown by Boltzmann and Gibbs, the average energy of a simple harmonic oscillator at temperature T is

$$\langle E \rangle = \frac{1}{Z} \sum_{n=0}^{\infty} E_n e^{-\beta E_n}, \tag{4.11}$$

where $\beta = 1/(k_B T)$ with k_B being the Boltzmann constant, and $Z = \sum_{n=0}^{\infty} e^{-\beta E_n}$. Suppose we want to calculate, approximately, the value of $\langle E \rangle$ when $k_B T = 100$. Since the terms in the sums for $\langle E \rangle$ and Z dwindle in size quite quickly as n becomes large, we can get a reasonable approximation by taking just the first 1000 terms in each sum. Working in units where $\hbar = \omega = 1$, here's a program to do the calculation:

File: qsho.py

```
from math import exp

terms = 1000
beta = 1/100
S = 0.0
Z = 0.0
for n in range(terms):
    E = n + 0.5
    weight = exp(-beta*E)
    S += weight*E
    Z += weight

print(S/Z)
```

Note a few features of this program:

1. Constants like the number of terms and the value of β are assigned to variables at the beginning of the program. As discussed in Section 2.7, this is good programming style because it makes them easy to find and modify and makes the rest of the program more readable.

2. We used just one for loop to calculate both sums. This saves time, making the program run faster.

3. Although the exponential $e^{-\beta E_n}$ occurs separately in both sums, we calculate it only once each time around the loop and save its value in the variable weight. This also saves time: exponentials take significantly longer to calculate than, for example, additions or multiplications. (Of course

"longer" is relative—the times involved are probably still less than a microsecond. But if one has to go many times around the loop even those short times can add up.)

If we run the program we get this result:

99.9554313409

The calculation (on my desktop computer) takes 0.01 seconds. Now let us try increasing the number of terms in the sums (which just means increasing the value of the variable terms at the top of the program). This will make our approximation more accurate and give us a better estimate of our answer, at the expense of taking more time to complete the calculation. If we increase the number of terms to a million then it does change our answer somewhat:

100.000833332

The calculation now takes 1.4 seconds, which is significantly longer, but still a short time in absolute terms.

Now let's increase the number of terms to a billion. When we do this the calculation takes 22 minutes to finish, but the result does not change at all:

100.000833332

There are three morals to this story. First, a billion operations is indeed doable—if a calculation is important to us we can probably wait twenty minutes for an answer. But it's approaching the limit of what is reasonable. If we increased the number of terms in our sum by another factor of ten the calculation would take 220 minutes, or nearly four hours. A factor of ten beyond that and we'd be waiting a couple of days for an answer.

Second, there is a balance to be struck between time spent and accuracy. In this case it was probably worthwhile to calculate a million terms of the sum—it didn't take long and the result was noticeably, though not wildly, different from the result for a thousand terms. But the change to a billion terms was clearly not worth the effort—the calculation took much longer to complete but the answer was exactly the same as before. We will see plenty of further examples in this book of calculations where we need to find an appropriate balance between speed and accuracy.

Third, it's pretty easy to write a program that will take forever to finish. If we set the program above to calculate a trillion terms, it would take weeks to run. So it's worth taking a moment, before you spend a whole lot of time writing and running a program, to do a quick estimate of how long you expect your calculation to take. If it's going to take a year then it's not worth it: you need to find a faster way to do the calculation, or settle for a quicker but less accurate answer. The simplest way to estimate running time is to make a rough count of the number of mathematical operations the calculation will involve; if the number is significantly greater than a billion, you have a problem.

EXAMPLE 4.3: MATRIX MULTIPLICATION

Suppose we have two $N \times N$ matrices represented as arrays A and B on the computer and we want to multiply them together to calculate their matrix product. Here is a fragment of code to do the multiplication and place the result in a new array called C:

```
from numpy import zeros
N = 1000
C = zeros([N,N],float)

for i in range(N):
    for j in range(N):
        for k in range(N):
            C[i,j] += A[i,k]*B[k,j]
```

We could use this code, for example, as the basis for a user-defined function to multiply arrays together. (As we saw in Section 2.4.4, Python already provides the function "dot" for calculating matrix products, but it's a useful exercise to write our own code for the calculation. Among other things, it helps us understand how many operations are involved in calculating such a product.)

How large a pair of matrices could we multiply together in this way if the calculation is to take a reasonable amount of time? The program has three nested for loops in it. The innermost loop, which runs through values of the variable k, goes around N times doing one multiplication operation each time and one addition, for a total of $2N$ operations. That whole loop is itself executed N times, once for each value of j in the middle loop, giving $2N^2$ operations. And those $2N^2$ operations are themselves performed N times as we go through the values of i in the outermost loop. The end result is that the matrix multiplication takes $2N^3$ operations overall. Thus if $N = 1000$, as above, the whole calculation would involve two billion operations, which is feasible in a few minutes of running time. Larger values of N, however, will rapidly become intractable. For $N = 2000$, for instance, we would have 16 billion op-

137

erations, which could take hours to complete. Thus the largest matrices we can multiply are about 1000×1000 in size.[4]

Exercise 4.4: Calculating integrals

Suppose we want to calculate the value of the integral

$$I = \int_{-1}^{1} \sqrt{1 - x^2}\, dx.$$

The integrand looks like a semicircle of radius 1:

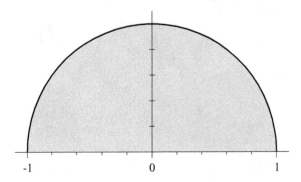

and hence the value of the integral—the area under the curve—must be equal to $\frac{1}{2}\pi = 1.57079632679\ldots$

Alternatively, we can evaluate the integral on the computer by dividing the domain of integration into a large number N of slices of width $h = 2/N$ each and then using the Riemann definition of the integral:

$$I = \lim_{N \to \infty} \sum_{k=1}^{N} h y_k,$$

---

[4]Interestingly, the direct matrix multiplication represented by the code given here is not the fastest way to multiply two matrices on a computer. *Strassen's algorithm* is an iterative method for multiplying matrices that uses some clever shortcuts to reduce the number of operations needed so that the total number is proportional to about $N^{2.8}$ rather than N^3. For very large matrices this can result in significantly faster computations. Unfortunately, Strassen's algorithm suffers from large numerical errors because of problems with subtraction of nearly equal numbers (see Section 4.2) and for this reason it is rarely used. On paper, an even faster method for matrix multiplication is the *Coppersmith–Winograd algorithm*, which requires a number of operations proportional to only about $N^{2.4}$, but in practice this method is so complex to program as to be essentially worthless—the extra complexity means that in real applications the method is always slower than direct multiplication.

where

$$y_k = \sqrt{1 - x_k^2} \qquad \text{and} \qquad x_k = -1 + hk.$$

We cannot in practice take the limit $N \to \infty$, but we can make a reasonable approximation by just making N large.

a) Write a program to evaluate the integral above with $N = 100$ and compare the result with the exact value. The two will not agree very well, because $N = 100$ is not a sufficiently large number of slices.

b) Increase the value of N to get a more accurate value for the integral. If we require that the program runs in about one second or less, how accurate a value can you get?

Evaluating integrals is a common task in computational physics calculations. We will study techniques for doing integrals in detail in the next chapter. As we will see, there are substantially quicker and more accurate methods than the simple one we have used here.

CHAPTER 5

INTEGRALS AND DERIVATIVES

IN THE preceding chapters we looked at the basics of computer programming using Python and solved some simple physics problems using what we learned. You will get plenty of further opportunities to polish your programming skills, but our main task from here on is to learn about the ideas and techniques of computational physics, the physical and mathematical insights that allow us to perform accurate calculations of physical quantities on the computer.

One of the most basic but also most important applications of computers in physics is the evaluation of integrals and derivatives. Numerical evaluation of integrals is a particularly crucial topic because integrals occur widely in physics calculations and, while some integrals can be done analytically in closed form, most cannot. They can, however, almost always be done on a computer. In this chapter we examine a number of different techniques for evaluating integrals and derivatives, as well as taking a brief look at the related operation of interpolation.

5.1 FUNDAMENTAL METHODS FOR EVALUATING INTEGRALS

Suppose we wish to evaluate the integral of a given function. Let us consider initially the simplest case, the integral of a function of a single variable over a finite range. We will study a range of techniques for the numerical evaluation of such integrals, but we start with the most basic—and also most widely used—the trapezoidal rule.[1]

---

[1] Also called the trapezium rule in British English.

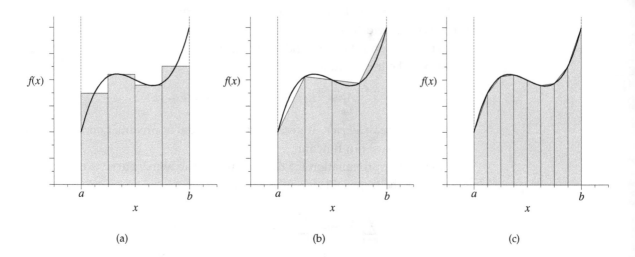

(a)   (b)   (c)

Figure 5.1: Estimating the area under a curve. (a) A simple scheme for estimating the area under a curve by dividing the area into rectangular slices. The gray shaded area approximates the area under the curve, though not very well. (b) The trapezoidal rule approximates the area as a set of trapezoids, and is usually more accurate. (c) With a larger number of slices, the shaded area is a more accurate approximation to the true area under the curve.

5.1.1 THE TRAPEZOIDAL RULE

Suppose we have a function $f(x)$ and we want to calculate its integral with respect to x from $x = a$ to $x = b$, which we denote $I(a, b)$:

$$I(a, b) = \int_a^b f(x)\, dx. \qquad (5.1)$$

This is equivalent to calculating the area under the curve of $f(x)$ from a to b. There is no known way to calculate such an area exactly in all cases on a computer, but we can do it approximately by the method shown in Fig. 5.1a: we divide the area up into rectangular slices, calculate the area of each one, and then add them up. This, however, is a pretty poor approximation. The area under the rectangles is not very close to the area under the curve.

A better approach, which involves very little extra work, is that shown in Fig. 5.1b, where the area is divided into trapezoids rather than rectangles. The area under the trapezoids is a considerably better approximation to the area under the curve, and this approach, though simple, often gives perfectly adequate results.

Suppose we divide the interval from a to b into N slices or steps, so that each slice has width $h = (b - a)/N$. Then the right-hand side of the kth slice falls at $a + kh$, and the left-hand side falls at $a + kh - h = a + (k - 1)h$. Thus the area of the trapezoid for this slice is

$$A_k = \tfrac{1}{2}h\big[f(a + (k - 1)h) + f(a + kh)\big]. \tag{5.2}$$

This is the *trapezoidal rule*. It gives us a trapezoidal approximation to the area under one slice of our function.

Now our approximation for the area under the whole curve is the sum of the areas of the trapezoids for all N slices:

$$
\begin{aligned}
I(a, b) \simeq \sum_{k=1}^{N} A_k &= \tfrac{1}{2}h \sum_{k=1}^{N} \big[f(a + (k - 1)h) + f(a + kh)\big] \\
&= h\big[\tfrac{1}{2}f(a) + f(a + h) + f(a + 2h) + \ldots + \tfrac{1}{2}f(b)\big] \\
&= h\bigg[\tfrac{1}{2}f(a) + \tfrac{1}{2}f(b) + \sum_{k=1}^{N-1} f(a + kh)\bigg].
\end{aligned}
\tag{5.3}
$$

This is the *extended trapezoidal rule*—it is the extension to many slices of the basic trapezoidal rule of Eq. (5.2). Being slightly sloppy in our usage, however, we will often refer to it simply as the trapezoidal rule. Note the structure of the formula: the quantity inside the square brackets is a sum over values of $f(x)$ measured at equally spaced points in the integration domain, and we take a half of the values at the start and end points but one times the value at all the interior points.

EXAMPLE 5.1: INTEGRATING A FUNCTION

Let us use the trapezoidal rule to calculate the integral of $x^4 - 2x + 1$ from $x = 0$ to $x = 2$. This is actually an integral we can do by hand, which means we don't really need to do it using the computer in this case, but it's a good first example because we can check easily if our program is working and how accurate an answer it gives.

Here is a program to do the integration using the trapezoidal rule with $N = 10$ slices:

File:
trapezoidal.py

```
def f(x):
    return x**4 - 2*x + 1
```

```
N = 10
a = 0.0
b = 2.0
h = (b-a)/N

s = 0.5*f(a) + 0.5*f(b)
for k in range(1,N):
    s += f(a+k*h)

print(h*s)
```

This is a straightforward translation of the trapezoidal rule formula into computer code: we create a function that calculates the integrand, set up all the constants used, evaluate the sum for the integral $I(a,b)$ term by term, and then multiply it by h and print it out.

If we run the program it prints

    4.50656

The correct answer is

$$\int_0^2 (x^4 - 2x + 1)\mathrm{d}x = \left[\tfrac{1}{5}x^5 - x^2 + x \right]_0^2 = 4.4. \tag{5.4}$$

So our calculation is moderately but not exceptionally accurate—the answer is off by about 2%.

We can make the calculation more accurate by increasing the number of slices. As shown in Fig. 5.1c, we approximate the area under the curve better when N is larger, though the program will also take longer to reach an answer because there are more terms in the sum to evaluate. If we increase the number of slices to $N = 100$ and run the program again we get 4.40107, which is now accurate to 0.02%, which is pretty good. And if we use $N = 1000$ we get 4.40001, which is accurate to 0.0002%. In Section 5.2 we will study in more detail the accuracy of the trapezoidal rule.

Exercise 5.1: In the on-line resources you will find a file called `velocities.txt`, which contains two columns of numbers, the first representing time t in seconds and the second the x-velocity in meters per second of a particle, measured once every second from time $t = 0$ to $t = 100$. The first few lines look like this:

0	0
1	0.069478
2	0.137694
3	0.204332
4	0.269083
5	0.331656

Write a program to do the following:

a) Read in the data and, using the trapezoidal rule, calculate from them the approximate distance traveled by the particle in the x direction as a function of time. See Section 2.4.3 on page 57 if you want a reminder of how to read data from a file.

b) Extend your program to make a graph that shows, on the same plot, both the original velocity curve and the distance traveled as a function of time.

5.1.2 SIMPSON'S RULE

The trapezoidal rule is the simplest of numerical integration methods, taking only a few lines of code as we have seen, but it is often perfectly adequate for calculations where no great accuracy is required. It happens frequently in physics calculations that we don't need an answer accurate to many significant figures and in such cases the ease and simplicity of the trapezoidal rule can make it the method of choice. One should not turn up one's nose at simple methods like this; they play an important role and are used widely. Moreover, the trapezoidal rule is the basis for several other more sophisticated methods of evaluating integrals, including the adaptive methods that we will study in Section 5.3 and the Romberg integration method of Section 5.4.

However, there are also cases where greater accuracy is required. As we have seen we can increase the accuracy of the trapezoidal rule by increasing the number N of steps used in the calculation. But in some cases, particularly for integrands that are rapidly varying, a very large number of steps may be needed to achieve the desired accuracy, which means the calculation can become slow. There are other, more advanced schemes for calculating integrals that can achieve high accuracy while still arriving at an answer quickly. In this section we study one such scheme, Simpson's rule.

In effect, the trapezoidal rule estimates the area under a curve by approximating the curve with straight-line segments—see Fig. 5.1b. We can often get a better result if we approximate the function instead with curves of some kind. Simpson's rule does exactly this, using quadratic curves, as shown in Fig. 5.2. In order to specify a quadratic completely one needs three points, not just two as with a straight line. So in this method we take a pair of adjacent slices and fit

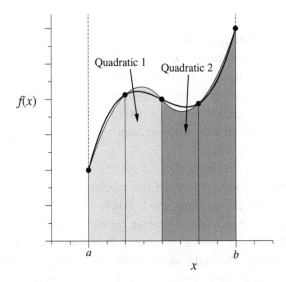

Figure 5.2: Simpson's rule. Simpson's rule involves fitting quadratic curves to pairs of slices and then calculating the area under the quadratics.

a quadratic through the three points that mark the boundaries of those slices. In Fig. 5.2 there are two quadratics, fitted to four slices. Simpson's rule involves approximating the integrand with quadratics in this way, then calculating the area under those quadratics, which gives an approximation to the area under the true curve.

Suppose, as before, that our integrand is denoted $f(x)$ and the spacing of adjacent points is h. And suppose for the purposes of argument that we have three points at $x = -h, 0,$ and $+h$. If we fit a quadratic $Ax^2 + Bx + C$ through these points, then by definition we will have:

$$f(-h) = Ah^2 - Bh + C, \qquad f(0) = C, \qquad f(h) = Ah^2 + Bh + C. \qquad (5.5)$$

Solving these equations simultaneously for A, B, and C gives

$$A = \frac{1}{h^2}\left[\tfrac{1}{2}f(-h) - f(0) + \tfrac{1}{2}f(h)\right], \quad B = \frac{1}{2h}\left[f(h) - f(-h)\right], \quad C = f(0), \quad (5.6)$$

and the area under the curve of $f(x)$ from $-h$ to $+h$ is given approximately by the area under the quadratic:

$$\int_{-h}^{h} (Ax^2 + Bx + C)\, dx = \tfrac{2}{3}Ah^3 + 2Ch = \tfrac{1}{3}h\left[f(-h) + 4f(0) + f(h)\right]. \quad (5.7)$$

This is *Simpson's rule*. It gives us an approximation to the area under two adjacent slices of our function. Note that the final formula for the area involves only h and the value of the function at evenly spaced points, just as with the trapezoidal rule. So to use Simpson's rule we don't actually have to worry about the details of fitting a quadratic—we just plug numbers into this formula and it gives us an answer. This makes Simpson's rule almost as simple to use as the trapezoidal rule, and yet Simpson's rule often gives much more accurate results, as we will see.

To use Simpson's rule to perform a general integral we note that Eq. (5.7) does not depend on the fact that our three points lie at $x = -h$, 0, and $+h$. If we were to slide the curve along the x-axis to either higher or lower values, the area underneath it would not change. So we can use the same rule for any three uniformly spaced points. Applying Simpson's rule involves dividing the domain of integration into many slices and using the rule to separately estimate the area under successive pairs of slices, then adding the estimates for all pairs to get the final answer. If, as before, we are integrating from $x = a$ to $x = b$ in slices of width h then the three points bounding the first pair of slices fall at $x = a$, $a + h$ and $a + 2h$, those bounding the second pair at $a + 2h$, $a + 3h$, $a + 4h$, and so forth. Then the approximate value of the entire integral is given by

$$I(a,b) \simeq \tfrac{1}{3}h\big[f(a) + 4f(a+h) + f(a+2h)\big]$$
$$+ \tfrac{1}{3}h\big[f(a+2h) + 4f(a+3h) + f(a+4h)\big] + \dots$$
$$+ \tfrac{1}{3}h\big[f(a+(N-2)h) + 4f(a+(N-1)h) + f(b)\big]. \qquad (5.8)$$

Note that the total number of slices must be even for this to work. Collecting terms together, we now have

$$I(a,b) \simeq \tfrac{1}{3}h\big[f(a) + 4f(a+h) + 2f(a+2h) + 4f(a+3h) + \dots + f(b)\big]$$
$$= \tfrac{1}{3}h\left[f(a) + f(b) + 4 \sum_{\substack{k\,\text{odd} \\ 1\dots N-1}} f(a+kh) + 2 \sum_{\substack{k\,\text{even} \\ 2\dots N-2}} f(a+kh)\right]. \qquad (5.9)$$

This formula is sometimes called the *extended Simpson's rule*, by analogy with the extended trapezoidal rule of Section 5.1.1, although for the sake of brevity we will just refer to it as Simpson's rule.

The sums over odd and even values of k can be conveniently accomplished in Python using a for loop of the form "for k in range(1,N,2)" for the odd terms or "for k in range(2,N,2)" for the even terms. Alternatively, we can rewrite Eq. (5.9) as

$$I(a,b) \simeq \tfrac{1}{3}h\left[f(a) + f(b) + 4 \sum_{k=1}^{N/2} f(a + (2k-1)h) + 2 \sum_{k=1}^{N/2-1} f(a+2kh)\right], \quad (5.10)$$

and just use an ordinary for loop (although this form is usually less convenient).

Comparing these equations to Eq. (5.3) we see that Simpson's rule is modestly more complicated than the trapezoidal rule, but not enormously so. Programs using it are still straightforward to create.

As an example, suppose we apply Simpson's rule with $N = 10$ slices to the integral from Example 5.1, $\int_0^2 (x^4 - 2x + 1)dx$, whose true value, as we saw, is 4.4. As shown in Exercise 5.2, this gives an answer of 4.400427, which is already accurate to better than 0.01%, orders of magnitude better than the trapezoidal rule with $N = 10$. Results for $N = 100$ and $N = 1000$ are better still—see the exercise.

If you need an accurate answer for an integral, Simpson's rule is a good choice in many cases, giving precise results with relatively little effort. Alternatively, if you need to evaluate an integral quickly—perhaps because you will be evaluating many integrals as part of a larger calculation—then Simpson's rule may again be a good choice, since it can give moderately accurate answers even with only a small number of steps.

Exercise 5.2:

a) Write a program to calculate an approximate value for the integral $\int_0^2 (x^4 - 2x + 1)\,dx$ from Example 5.1, but using Simpson's rule with 10 slices instead of the trapezoidal rule. You may wish to base your program on the trapezoidal rule program on page 142.

b) Run the program and compare your result to the known correct value of 4.4. What is the fractional error on your calculation?

c) Modify the program to use a hundred slices instead, then a thousand. Note the improvement in the result. How do the results compare with those from Example 5.1 for the trapezoidal rule with the same numbers of slices?

Exercise 5.3: Consider the integral

$$E(x) = \int_0^x e^{-t^2}\,dt.$$

a) Write a program to calculate $E(x)$ for values of x from 0 to 3 in steps of 0.1. Choose for yourself what method you will use for performing the integral and a suitable number of slices.

b) When you are convinced your program is working, extend it further to make a graph of $E(x)$ as a function of x. If you want to remind yourself of how to make a graph, you should consult Section 3.1, starting on page 88.

Note that there is no known way to perform this particular integral analytically, so numerical approaches are the only way forward.

Exercise 5.4: The diffraction limit of a telescope

Our ability to resolve detail in astronomical observations is limited by the diffraction of light in our telescopes. Light from stars can be treated effectively as coming from a point source at infinity. When such light, with wavelength λ, passes through the circular aperture of a telescope (which we'll assume to have unit radius) and is focused by the telescope in the focal plane, it produces not a single dot, but a circular diffraction pattern consisting of central spot surrounded by a series of concentric rings. The intensity of the light in this diffraction pattern is given by

$$I(r) = \left(\frac{J_1(kr)}{kr} \right)^2,$$

where r is the distance in the focal plane from the center of the diffraction pattern, $k = 2\pi/\lambda$, and $J_1(x)$ is a Bessel function. The Bessel functions $J_m(x)$ are given by

$$J_m(x) = \frac{1}{\pi} \int_0^\pi \cos(m\theta - x\sin\theta)\, d\theta,$$

where m is a nonnegative integer and $x \geq 0$.

The diffraction pattern produced by a point source of light when viewed through a telescope.

a) Write a Python function J(m,x) that calculates the value of $J_m(x)$ using Simpson's rule with $N = 1000$ points. Use your function in a program to make a plot, on a single graph, of the Bessel functions J_0, J_1, and J_2 as a function of x from $x = 0$ to $x = 20$.

b) Make a second program that makes a density plot of the intensity of the circular diffraction pattern of a point light source with $\lambda = 500\,\text{nm}$, in a square region of the focal plane, using the formula given above. Your picture should cover values of r from zero up to about $1\,\mu\text{m}$.

Hint 1: You may find it useful to know that $\lim_{x\to 0} J_1(x)/x = \frac{1}{2}$. Hint 2: The central spot in the diffraction pattern is so bright that it may be difficult to see the rings around it on the computer screen. If you run into this problem a simple way to deal with it is to use one of the other color schemes for density plots described in Section 3.3. The "hot" scheme works well. For a more sophisticated solution to the problem, the imshow function has an additional argument vmax that allows you to set the value that corresponds to the brightest point in the plot. For instance, if you say "imshow(x,vmax=0.1)", then elements in x with value 0.1, or any greater value, will produce the brightest (most positive) color on the screen. By lowering the `vmax` value, you can reduce the total range of values between the minimum and maximum brightness, and hence increase the sensitivity of the plot, making subtle details visible. (There is also a `vmin` argument that can be used to set the value that corresponds to the dimmest (most negative) color.) For this exercise a value of `vmax=0.01` appears to work well.

5.2 ERRORS ON INTEGRALS

Our numerical integrals are only approximations. As with most numerical calculations there is usually a rounding error when we calculate an integral, as described in Section 4.2, but this is not the main source of error. The main source of error is the so-called *approximation error*—the fact that our integration rules themselves are only approximations to the true integral. Both the trapezoidal and Simpson rules calculate the area under an approximation (either linear or quadratic) to the integrand, not the integrand itself. How big an error does this approximation introduce?

Consider again an integral $\int_a^b f(x)\,dx$, and let us first look at the trapezoidal rule of Eq. (5.3). To simplify our notation a little, let us define $x_k = a + kh$ as a shorthand for the positions at which we evaluate the integrand $f(x)$. We will refer to these positions as *sample points*. Now consider one particular slice of the integral, the one that falls between x_{k-1} and x_k, and let us perform a Taylor expansion of $f(x)$ about x_{k-1} thus:

$$f(x) = f(x_{k-1}) + (x - x_{k-1})f'(x_{k-1}) + \tfrac{1}{2}(x - x_{k-1})^2 f''(x_{k-1}) + \ldots \quad (5.11)$$

where f' and f'' denote the first and second derivatives of f respectively. Integrating this expression from x_{k-1} to x_k gives

$$\int_{x_{k-1}}^{x_k} f(x)\,dx = f(x_{k-1}) \int_{x_{k-1}}^{x_k} dx + f'(x_{k-1}) \int_{x_{k-1}}^{x_k} (x - x_{k-1})\,dx$$
$$+ \tfrac{1}{2} f''(x_{k-1}) \int_{x_{k-1}}^{x_k} (x - x_{k-1})^2\,dx + \ldots \quad (5.12)$$

Now we make the substitution $u = x - x_{k-1}$, which gives

$$\int_{x_{k-1}}^{x_k} f(x)\,dx = f(x_{k-1}) \int_0^h du + f'(x_{k-1}) \int_0^h u\,du + \tfrac{1}{2} f''(x_{k-1}) \int_0^h u^2\,du + \ldots$$
$$= hf(x_{k-1}) + \tfrac{1}{2}h^2 f'(x_{k-1}) + \tfrac{1}{6}h^3 f''(x_{k-1}) + O(h^4), \quad (5.13)$$

where $O(h^4)$ denotes the rest of the terms in the series, those in h^4 and higher, which we are neglecting.

We can do a similar expansion around $x = x_k$ and again integrate from x_{k-1} to x_k to get

$$\int_{x_{k-1}}^{x_k} f(x)\,dx = hf(x_k) - \tfrac{1}{2}h^2 f'(x_k) + \tfrac{1}{6}h^3 f''(x_k) - O(h^4). \qquad (5.14)$$

Then, taking the average of Eqs. (5.13) and (5.14), we get

$$\int_{x_{k-1}}^{x_k} f(x)\,dx = \tfrac{1}{2}h[f(x_{k-1}) + f(x_k)] + \tfrac{1}{4}h^2[f'(x_{k-1}) - f'(x_k)]$$

$$+ \tfrac{1}{12}h^3[f''(x_{k-1}) + f''(x_k)] + O(h^4). \qquad (5.15)$$

Finally, we sum this expression over all slices k to get the full integral that we want:

$$\int_a^b f(x)\,dx = \sum_{k=1}^{N} \int_{x_{k-1}}^{x_k} f(x)\,dx$$

$$= \tfrac{1}{2}h \sum_{k=1}^{N}[f(x_{k-1}) + f(x_k)] + \tfrac{1}{4}h^2[f'(a) - f'(b)]$$

$$+ \tfrac{1}{12}h^3 \sum_{k=1}^{N}[f''(x_{k-1}) + f''(x_k)] + O(h^4). \qquad (5.16)$$

Let's take a close look at this expression to see what's going on.

The first sum on the right-hand side of the equation is precisely equal to the trapezoidal rule, Eq. (5.3). When we use the trapezoidal rule, we evaluate only this sum and discard all the terms following. The size of the discarded terms—the rest of the series—measures the amount we would have to add to the trapezoidal rule value to get the true value of the integral. In other words it is equal to the error we incur when we use the trapezoidal rule, the so-called approximation error.

In the second term, the term in h^2, notice that almost all of the terms have canceled out of the sum, leaving only the first and last terms, the ones evaluated at a and b. Although we haven't shown it, a similar cancellation happens for the terms in h^4, h^6, and all even powers of h.

Now take a look at the term in h^3 and notice the following useful fact: the sum in this term is itself, to within an overall constant, just the trapezoidal rule approximation to the integral of $f''(x)$ over the interval from a to b. Specifically, if we take Eq. (5.16) and make the substitution $f(x) \to f''(x)$ on the left-hand side, we get

$$\int_a^b f''(x)\,dx = \tfrac{1}{2}h \sum_{k=1}^{N}[f''(x_{k-1}) + f''(x_k)] + O(h^2). \qquad (5.17)$$

Multiplying by $\frac{1}{6}h^2$ and rearranging, we then get

$$\frac{1}{12}h^3 \sum_{k=1}^{N} [f''(x_{k-1}) + f''(x_k)] = \frac{1}{6}h^2 \int_a^b f''(x)\,\mathrm{d}x + \mathrm{O}(h^4)$$

$$= \frac{1}{6}h^2[f'(b) - f'(a)] + \mathrm{O}(h^4), \qquad (5.18)$$

since the integral of $f''(x)$ is just $f'(x)$. Substituting this result into Eq. (5.16) and canceling some terms, we find that

$$\int_a^b f(x)\,\mathrm{d}x = \frac{1}{2}h \sum_{k=1}^{N} [f(x_{k-1}) + f(x_k)] + \frac{1}{12}h^2[f'(a) - f'(b)] + \mathrm{O}(h^4). \quad (5.19)$$

Thus, to leading order in h, the value of the terms dropped when we use the trapezoidal rule, which equals the approximation error ϵ on the integral, is

$$\epsilon = \frac{1}{12}h^2[f'(a) - f'(b)]. \qquad (5.20)$$

This is the *Euler–Maclaurin formula* for the error on the trapezoidal rule. More correctly it is the first term in the Euler–Maclaurin formula; the full formula keeps the terms to all orders in h. We can see from Eq. (5.19) that the next term in the series is of order h^4. We might imagine it would be of order h^3, but the h^3 term cancels out, and in fact it's fairly straightforward to show that only even powers of h survive in the full formula at all orders, so the next term after h^4 is h^6, then h^8, and so forth. So long as h is small, however, we can neglect the h^4 and higher terms—the leading term, Eq. (5.20), is usually enough.

Equation (5.19) tells us that the trapezoidal rule is a *first-order* integration rule, meaning it is accurate up to and including terms proportional to h and the leading-order approximation error is of order h^2. That is, a first-order rule is *accurate* to $\mathrm{O}(h)$ and has an *error* $\mathrm{O}(h^2)$.

In addition to approximation error, there is also a rounding error on our calculation. As discussed in Section 4.2, this rounding error will have approximate size C times the value of the integral, where C is the error constant, which is about 10^{-16} in current versions of Python.[2] Equation (5.20) tells us that the

---

[2]One might imagine that the rounding error would be larger than this because the trapezoidal rule involves a sum of terms in Eq. (5.3) and each term will incur its own rounding error, the individual errors accumulating over the course of the calculation. As shown in Section 4.2 and Eq. (4.7), however, the size of such cumulative errors goes up only as $\sqrt{N}$, while the trapezoidal rule equation (5.3) includes a factor of h, which falls off as $1/N$. The net result is that the theoretical cumulative error on the trapezoidal rule actually decreases as $1/\sqrt{N}$, rather than increasing, so it is safe to say that the final error is no greater than the error incurred by the final operation in the calculation, which will have size C times the final value.

approximation error gets smaller as h gets smaller, so we can make our integral more accurate by using smaller h or, equivalently, a larger number N of slices. However, there is little point in making h so small that the approximation error becomes much smaller than the rounding error. Further decreases in h beyond this point will only make our program slower, by increasing the number of terms in the sum for Eq. (5.3), without improving the accuracy of our calculation significantly, since accuracy will be dominated by the rounding error.

Thus decreases in h will only help us up to the point at which the approximation and rounding errors are roughly equal, which is the point where

$$\tfrac{1}{12}h^2\left[f'(a) - f'(b)\right] \simeq C \int_a^b f(x)\,\mathrm{d}x. \tag{5.21}$$

Rearranging for h we get

$$h \simeq \sqrt{\frac{12 \int_a^b f(x)\,\mathrm{d}x}{f'(a) - f'(b)}}\, C^{1/2}. \tag{5.22}$$

Or we can set $h = (b - a)/N$ to get

$$N \simeq (b - a)\sqrt{\frac{f'(a) - f'(b)}{12 \int_a^b f(x)\,\mathrm{d}x}}\, C^{-1/2}. \tag{5.23}$$

Thus if, for example, all the factors except the last are of order unity, then rounding error will become important when $N \simeq 10^8$. Looked at another way, this is the point at which the accuracy of the trapezoidal rule reaches the "machine precision," the maximum accuracy with which the computer can represent the result. There is no point increasing the number of integration slices beyond this point; the calculation will not become any more accurate. However, $N = 10^8$ would be an unusually large number of slices for the trapezoidal rule—it would be rare to use such a large number when equivalent accuracy can be achieved using much smaller N with a more accurate rule such as Simpson's rule. In most practical situations, therefore, we will be in the regime where approximation error is the dominant source of inaccuracy and it is safe to assume that rounding error can be ignored.

We can do an analogous error analysis for Simpson's rule. The algebra is similar but more tedious. Here we'll just quote the results. For an integral over the interval from a to b the approximation error is given to leading order by

$$\epsilon = \tfrac{1}{180}h^4\left[f'''(a) - f'''(b)\right]. \tag{5.24}$$

Thus Simpson's rule is a third-order integration rule—two orders better than the trapezoidal rule—with a fourth-order approximation error. For small values of h this means that the error on Simpson's rule will typically be much smaller than the error on the trapezoidal rule and it explains why Simpson's rule gave such superior results in our example calculations (see Section 5.1.2).

The rounding error for Simpson's rule is again of order $C \int_a^b f(x)\,dx$ and the equivalent of Eq. (5.23) is

$$N = (b-a)\sqrt[4]{\frac{f'''(a) - f'''(b)}{180 \int_a^b f(x)\,dx}}\, C^{-1/4}. \tag{5.25}$$

If, again, the leading factors are roughly of order unity, this implies that rounding error will become important when $N \simeq 10\,000$. Beyond this point Simpson's rule is so accurate that its accuracy exceeds the machine precision of the computer and there is no point using larger values of N. By contrast with the case for the trapezoidal rule, $N = 10\,000$ is not an unusually large number of slices to use in a calculation. Calculations with ten thousand slices can be done easily in a fraction of a second. Thus it is worth bearing this result in mind: there is no point using more than a few thousand slices with Simpson's rule because the calculation will reach the limits of precision of the computer and larger values of N will do no further good.

Finally in this section, let us note that while Simpson's rule does in general give superior accuracy, it is not always guaranteed to do better than the trapezoidal rule, since the errors on the trapezoidal and Simpson rules also depend on derivatives of the integrand function via Eqs. (5.20) and (5.24). It would be possible, for instance, for $f'''(a)$ by bad luck to be large in some particular instance, making the error in Eq. (5.24) similarly large, and possibly worse than the error for the trapezoidal rule. It would be fair to say that Simpson's rule usually gives better results than the trapezoidal rule, but the prudent scientist will bear in mind that it can do worse on occasion.

5.2.1 PRACTICAL ESTIMATION OF ERRORS

The Euler–Maclaurin formula of Eq. (5.20), or its equivalent for Simpson's rule, Eq. (5.24), allows us to calculate the error on our integrals provided we have a known closed-form expression for the integrand $f(x)$, so that we can calculate the derivatives that appear in the formulas. Unfortunately, in many cases—perhaps most—we have no such expression. For instance, the integrand may not be a mathematical function at all but a set of measurements made in the laboratory, or it might itself be the output of another computer program. In such cases we cannot differentiate the function and Eq. (5.20) or (5.24) will not work. There is, however, still a way to calculate the error.

Suppose, as before, that we are evaluating an integral over the interval from $x = a$ to $x = b$ and let's assume that we are using the trapezoidal rule, since it makes the argument simpler, although the method described here extends to Simpson's rule too. Let us perform the integral with some number of steps N_1, so that the step size is $h_1 = (b - a)/N_1$, and let us denote by I_1 the value of the integral that we calculate.

Then here's the trick: we now double the number of steps and perform the integral again. That is we define a new number of steps $N_2 = 2N_1$ and a new step size $h_2 = (b - a)/N_2 = \frac{1}{2}h_1$ and we reevaluate the integral using the trapezoidal rule, giving a new answer I_2, which will normally be more accurate than the previous one. As we have seen, the trapezoidal rule introduces an error of order $O(h^2)$, which means when we half the value of h we *quarter* the size of our error. Knowing this fact allows us to estimate how big the error is.

Suppose that the true value of our integral is I. The difference between the true value and our first estimate I_1 is equal by definition to the error on that estimate, which as we have said is proportional to h^2, so let us write it as ch^2, where c is a constant. Then I and I_1 are related by $I = I_1 + ch_1^2$, neglecting higher-order terms.

We can also write a similar formula for our second estimate I_2 of the integral, with N_2 steps: $I = I_2 + ch_2^2$. Equating the two expressions for I we then get

$$I_1 + ch_1^2 = I_2 + ch_2^2, \tag{5.26}$$

or

$$I_2 - I_1 = ch_1^2 - ch_2^2 = 3ch_2^2, \tag{5.27}$$

where we have made use of the fact that $h_1 = 2h_2$. Rearranging this expression then gives the error ϵ_2 on the second estimate of the integral to be

$$\epsilon_2 = ch_2^2 = \tfrac{1}{3}(I_2 - I_1). \tag{5.28}$$

As we have written it, this expression can be either positive or negative, depending on which way the error happens to go. If we want only the absolute size of the error then we can take the absolute value $\frac{1}{3}|I_2 - I_1|$, which in Python would be done using the built-in function abs.

This method gives us a simple way to estimate the error on the trapezoidal rule without using the Euler–Maclaurin formula. Indeed, even in cases where we could in principle use the Euler–Maclaurin formula because we know the mathematical form of the integrand, it is often simpler in practice to use the method of Eq. (5.28) instead—it is easy to program and gives reliable answers.

The same principle can be applied to integrals evaluated using Simpson's rule too. The equivalent of Eq. (5.28) in that case turns out to be

$$\epsilon_2 = \tfrac{1}{15}(I_2 - I_1). \tag{5.29}$$

The derivation is left to the reader (see Exercise 5.5).

Exercise 5.5: Error on Simpson's rule
Following the same line of argument that led to Eq. (5.28), show that the error on an integral evaluated using Simpson's rule is given, to leading order in h, by Eq. (5.29).

Exercise 5.6: Write a program, or modify an earlier one, to once more calculate the value of the integral $\int_0^2 (x^4 - 2x + 1)\,dx$ from Example 5.1, using the trapezoidal rule with 20 slices, but this time have the program also print an estimate of the error on the result, calculated using the method of Eq. (5.28). To do this you will need to evaluate the integral twice, once with $N_1 = 10$ slices and then again with $N_2 = 20$ slices. Then Eq. (5.28) gives the error. How does the error calculated in this manner compare with a direct computation of the error as the difference between your value for the integral and the true value of 4.4? Why do the two not agree perfectly?

5.3 CHOOSING THE NUMBER OF STEPS

So far we have not specified how the number N of steps used in our integrals is to be chosen. In our example calculations we just chose round numbers and looked to see if the results seemed reasonable. This is fine for quick calculations, but for serious physics we want a more principled approach. In some calculations we may know in advance how many steps we want to use. Sometimes we have a "budget," a certain amount of computer time that we can spend on a calculation and our goal is simply to make the most accurate calculation we can in the given amount of time. If we know, for instance, that we have time to do a thousand steps, then that's what we do.

But a more common situation is that we want to calculate the value of an integral to a given accuracy, such as four decimal places, and we would like to know how many steps will be needed. So long as the desired accuracy does not exceed the fundamental limit set by the machine precision of our computer—the rounding error that limits all calculations—then it should always be possible to meet our goal by using a large enough number of steps. At the same time, we want to avoid using more steps than are necessary, since more steps take more time and our calculation will be slower. Ideally we would like an N that gives us the accuracy we want and no more.

A simple way to achieve this is to start with a small value of N and repeatedly double it until we achieve the accuracy we want. As we saw in Section 5.2.1, there is a simple formula, Eq. (5.28), for calculating the error on an integral when we double the number of steps. By using this formula with repeated doublings we can evaluate an integral to exactly the accuracy we want.

The procedure is straightforward. We start off by evaluating the integral with some small number of steps N_1. For instance, we might choose $N_1 = 10$. Then we double the number to $N_2 = 2N_1$, evaluate the integral again, and apply Eq. (5.28) to calculate the error. If the error is small enough to satisfy our accuracy requirements, then we're done—we have our answer. If not, we double again to $N_3 = 2N_2$ and we keep on doubling until we achieve the required accuracy. The error on the ith step of the process is given by the obvious generalization of Eq. (5.28):

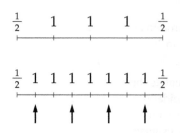

$$\epsilon_i = \tfrac{1}{3}(I_i - I_{i-1}), \tag{5.30}$$

where I_i is the ith estimate of the integral. This method is an example of an *adaptive integration* method, one that varies its own parameters to get a desired answer.

A particularly nice feature of this method is that when we double the number of steps we don't actually have to recalculate the entire integral again. We can reuse our previous calculation rather than just throwing it away. To see this, take a look at Fig. 5.3. The top part of the figure depicts the locations of the sample points, the values of x at which the integrand is evaluated in the trapezoidal rule. The sample points are regularly spaced, and bear in mind that the first and last points are treated differently from the others—the trapezoidal rule formula, Eq. (5.3), specifies that the values of $f(x)$ at these points are multiplied by a factor of $\tfrac{1}{2}$ where the values at the interior points are multiplied by 1.

The lower part of the figure shows what happens when we double the number of slices. This adds an additional set of sample points half way between

Figure 5.3: Doubling the number of steps in the trapezoidal rule. Top: We evaluate the integrand at evenly spaced points as shown, with the value at each point being multiplied by the appropriate factor. Bottom: when we double the number of steps, we effectively add a new set of points, half way between the previous points, as indicated by the arrows.

the old ones, as indicated by the arrows. Note that the original points are still included in the calculation and still carry the same multiplying factors as before—$\frac{1}{2}$ at the ends and 1 in the middle—while the new points are all multiplied by a simple factor of 1. Thus we have all of the same terms in our trapezoidal rule sum that we had before, terms that we have already evaluated, but we also have a set of new ones, which we have to add into the sum to calculate its full value. In the jargon of computational physics we say that the sample points for the first estimate of the integral are *nested* inside the points for the second estimate.

To put this in mathematical terms, consider the trapezoidal rule at the *i*th step of the calculation. Let the number of slices at this step be N_i and the width of a slice be $h_i = (b - a) / N_i$, and note that on the previous step there were half as many slices of twice the width, so that $N_{i-1} = \frac{1}{2} N_i$ and $h_{i-1} = 2h_i$. Then

$$I_i = h_i \left[\tfrac{1}{2} f(a) + \tfrac{1}{2} f(b) + \sum_{k=1}^{N_i - 1} f(a + kh_i) \right]$$

$$= h_i \left[\tfrac{1}{2} f(a) + \tfrac{1}{2} f(b) + \sum_{\substack{k \text{ even} \\ 2...N_i - 2}} f(a + kh_i) + \sum_{\substack{k \text{ odd} \\ 1...N_i - 1}} f(a + kh_i) \right]. \quad (5.31)$$

But

$$\sum_{\substack{k \text{ even} \\ 2...N_i - 2}} f(a + kh_i) = \sum_{k=1}^{N_i / 2 - 1} f(a + 2kh_i) = \sum_{k=1}^{N_{i-1} - 1} f(a + kh_{i-1}), \quad (5.32)$$

and hence

$$I_i = \tfrac{1}{2} h_{i-1} \left[\tfrac{1}{2} f(a) + \tfrac{1}{2} f(b) + \sum_{k=1}^{N_{i-1} - 1} f(a + kh_{i-1}) \right] + h_i \sum_{\substack{k \text{ odd} \\ 1...N_i - 1}} f(a + kh_i). \quad (5.33)$$

But the term $h_{i-1}[\ldots]$ in this equation is precisely the trapezoidal rule estimate I_{i-1} of the integral on the previous iteration of the process, so

$$I_i = \tfrac{1}{2} I_{i-1} + h_i \sum_{\substack{k \text{ odd} \\ 1...N_i - 1}} f(a + kh_i). \quad (5.34)$$

In effect, our old estimate gives us half of the terms in our trapezoidal rule sum and we only have to calculate the other half. In this way we avoid ever recalculating any term that has already been calculated, meaning that each term in our sums is calculated only once, regardless of how many levels of the calculation it's used in. This means it takes only about as much work to calculate I_i

going through all the successive levels $I_1, I_2, I_3, \ldots$ as it does to calculate I_i outright using the ordinary trapezoidal rule. Thus we pay very little extra price in terms of the running time of our program to use this adaptive method and we gain the significant advantage of a guarantee in the accuracy of the integral.

The entire process is as follows:

1. Choose an initial number of steps N_1 and decide on the target accuracy for the value of the integral. Calculate the first approximation I_1 to the integral using the chosen value of N_1 with the standard trapezoidal rule formula, Eq. (5.3).
2. Double the number of steps and use Eq. (5.34) to calculate an improved estimate of the integral. Also calculate the error on that estimate from Eq. (5.30).
3. If the absolute magnitude of the error is less than the target accuracy for the integral, stop. Otherwise repeat from step 2.

The sum over odd values of k in Eq. (5.34) can be conveniently performed in Python with a for loop of the form "for k in range(1,N,2)".

We can also derive a similar method for integrals evaluated using Simpson's rule. Again we double the number of steps on each iteration of the process and the equivalent of Eq. (5.30) is

$$\epsilon_i = \tfrac{1}{15}(I_i - I_{i-1}). \tag{5.35}$$

The equivalent of Eq. (5.34) is a little more complicated. We define

$$S_i = \tfrac{1}{3}\left[f(a) + f(b) + 2 \sum_{\substack{k \text{ even} \\ 2 \ldots N_i-2}} f(a + kh_i)\right], \tag{5.36}$$

and

$$T_i = \tfrac{2}{3} \sum_{\substack{k \text{ odd} \\ 1 \ldots N_i-1}} f(a + kh_i). \tag{5.37}$$

Then we can show that

$$S_i = S_{i-1} + T_{i-1}, \tag{5.38}$$

and

$$I_i = h_i(S_i + 2T_i). \tag{5.39}$$

Thus for Simpson's rule the complete process is:

1. Choose an initial number of steps and a target accuracy, and calculate the sums S_1 and T_1 from Eqs. (5.36) and (5.37) and the initial value I_1 of the integral from Eq. (5.39).

2. Double the number of steps then use Eqs. (5.37), (5.38), and (5.39) to calculate the new values of S_i and T_i and the new estimate of the integral. Also calculate the error on that estimate from Eq. (5.35).

3. If the absolute magnitude of the error is less than the target accuracy for the integral, stop. Otherwise repeat from step 2.

Again notice that on each iteration of the process you only have to calculate one sum, Eq. (5.37), which includes only those terms in the Simpson's rule formula that have not previously been calculated. As a result, the complete calculation of I_i takes very little more computer time than the basic Simpson rule.

5.4   ROMBERG INTEGRATION

We can do even better than the adaptive method of the last section with only a little more effort. Let us go back to the trapezoidal rule again. We have seen that the leading-order error on the trapezoidal rule, at the ith step of the adaptive method, can be written as ch_i^2 for some constant c and is given by Eq. (5.30) to be

$$ch_i^2 = \tfrac{1}{3}(I_i - I_{i-1}). \tag{5.40}$$

But by definition the true value of the integral is $I = I_i + ch_i^2 + O(h_i^4)$, where we are including the $O(h_i^4)$ term to remind us of the next term in the series—see Eq. (5.19). (Remember that there are only even-order terms in this series.) So in other words

$$I = I_i + \tfrac{1}{3}(I_i - I_{i-1}) + O(h_i^4). \tag{5.41}$$

But this expression is now accurate to third order, and has only a fourth order error, which is as accurate as Simpson's rule, and yet we calculated it using only our results from the trapezoidal rule, with hardly any extra work; we are just reusing numbers we already calculated while carrying out the repeated doubling procedure of Section 5.3.

We can take this process further. Let us refine our notation a little and define

$$R_{i,1} = I_i, \qquad R_{i,2} = I_i + \tfrac{1}{3}(I_i - I_{i-1}) = R_{i,1} + \tfrac{1}{3}(R_{i,1} - R_{i-1,1}). \tag{5.42}$$

Then, from Eq. (5.41),

$$I = R_{i,2} + c_2 h_i^4 + O(h_i^6), \tag{5.43}$$

where c_2 is another constant and we have made use of the fact that the series for I contains only even powers of h_i. Analogously,

$$I = R_{i-1,2} + c_2 h_{i-1}^4 + O(h_{i-1}^6) = R_{i-1,2} + 16c_2 h_i^4 + O(h_i^6). \qquad (5.44)$$

Since these last two equations both give expressions for I we can equate them and rearrange to get

$$c_2 h_i^4 = \tfrac{1}{15}(R_{i,2} - R_{i-1,2}) + O(h_i^6). \qquad (5.45)$$

Substituting this expression back into (5.43) gives

$$I = R_{i,2} + \tfrac{1}{15}(R_{i,2} - R_{i-1,2}) + O(h_i^6). \qquad (5.46)$$

Now we have eliminated the h_i^4 term and generated an estimate accurate to fifth order, with a sixth-order error!

We can continue this process, canceling out higher and higher order error terms and getting more and more accurate results. In general, if $R_{i,m}$ is an estimate calculated at the ith round of the doubling procedure and accurate to order h^{2m-1}, with an error of order h^{2m}, then

$$I = R_{i,m} + c_m h_i^{2m} + O(h_i^{2m+2}), \qquad (5.47)$$

$$I = R_{i-1,m} + c_m h_{i-1}^{2m} + O(h_{i-1}^{2m+2}) = R_{i-1,m} + 4^m c_m h_i^{2m} + O(h_i^{2m+2}). \qquad (5.48)$$

Equating the two and rearranging we have

$$c_m h_i^{2m} = \frac{1}{4^m - 1}(R_{i,m} - R_{i-1,m}) + O(h_i^{2m+2}), \qquad (5.49)$$

and substituting this into Eq. (5.47) gives

$$I = R_{i,m+1} + O(h_i^{2m+2}), \qquad (5.50)$$

where

$$R_{i,m+1} = R_{i,m} + \frac{1}{4^m - 1}(R_{i,m} - R_{i-1,m}), \qquad (5.51)$$

which is accurate to order h^{2m+1} with an error of order h^{2m+2}. The calculation also gives us an estimate of the error—Eq. (5.49) is precisely the error on $R_{i,m}$ (see Eq. (5.47))—and hence we can say how accurate our results are.

To make use of these results in practice we do the following:

1. We calculate our first two estimates of the integral using the regular trapezoidal rule: $I_1 \equiv R_{1,1}$ and $I_2 \equiv R_{2,1}$.

2. From these we calculate the more accurate estimate $R_{2,2}$ using Eq. (5.51). This is as much as we can do with only the two starting estimates.
3. Now we calculate the next trapezoidal rule estimate $I_3 \equiv R_{3,1}$ and from this, with Eq. (5.51), we calculate $R_{3,2}$, and then $R_{3,3}$.
4. At each successive stage we compute one more trapezoidal rule estimate $I_i \equiv R_{i,1}$, and from it, with very little extra effort, we can calculate $R_{i,2} \dots R_{i,i}$.
5. For each estimate we can also calculate the error, Eq. (5.49), which allows us to halt the calculation when the error on our estimate of the integral meets some desired target.

Perhaps a picture will help make the process clearer. This diagram shows which values $R_{i,m}$ are needed to calculate further Rs:

$$I_1 \equiv R_{1,1}$$
$$\searrow$$
$$I_2 \equiv R_{2,1} \rightarrow R_{2,2}$$
$$\searrow \qquad \searrow$$
$$I_3 \equiv R_{3,1} \rightarrow R_{3,2} \rightarrow R_{3,3}$$
$$\searrow \qquad \searrow \qquad \searrow$$
$$I_4 \equiv R_{4,1} \rightarrow R_{4,2} \rightarrow R_{4,3} \rightarrow R_{4,4}$$
$$\searrow \qquad \searrow \qquad \searrow \qquad \searrow$$

Each row here lists one trapezoidal rule estimate I_i followed by the other higher-order estimates it allows us to make. The arrows show which previous estimates go into the calculation of each new one via Eq. (5.51).

Note how each fundamental trapezoidal rule estimate I_i allows us to go one step further with calculating the $R_{i,m}$. The most accurate estimate we get from the whole process is the very last one: if we do n levels of the process, then the last estimate is $R_{n,n}$ and is accurate to order h_n^{2n}.

Errors on our estimates are given by Eq. (5.49). If we're being picky, however, we should point out that the equation gives us the error on every estimate *except* the last one in each row (which is the one we really care about). The equation says that the error on $R_{n,n}$ would be $(R_{n,n} - R_{n-1,n})/(4^n - 1)$ but there is no $R_{n-1,n}$ so we cannot use the formula in this case. In practice this means we have to content ourselves with the error estimate for the penultimate entry in each row, which is normally bigger than the error on the final entry. The best we can say is that the final entry in the row is our most accurate estimate of the integral and that its error is at least as good as the error for the entry that precedes it, which is given by Eq. (5.49). This is not ideal, but in practice it's usually good enough.

This whole procedure is called *Romberg integration*. It's essentially an "add-on" to our earlier trapezoidal rule scheme: all the tough work is done in the trapezoidal rule calculations and the Romberg integration takes almost no extra computer time (although it does involve extra programming). The payoff is a value for the integral that is accurate to much higher order in h than the simple trapezoidal rule value (or even than Simpson's rule). And when used in an adaptive scheme that halts the calculation once the required accuracy is reached, it can significantly reduce the time needed to evaluate integrals because it reduces the number of trapezoidal rule steps we have to do.

The method does have its limitations. We are in essence calculating the value of our integral by making a series expansion in powers of the step size h. This means that the method works best in cases where such power series converge rapidly. If one needs hundreds of terms in the series to get good convergence then the method is not going to give us any advantage over the simple trapezoidal rule. This can happen if the integrand $f(x)$ is poorly behaved, containing wild fluctuations, for instance, or singularities, or if it is noisy. If your integrand displays these types of pathologies then Romberg integration is not a good choice. The simpler adaptive trapezoidal method of Section 5.3 will give better results. In cases where the integrand is smooth and well-behaved, however, Romberg integration can give significantly more accurate results significantly faster than either the trapezoidal or Simpson rules.

Romberg integration is an example of the more general technique of *Richardson extrapolation*, in which high-order estimates of quantities are calculated iteratively from lower-order ones. We will see another example of Richardson extrapolation in Section 8.5.5, when we apply it to the solution of differential equations.

Exercise 5.7: Consider the integral

$$I = \int_0^1 \sin^2 \sqrt{100x}\, dx.$$

a) Write a program that uses the adaptive trapezoidal rule method of Section 5.3 and Eq. (5.34) to calculate the value of this integral to an approximate accuracy of $\epsilon = 10^{-6}$ (i.e., correct to six digits after the decimal point). Start with one single integration slice and work up from there to two, four, eight, and so forth. Have your program print out the number of slices, its estimate of the integral, and its estimate of the error on the integral, for each value of the number of slices N, until the target accuracy is reached. (Hint: You should find the result is around $I = 0.45$.)

b) Now modify your program to evaluate the same integral using the Romberg integration technique described in this section. Have your program print out a triangular table of values, as on page 161, of all the Romberg estimates of the integral. Calculate the error on your estimates using Eq. (5.49) and again continue the calculation until you reach an accuracy of $\epsilon = 10^{-6}$. You should find that the Romberg method reaches the required accuracy considerably faster than the trapezoidal rule alone.

Exercise 5.8: Write a program that uses the adaptive Simpson's rule method of Section 5.3 and Eqs. (5.35) to (5.39) to calculate the same integral as in Exercise 5.7, again to an approximate accuracy of $\epsilon = 10^{-6}$. Starting this time with two integration slices, work up from there to four, eight, and so forth, printing out the results at each step until the required accuracy is reached. You should find you reach that accuracy for a significantly smaller number of slices than with the trapezoidal rule calculation in part (a) of Exercise 5.7, but a somewhat larger number than with the Romberg integration of part (b).

5.5 HIGHER-ORDER INTEGRATION METHODS

As we have seen, the trapezoidal rule is based on approximating an integrand $f(x)$ with straight-line segments, while Simpson's rule uses quadratics. We can create higher-order (and hence potentially more accurate) rules by using higher-order polynomials, fitting $f(x)$ with cubics, quartics, and so forth. The general form of the trapezoidal and Simpson rules is

$$\int_a^b f(x)\,dx \simeq \sum_{k=1}^{N} w_k f(x_k), \tag{5.52}$$

where the x_k are the positions of the sample points at which we calculate the integrand and the w_k are some set of weights. In the trapezoidal rule, Eq. (5.3), the first and last weights are $\frac{1}{2}$ and the others are all 1, while in Simpson's rule the weights are $\frac{1}{3}$ for the first and last slices and alternate between $\frac{4}{3}$ and $\frac{2}{3}$ for the other slices—see Eq. (5.9).

For higher-order rules the basic form is the same: after fitting to the appropriate polynomial and integrating we end up with a set of weights that multiply the values $f(x_k)$ of the integrand at evenly spaced sample points. Here are the weights up to quartic order:

Degree	Polynomial	Coefficients
1 (trapezoidal rule)	Straight line	$\frac{1}{2}, 1, 1, \ldots, 1, \frac{1}{2}$
2 (Simpson's rule)	Quadratic	$\frac{1}{3}, \frac{4}{3}, \frac{2}{3}, \frac{4}{3}, \ldots, \frac{4}{3}, \frac{1}{3}$
3	Cubic	$\frac{3}{8}, \frac{9}{8}, \frac{9}{8}, \frac{3}{4}, \frac{9}{8}, \frac{9}{8}, \frac{3}{4}, \ldots, \frac{9}{8}, \frac{3}{8}$
4	Quartic	$\frac{14}{45}, \frac{64}{45}, \frac{8}{15}, \frac{64}{45}, \frac{28}{45}, \frac{64}{45}, \frac{8}{15}, \frac{64}{45}, \ldots, \frac{64}{45}, \frac{14}{45}$

Higher-order integration rules of this kind are called *Newton–Cotes formulas* and in principle they can be extended to any order we like.

However, we can do better still. The point to notice is that the trapezoidal rule is *exact* if the function being integrated is actually a straight line, because then the straight-line approximation isn't an approximation at all. Similarly, Simpson's rule is exact if the function being integrated is a quadratic, and the kth Newton–Cotes rule is exact if the function being integrated is a degree-k polynomial.

But if we have N sample points, then presumably that means we could just fit one $(N-1)$th-degree polynomial to the whole integration interval, and get an integration method that is exact for $(N-1)$th-degree polynomials—and for any lower-degree polynomials as well. (Note that it's $N-1$ because you need three points to fit a quadratic, four for a cubic, and so forth.)

But we can do even better than this. We have been assuming here that the sample points are evenly spaced. Methods with evenly spaced points are relatively simple to program, and it's easy to increase the number of points by adding new points half way between the old ones, as we saw in Section 5.3. However, it is also possible to derive integration methods that use unevenly spaced points and, while they lack some of the advantages above, they have others of their own. In particular, they can give very accurate answers with only small numbers of points, making them particularly suitable for cases where we need to do integrals very fast, or where evaluation of the integrand itself takes a long time.

Suppose then that we broaden our outlook to include rules of the form of Eq. (5.52), but where we are allowed to vary not only the weights w_k but also the positions x_k of the sample points. Any choice of positions is allowed, and particularly ones that are not equally spaced. As we have said, it is possible to create an integration method that is exact for polynomials up to degree $N-1$ with N equally spaced points. Varying the positions of the points gives us N extra degrees of freedom, which suggests that it then might be possible to create an integration rule that is exact for polynomials up to degree $2N-1$ if all of those degrees of freedom are chosen correctly. For large values of N this could give us the power to fit functions very accurately indeed, and hence to do very accurate integrals. It turns out indeed that it is possible to do this and the developments lead to the superbly accurate integration method known as *Gaussian quadrature*, which we describe in the next section.

5.6 GAUSSIAN QUADRATURE

The derivation of the Gaussian quadrature method has two parts. First, we will see how to derive integration rules with nonuniform sample points x_k. Then we will choose the particular set of nonuniform points that give the optimal integration rule.

5.6.1 NONUNIFORM SAMPLE POINTS

Suppose we are given a nonuniform set of N points x_k and we wish to create an integration rule of the form (5.52) that calculates integrals over a given interval from a to b, based only on the values $f(x_k)$ of the integrand at those points. In other words, we want to choose weights w_k so that Eq. (5.52) works for general $f(x)$. To do this, we will fit a single polynomial through the values $f(x_k)$ and then integrate that polynomial from a to b to calculate an approximation to the true integral. To fit N points we need to use a polynomial of degree $N - 1$. The fitting can be done using the *method of interpolating polynomials*.

Consider the following quantity:

$$
\phi_k(x) = \prod_{\substack{m=1...N \\ m \neq k}} \frac{(x - x_m)}{(x_k - x_m)}
$$

$$
= \frac{(x - x_1)}{(x_k - x_1)} \times \ldots \times \frac{(x - x_{k-1})}{(x_k - x_{k-1})} \times \frac{(x - x_{k+1})}{(x_k - x_{k+1})} \times \ldots \times \frac{(x - x_N)}{(x_k - x_N)},
$$

(5.53)

which is called an *interpolating polynomial*. Note that the numerator contains one factor for each sample point except the point x_k. Thus $\phi_k(x)$ is a polynomial in x of degree $N - 1$. For values of k from 1 to N, Eq. (5.53) defines N different such polynomials.

You can confirm for yourself that if we evaluate $\phi_k(x)$ at one of the sample points $x = x_m$ we get

$$
\phi_k(x_m) = \begin{cases} 1 & \text{if } m = k, \\ 0 & \text{if } m \neq k, \end{cases}
$$

(5.54)

165

or, to be more concise,

$$\phi_k(x_m) = \delta_{km}, \tag{5.55}$$

where δ_{km} is the Kronecker delta—the quantity that is 1 when $k = m$ and zero otherwise.

So now consider the following expression:

$$\Phi(x) = \sum_{k=1}^{N} f(x_k)\,\phi_k(x). \tag{5.56}$$

Since it is a linear combination of polynomials of degree $N - 1$, this entire quantity is also a polynomial of degree $N - 1$. And if we evaluate it at any one of the sample points $x = x_m$ we get

$$\Phi(x_m) = \sum_{k=1}^{N} f(x_k)\,\phi_k(x_m) = \sum_{k=1}^{N} f(x_k)\,\delta_{km} = f(x_m), \tag{5.57}$$

where we have used Eq. (5.55).

In other words $\Phi(x)$ is a polynomial of degree $N - 1$ that fits the integrand $f(x)$ at all of the sample points. This is exactly the quantity we were looking for to create our integration rule. Moreover, the polynomial of degree $N - 1$ that fits a given N points is unique: it has N free coefficients and our points give us N constraints, so the coefficients are completely determined. Hence $\Phi(x)$ is not merely *a* polynomial that fits our points, it is *the* polynomial. There are no others.

To calculate an approximation to our integral, all we have to do now is integrate $\Phi(x)$ from a to b thus:

$$\int_a^b f(x)\,dx \simeq \int_a^b \Phi(x)\,dx = \int_a^b \sum_{k=1}^{N} f(x_k)\phi_k(x)\,dx$$

$$= \sum_{k=1}^{N} f(x_k) \int_a^b \phi_k(x)\,dx, \tag{5.58}$$

where we have interchanged the order of the sum and integral in the second line. Comparing this expression with Eq. (5.52) we now see that the weights we need for our integration rule are given by

$$w_k = \int_a^b \phi_k(x)\,dx. \tag{5.59}$$

In other words we have found a general method for creating an integration rule of the form (5.52) for any set of sample points x_k: we simply set the weights w_k

equal to the integrals of the interpolating polynomials, Eq. (5.53), over the domain of integration.

There is no general closed-form formula for the integrals of the interpolating polynomials.[3] In some special cases it is possible to perform the integrals exactly, but often it is not, in which case we may have to perform them on the computer, using one of our other integration methods, such as Simpson's rule or Romberg integration. This may seem to defeat the point of our calculation, which was to find an integration method that didn't rely on uniformly spaced sample points, and here we are using Simpson's rule, which has uniformly spaced points! But in fact the exercise is not as self-defeating as it may appear. The important point to notice is that we only have to calculate the weights w_k once, and then we can use them in Eq. (5.52) to integrate as many different functions over the given integration domain as we like. So we may have to put some effort into the calculation of the weights, using, say, Simpson's rule with very many slices to get as accurate an answer as possible. But we only have to do it once, and thereafter other integrals can be done rapidly and accurately using Eq. (5.52).

In fact, it's better than this. Once one has calculated the weights for a particular set of sample points and domain of integration, it's possible to map those weights and points onto any other domain and get an integration rule of the form (5.52) without having to recalculate the weights. Typically one gives sample points and weights arranged in a standard interval, which for historical reasons is usually taken to be the interval from $x = -1$ to $x = +1$. Thus to specify an integration rule one gives a set of sample points in the range $-1 \leq x_k \leq 1$ and a set of weights

$$w_k = \int_{-1}^{1} \phi_k(x) \, \mathrm{d}x. \tag{5.60}$$

If we want to use any integration domain other than the one from -1 to $+1$, we map these values to that other domain. Since the area under a curve doesn't depend on where that curve is along the x line, the sample points can be slid up and down the x line *en masse* and the integration rule will still work fine. If the desired domain is wider or narrower than the interval from -1 to $+1$ then we also need to spread the points out or squeeze them together. The correct rule

---

[3]One can in principle expand Eq. (5.53) and then integrate the resulting expression term by term, since powers of x can be integrated in closed form. However, the result would be a sum of 2^{N-1} different terms, which would be intractable even for the fastest computers, for relatively modest values of N.

for mapping the points to a general domain that runs from $x = a$ to $x = b$ is:

$$x'_k = \tfrac{1}{2}(b - a)x_k + \tfrac{1}{2}(b + a).$$ (5.61)

Similarly the weights do not change if we are simply sliding the sample points up or down the x line, but if the width of the integration domain changes then the value of the integral will increase or decrease by a corresponding factor, and hence the weights have to be rescaled thus:

$$w'_k = \tfrac{1}{2}(b - a)w_k.$$ (5.62)

Once we have calculated the rescaled positions and weights then the integral itself is given by

$$\int_a^b f(x)\,dx \simeq \sum_{k=1}^n w'_k f(x'_k).$$ (5.63)

5.6.2 Sample points for Gaussian quadrature

The developments of the previous section solve half our problem. Given the positions of the sample points x_k they tell us how to choose the weights w_k, but we still need to choose the sample points. As we argued in Section 5.5, in the best case it should be possible to choose our N points so that our integration rule is exact for all polynomial integrands up to and including polynomials of degree $2N - 1$. The proof that this is indeed possible, and the accompanying derivation of the positions, is not difficult, but it is quite long and it's not really important for our purposes. If you want to see it, it's given in Appendix C on page 514. Here we'll just look at the results, which definitely are important and useful.

The bottom line is this: to get an integration rule accurate up to the highest possible degree of $2N - 1$, the sample points x_k should be chosen to coincide with the zeros of the Nth Legendre polynomial $P_N(x)$, rescaled if necessary to the window of integration using Eq. (5.61), and the corresponding weights w_k are

$$w_k = \left[\frac{2}{(1 - x^2)} \left(\frac{dP_N}{dx} \right)^{-2} \right]_{x = x_k},$$ (5.64)

also rescaled if necessary, using Eq. (5.62).

This method is called *Gaussian quadrature*[4] and although it might sound rather formidable from the description above, in practice it's beautifully simple: given the values x_k and w_k for your chosen N, all you have to do is rescale

---

[4]It's called "Gaussian" because it was pioneered by the legendary mathematician Carl

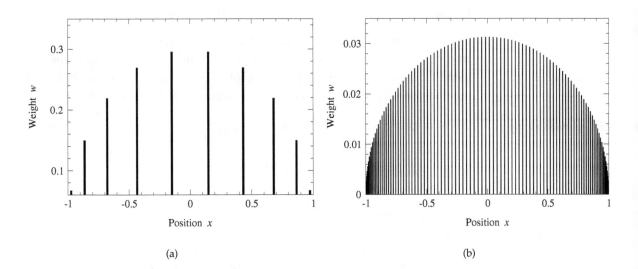

Figure 5.4: Sample points and weights for Gaussian quadrature. The positions and heights of the bars represent the sample points and their associated weights for Gaussian quadrature with (a) $N = 10$ and (b) $N = 100$.

them if necessary using Eqs. (5.61) and (5.62) and then perform the sum in Eq. (5.63).

The only catch is finding the values in the first place. In principle the results quoted above tell us everything we need to know but in practice the zeros of the Legendre polynomials are not trivial to compute. Tables containing values of x_k and w_k up to about $N = 20$ can be found in books or on-line,[5] or they can be calculated for any N using a suitable computer program. Python functions to perform the calculation are given in Appendix E and also in the on-line resources in the file gaussxw.py—Example 5.2 below shows how to use them. Figure 5.4 shows what the sample points and weights look like for the cases $N = 10$ and $N = 100$. Note how the points get closer together at the edges while at the same time the weights get smaller.

Friedrich Gauss. "Quadrature" is an old (19th century) name for numerical integration—Gauss's work predates the invention of computers, to a time when people did numerical integrals by hand, meaning they were *very* concerned about getting the best answers when N is small. When you're doing calculations by hand, Simpson's rule with $N = 1000$ is not an option.

[5]See for example Abramowitz, M. and Stegun, I. A., eds., *Handbook of Mathematical Functions*, Dover Publishing, New York (1974).

EXAMPLE 5.2: GAUSSIAN INTEGRAL OF A SIMPLE FUNCTION

Consider again the integral we did in Example 5.1, $\int_0^2 (x^4 - 2x + 1)\, dx$, whose true value, as we saw, is 4.4. Here's a program to evaluate the same integral using Gaussian quadrature. Just to emphasize the impressive power of the method, we will perform the calculation with only $N = 3$ sample points:

File: gaussint.py

```python
from gaussxw import gaussxw

def f(x):
    return x**4 - 2*x + 1

N = 3
a = 0.0
b = 2.0

# Calculate the sample points and weights, then map them
# to the required integration domain
x,w = gaussxw(N)
xp = 0.5*(b-a)*x + 0.5*(b+a)
wp = 0.5*(b-a)*w

# Perform the integration
s = 0.0
for k in range(N):
    s += wp[k]*f(xp[k])

print(s)
```

For this program to work you must have a copy of the file gaussxw.py in the same folder as the program itself.

Note how the function gaussxw(N) returns two variables, not just one. We discussed functions of this type in Section 2.6 but this is the first time we've seen one in use. In this case the variables are arrays, x and w, containing the sample points and weights for Gaussian quadrature on N points over the standard interval from -1 to $+1$. Notice also how we mapped the points and the weights from the standard interval to our desired integration domain: we have used Python's ability to perform calculations with entire arrays to achieve the mapping in just two lines.

There is also an alternative function gaussxwab(N,a,b) that calculates the positions and weights and then does the mapping for you. To use this function, we would say "from gaussxw import gaussxwab", then

```
x,w = gaussxwab(N,a,b)
s = 0.0
for k in range(N):
    s += w[k]*f(x[k])
```

It's worth noting that the calculation of the sample points and weights takes quite a lot of work—the functions above may take a second or so to complete the calculation. That's fine if you call them only once in your program, but you should avoid calling them many times or you may find your program runs slowly. Thus, for instance, if you need to do many integrals over different domains of integration, you should call the function gaussxw once to calculate the sample points over the standard interval from -1 to $+1$ and then map the points yourself to the other integration domains you need. Calling gaussxwab separately for each different integration domain would be slow and wasteful, since it would needlessly recalculate the zeros of the Legendre polynomial each time.

Our Gaussian quadrature program is quite simple—only a little more complicated than the program for the trapezoidal rule in Example 5.1. Yet when we run it, it prints the following:

4.4

The program has calculated the answer exactly, with just three sample points! This is not a mistake, or luck, or a coincidence. It's exactly what we expect. Gaussian integration on N points gives exact answers for the integrals of polynomial functions up to and including polynomials of degree $2N - 1$, which means degree five when $N = 3$. The function $x^4 - 2x + 1$ that we are integrating here is a degree-four polynomial, so we expect the method to return the exact answer of 4.4, and indeed it does. Nonetheless, the performance of the program does seem almost magical in this case: the program has evaluated the integrand at just three points and from those three values alone it is, amazingly, able to deduce the integral of the entire function exactly.

This is the strength of Gaussian quadrature: it can give remarkably accurate answers, even with small numbers of sample points. This makes it especially useful in situations where you cannot afford to use large numbers of points, either because you need to be able to calculate an answer very quickly or because evaluating your integrand takes a long time even for just a few points.

The method does have its disadvantages. In particular, because the sample points are not uniformly distributed it takes more work if we want to employ the trick of repeatedly doubling N, as we did in Section 5.3, to successively improve the accuracy of the integral—if we change the value of N then all the sample points and weights have to be recalculated, and the entire sum over points, Eq. (5.52), has to be redone. We cannot reuse the calculations for old sample points as we did with the trapezoidal rule; in the language of computational physics we would say that the sample points are not nested.[6]

Exercise 5.9: Heat capacity of a solid

Debye's theory of solids gives the heat capacity of a solid at temperature T to be

$$C_V = 9V\rho k_B \left(\frac{T}{\theta_D}\right)^3 \int_0^{\theta_D/T} \frac{x^4 e^x}{(e^x - 1)^2}\, dx,$$

where V is the volume of the solid, ρ is the number density of atoms, k_B is Boltzmann's constant, and θ_D is the so-called *Debye temperature*, a property of solids that depends on their density and speed of sound.

a) Write a Python function cv(T) that calculates C_V for a given value of the temperature, for a sample consisting of 1000 cubic centimeters of solid aluminum, which has a number density of $\rho = 6.022 \times 10^{28}\,\mathrm{m}^{-3}$ and a Debye temperature of $\theta_D = 428\,\mathrm{K}$. Use Gaussian quadrature to evaluate the integral, with $N = 50$ sample points.

b) Use your function to make a graph of the heat capacity as a function of temperature from $T = 5\,\mathrm{K}$ to $T = 500\,\mathrm{K}$.

---

[6]There are other methods, such as *Gauss–Kronrod quadrature* and *Clenshaw–Curtis quadrature*, which have nonuniformly distributed sample points and still permit nesting, although these methods have their own disadvantages. Gauss–Kronrod quadrature permits only one step of nesting: it provides two sets of integration points, one nested inside the other, but no way to generate subsequent points nested inside those. Two sets of points are enough to make error estimates, via a formula analogous to Eq. (5.28), but one cannot keep on doubling the number of points to reduce the error below a given target, as with the adaptive method of Section 5.3. Clenshaw–Curtis quadrature does permit nesting over an arbitrary number of steps, but is not based on an integration rule of the simple form (5.52). Instead the method uses a more complicated formula whose evaluation involves, among other steps, performing a Fourier transform, which is more computationally demanding, and hence slower, than the simple sum used in Gaussian quadrature. In addition, neither Gauss–Kronrod quadrature nor Clenshaw–Curtis quadrature achieves the level of accuracy provided by Gaussian quadrature, although both are highly accurate and probably good enough for most purposes. Gauss–Kronrod quadrature in particular is widely used in mathematical software to compute definite integrals, because of its ability to provide both good accuracy and error estimates. Gauss–Kronrod quadrature is discussed further in Appendix C.

Exercise 5.10: Period of an anharmonic oscillator

The simple harmonic oscillator crops up in many places. Its behavior can be studied readily using analytic methods and it has the important property that its period of oscillation is a constant, independent of its amplitude, making it useful, for instance, for keeping time in watches and clocks.

Frequently in physics, however, we also come across anharmonic oscillators, whose period varies with amplitude and whose behavior cannot usually be calculated analytically. A general classical oscillator can be thought of as a particle in a concave potential well. When disturbed, the particle will rock back and forth in the well:

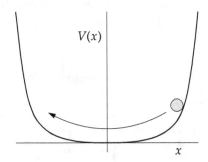

The harmonic oscillator corresponds to a quadratic potential $V(x) \propto x^2$. Any other form gives an anharmonic oscillator. (Thus there are many different kinds of anharmonic oscillator, depending on the exact form of the potential.)

One way to calculate the motion of an oscillator is to write down the equation for the conservation of energy in the system. If the particle has mass m and position x, then the total energy is equal to the sum of the kinetic and potential energies thus:

$$E = \tfrac{1}{2}m\left(\frac{dx}{dt}\right)^2 + V(x).$$

Since the energy must be constant over time, this equation is effectively a (nonlinear) differential equation linking x and t.

Let us assume that the potential $V(x)$ is symmetric about $x = 0$ and let us set our anharmonic oscillator going with amplitude a. That is, at $t = 0$ we release it from rest at position $x = a$ and it swings back towards the origin. Then at $t = 0$ we have $dx/dt = 0$ and the equation above reads $E = V(a)$, which gives us the total energy of the particle in terms of the amplitude.

a) When the particle reaches the origin for the first time, it has gone through one quarter of a period of the oscillator. By rearranging the equation above for dx/dt and then integrating with respect to t from 0 to $\tfrac{1}{4}T$, show that the period T is given by

$$T = \sqrt{8m} \int_0^a \frac{dx}{\sqrt{V(a) - V(x)}}.$$

173

b) Suppose the potential is $V(x) = x^4$ and the mass of the particle is $m = 1$. Write a Python function that calculates the period of the oscillator for given amplitude a using Gaussian quadrature with $N = 20$ points, then use your function to make a graph of the period for amplitudes ranging from $a = 0$ to $a = 2$.

c) You should find that the oscillator gets faster as the amplitude increases, even though the particle has further to travel for larger amplitude. And you should find that the period diverges as the amplitude goes to zero. How do you explain these results?

Exercise 5.11: Suppose a plane wave of wavelength λ, such as light or a sound wave, is blocked by an object with a straight edge, represented by the solid line at the bottom of this figure:

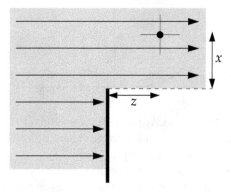

The wave will be diffracted at the edge and the resulting intensity at the position (x, z) marked by the dot is given by near-field diffraction theory to be

$$I = \frac{I_0}{8} \left([2C(u) + 1]^2 + [2S(u) + 1]^2 \right),$$

where I_0 is the intensity of the wave before diffraction and

$$u = x\sqrt{\frac{2}{\lambda z}}, \qquad C(u) = \int_0^u \cos \tfrac{1}{2} \pi t^2 \, dt, \qquad S(u) = \int_0^u \sin \tfrac{1}{2} \pi t^2 \, dt.$$

Write a program to calculate I/I_0 and make a plot of it as a function of x in the range -5 m to 5 m for the case of a sound wave with wavelength $\lambda = 1$ m, measured $z = 3$ m past the straight edge. Calculate the integrals using Gaussian quadrature with $N = 50$ points. You should find significant variation in the intensity of the diffracted sound— enough that you could easily hear the effect if sound were diffracted, say, at the edge of a tall building.

5.6.3 ERRORS ON GAUSSIAN QUADRATURE

In our study of the trapezoidal rule we derived an expression, the Euler–Maclaurin formula of Eq. (5.20), for the approximation error on the value of an integral. There exists a corresponding expression for Gaussian quadrature but it is, unfortunately, ungainly and not easy to use in practice. What it does tell us, however, is that Gaussian quadrature is impressively accurate. Roughly speaking, the approximation error—the difference between the value of an integral calculated using Gaussian quadrature and the true value of the same integral, neglecting rounding error—improves by a factor of c/N^2 when we increase the number of samples by *just one*, where c is a constant whose value depends on the detailed shape of the integrand and the size of the domain of integration. Thus, for instance, if we go from $N = 10$ to $N = 11$ our estimate of the integral will improve by a factor of order a hundred. This means that we converge extremely quickly on the true value of the integral, and in practice it is rarely necessary to use more than a few tens of points, or at most perhaps a hundred, to get an estimate of an integral accurate to the limits of precision of the computer.

There are some caveats. An important one is that the function being integrated must be reasonably smooth. When one is calculating an integral using a relatively small number of sample points, the points will inevitably be far apart, which leaves room for the function to vary significantly between them. Since Gaussian quadrature looks only at the values of the function at the sample points and nowhere else, substantial variation between points is not taken into account in calculating the value of the integral. If, on the other hand, the function is relatively smooth, then the samples we take will give a good approximation of the function's behavior and Gaussian quadrature will work well. Thus for rapidly varying functions one needs to use enough sample points to capture the variation, and in such cases larger values of N may be warranted.

Another issue is that there is no direct equivalent of Eq. (5.28) for estimating the error in practice. As we have said, however, the error improves by a factor of c/N^2 when the number of samples is increased by one, which is typically a substantial improvement if N is reasonably large. And if we *double* the value of N then we compound many such improvements, giving an overall reduction in the error by a factor of something like N^{-2N}, which is typically a huge improvement.

If we make a Gaussian estimate I_N of the true value I of an integral using

N sample points, then $I = I_N + \epsilon_N$, where ϵ_N is the approximation error. And if we double the number of samples to $2N$ we have, $I = I_{2N} + \epsilon_{2N}$. Equating the two expressions for I and rearranging, we have

$$\epsilon_N - \epsilon_{2N} = I_{2N} - I_N. \tag{5.65}$$

But, as we have argued, the error is expected to improve by a large factor when we double the number of sample points, meaning that $\epsilon_{2N} \ll \epsilon_N$. So, to a good approximation,

$$\epsilon_N \simeq I_{2N} - I_N. \tag{5.66}$$

Another way of saying this is that I_{2N} is so much better an estimate of the true value of the integral than I_N that for the purposes of estimating the error we can treat it as if it were the true value, so that $I_{2N} - I_N$ is a good estimate of the error.

We can use Eq. (5.66) in an adaptive integration method where we double the number of sample points at each step, calculating the error and repeating until the desired target accuracy is reached. Such a method is not entirely satisfactory, for a couple of reasons. First, when we double the number of sample points from N to $2N$, Eq. (5.66) gives us only the error on the *previous* estimate of the integral I_N, not on the new estimate I_{2N}. This means that we always end up doubling N one more time than is strictly necessary to achieve the desired accuracy, and that the final value for the integral will probably be significantly more accurate than we really need it to be, which means we have wasted time on unnecessary calculations. Second, we have to perform the entire calculation of the integral anew for each new value of N. As mentioned earlier, and unlike the adaptive trapezoidal method of Section 5.3, we cannot reuse the results of earlier calculations to speed up the computation. So an adaptive calculation of this type would be slower than just a single instance of Gaussian quadrature. On the other hand, it's straightforward to show that the total number of terms in all the sums we perform, over all steps of the process, is never greater than twice the final value of N used, which means that the adaptive procedure costs us no more than about twice the effort required for the simple Gaussian quadrature. Moreover, as we have said, we rarely need to go beyond $N = 100$ to get a highly accurate answer, so the number of times we double N is typically rather small. If we start with, say, $N = 10$, we will probably only have to double three or four times. The net result is that, despite the extra work, Gaussian quadrature is often more efficient than methods like the trapezoidal rule or Simpson's rule in terms of overall time needed to get an answer to a desired degree of accuracy.

An alternative, though more complex, solution to the problem of estimating the error in Gaussian quadrature is to use *Gauss–Kronrod quadrature*, a variant of Gaussian quadrature based on the properties of Stieltjes polynomials, which provides not only an accurate estimate of our integral (though not quite as accurate as ordinary Gaussian quadrature) but also an estimate of the error. We will not use Gauss–Kronrod quadrature in this book, but the interested reader can find a short discussion, with some derivations, in Appendix C.

5.7   CHOOSING AN INTEGRATION METHOD

We have studied a number of different integration methods in this chapter: the trapezoidal rule and Simpson's rule as well as adaptive versions of both, Romberg integration, and Gaussian integration. You might well ask at this point which of all these methods is the best? Which one should I use, in practice, if I need to evaluate an integral?

There is no one answer to this question. Which method you should use depends on the particular problem confronting you. A good general principle, however, is that higher-order methods such as Romberg and Gaussian integration—methods that allow you to make accurate estimates of integrals using relatively few sample points—work best when applied to smooth, well-behaved functions. If your function is not smooth or is poorly behaved in some way, then simpler methods, and particularly the trapezoidal rule, are the way to go. The reason is that any integration method knows about the value of the integrand only at its sample points. If the integrand varies significantly in between the sample points, then that variation will not be reflected in the computed value of the integral, which can lead to inaccurate results. If you're evaluating an integral using only ten or twenty sample points, it's crucial that those points give a good picture of the integrand—if you join up the dots the result should capture most of the shape of the function. If it does not then high-order methods using few sample points will not do a good job.

Bearing this principle in mind, here is a guide to the kinds of problems each of our integration methods is good for.

The trapezoidal rule: The trapezoidal rule of Section 5.1.1 is trivial to program and hence is a good choice when you need a quick answer for an integral. It's not very accurate, but sometimes you don't need great accuracy. It uses equally spaced sample points, which is appropriate for problems such as integrating data from laboratory experiments that are sampled at uniform time intervals. The trapezoidal rule is also a good choice for poorly behaved functions—those that vary widely, contain singularities, or are noisy. It is usually a better choice for such functions than the other methods we have considered. In its adaptive form (Section 5.3) it can also give us a guaranteed accuracy for an integral, although it may take more computer time to achieve that accuracy than other methods.

Simpson's rule: Simpson's rule (Section 5.1.2) has many of the benefits of the trapezoidal rule, such as simplicity of programming and equally spaced sample points. It gives greater accuracy than the trapezoidal rule with the same number of sample points, or the same accuracy with fewer points, but relies on higher-order approximation of the integrand, which can lead to problems if the integrand is noisy or otherwise not smooth—use it with caution if you are unsure of the nature of your integrand. Its adaptive form (Section 5.3) provides a result of guaranteed accuracy, and does so faster than the equivalent trapezoidal rule calculation, but again may be less suitable for poorly behaved integrands.

Romberg integration: When using equally spaced sample points, Romberg integration (Section 5.4) is the quintessential higher-order integration method. It gives exceptionally accurate estimates of integrals with a minimum number of sample points, plus error estimates that allow you to halt the calculation once you have achieved a desired accuracy. Since it relies on extrapolating answers from measurements of the integrand at only a few points, however, Romberg integration will not work well for wildly varying integrands, noisy integrands, or integrands with mathematically pathological behaviors like singularities. It is best applied to smooth functions whose form can be determined accurately from only a small number of sample points.

Gaussian quadrature: Gaussian quadrature (Section 5.6) has many of the same advantages as Romberg integration (potentially highly accurate estimates from small numbers of sample points) as well as the same disadvantages (poor performance for badly behaved integrands). It is also simple to program, as simple as any of the other methods we have considered. The hard work of the method lies in the calculation of the integration points and weights, which is normally done for you by standard software, and the Gaussian integral itself requires only the evaluation of a single sum in the form of Eq. (5.52). It has the additional advantage over Romberg integration of still higher-order accuracy and indeed, in a certain formal sense, it is the highest-order, and hence potentially most accurate, integration rule available. The price you pay for this is that the integration points are unequally spaced. If you need equally spaced points, then Gaussian quadrature is not the method for you.

Armed with these guidelines, you should be able to choose a suitable integration method for most problems you come up against.

5.8 INTEGRALS OVER INFINITE RANGES

Often in physics we encounter integrals over infinite ranges, like $\int_0^\infty f(x) \, dx$. The techniques we have seen so far don't work for these integrals because we'd need an infinite number of sample points to span an infinite range. The solution to this problem is to change variables. For an integral over the range from 0 to ∞ the standard change of variables is

$$z = \frac{x}{1+x} \qquad \text{or equivalently} \qquad x = \frac{z}{1-z}. \tag{5.67}$$

Then $dx = dz/(1-z)^2$ and

$$\int_0^\infty f(x) \, dx = \int_0^1 \frac{1}{(1-z)^2} f\left(\frac{z}{1-z}\right) dz, \tag{5.68}$$

which can be done using any of the techniques earlier in the chapter, including the trapezoidal and Simpson rules, or Gaussian quadrature.

This is not the only change of variables that we can use, however. In fact, a change of the form

$$z = \frac{x}{c+x} \tag{5.69}$$

would work for any value of c, or $z = x^\gamma/(1+x^\gamma)$ for any γ, or any of a range of other possibilities. Some choices typically work better than others for particular integrals and sometimes you have to play around with things a little to find what works for a given problem, but Eq. (5.67) is often a good first guess. (See Exercise 5.17 for a counterexample.)

To do an integral over a range from some nonzero value a to ∞ we can use a similar approach, but make two changes of variables, first to $y = x - a$, which shifts the start of the integration range to 0, and then $z = y/(1+y)$ as in Eq. (5.67). Or we can combine both changes into a single one:

$$z = \frac{x-a}{1+x-a} \qquad \text{or} \qquad x = \frac{z}{1-z} + a, \tag{5.70}$$

and again $dx = dz/(1-z)^2$, so that

$$\int_a^\infty f(x) \, dx = \int_0^1 \frac{1}{(1-z)^2} f\left(\frac{z}{1-z} + a\right) dz. \tag{5.71}$$

179

Integrals from $-\infty$ to a can be done the same way—just substitute $z \to -z$.

For integrals that run from $-\infty$ to ∞ we can split the integral into two parts, one from $-\infty$ to 0 and one from 0 to ∞, and then use the tricks above for the two integrals separately. Or we can put the split at some other point a and perform separate integrals from $-\infty$ to a and from a to ∞. Alternatively, one could use a single change of variables, such as

$$x = \frac{z}{1-z^2}, \qquad dx = \frac{1+z^2}{(1-z^2)^2}\,dz, \tag{5.72}$$

which would give

$$\int_{-\infty}^{\infty} f(x)\,dx = \int_{-1}^{1} \frac{1+z^2}{(1-z^2)^2} f\left(\frac{z}{1-z^2}\right) dz. \tag{5.73}$$

Another possibility, perhaps simpler, is

$$x = \tan z, \qquad dx = \frac{dz}{\cos^2 z}, \tag{5.74}$$

which gives

$$\int_{-\infty}^{\infty} f(x)\,dx = \int_{-\pi/2}^{\pi/2} \frac{f(\tan z)}{\cos^2 z}\,dz. \tag{5.75}$$

EXAMPLE 5.3: INTEGRATING OVER AN INFINITE RANGE

Let us calculate the value of the following integral using Gaussian quadrature:

$$I = \int_0^{\infty} e^{-t^2}\,dt. \tag{5.76}$$

We make the change of variables given in Eq. (5.67) and the integral becomes

$$I = \int_0^1 \frac{e^{-z^2/(1-z)^2}}{(1-z)^2}\,dz. \tag{5.77}$$

We can modify our program from Example 5.2 to perform this integral using Gaussian quadrature with $N = 50$ sample points:

File: intinf.py

```
from gaussxw import gaussxwab
from math import exp

def f(z):
    return exp(-z**2/(1-z)**2)/(1-z)**2
```

```
N = 50
a = 0.0
b = 1.0
x,w = gaussxwab(N,a,b)
s = 0.0
for k in range(N):
    s += w[k]*f(x[k])
print(s)
```

If we run this program it prints

0.886226925453

In fact, the value of this particular integral is known exactly to be $\frac{1}{2}\sqrt{\pi} = 0.886226925453\ldots$ Again we see the impressive accuracy of the Gaussian quadrature method: with just 50 sample points, we have calculated an estimate of the integral that is correct to the limits of precision of the computer.

_____

Exercise 5.12: The Stefan–Boltzmann constant

The Planck theory of thermal radiation tells us that in the (angular) frequency interval ω to $\omega + d\omega$, a black body of unit area radiates electromagnetically an amount of thermal energy per second equal to $I(\omega)\,d\omega$, where

$$I(\omega) = \frac{\hbar}{4\pi^2 c^2}\,\frac{\omega^3}{(e^{\hbar\omega/k_B T} - 1)}.$$

Here $\hbar$ is Planck's constant over 2π, c is the speed of light, and k_B is Boltzmann's constant.

a) Show that the total rate at which energy is radiated by a black body per unit area, over all frequencies, is

$$W = \frac{k_B^4 T^4}{4\pi^2 c^2 \hbar^3}\int_0^\infty \frac{x^3}{e^x - 1}\,dx.$$

b) Write a program to evaluate the integral in this expression. Explain what method you used, and how accurate you think your answer is.

c) Even before Planck gave his theory of thermal radiation around the turn of the 20th century, it was known that the total energy W given off by a black body per unit area per second followed Stefan's law: $W = \sigma T^4$, where σ is the Stefan–Boltzmann constant. Use your value for the integral above to compute a value for the Stefan–Boltzmann constant (in SI units) to three significant figures. Check your result against the known value, which you can find in books or on-line. You should get good agreement.

Exercise 5.13: Quantum uncertainty in the harmonic oscillator

In units where all the constants are 1, the wavefunction of the nth energy level of the one-dimensional quantum harmonic oscillator—i.e., a spinless point particle in a quadratic potential well—is given by

$$\psi_n(x) = \frac{1}{\sqrt{2^n n! \sqrt{\pi}}} e^{-x^2/2} H_n(x),$$

for $n = 0 \ldots \infty$, where $H_n(x)$ is the nth Hermite polynomial. Hermite polynomials satisfy a relation somewhat similar to that for the Fibonacci numbers, although more complex:

$$H_{n+1}(x) = 2x H_n(x) - 2n H_{n-1}(x).$$

The first two Hermite polynomials are $H_0(x) = 1$ and $H_1(x) = 2x$.

a) Write a user-defined function `H(n,x)` that calculates $H_n(x)$ for given x and any integer $n \geq 0$. Use your function to make a plot that shows the harmonic oscillator wavefunctions for $n = 0, 1, 2,$ and 3, all on the same graph, in the range $x = -4$ to $x = 4$. Hint: There is a function `factorial` in the `math` package that calculates the factorial of an integer.

b) Make a separate plot of the wavefunction for $n = 30$ from $x = -10$ to $x = 10$. Hint: If your program takes too long to run in this case, then you're doing the calculation wrong—the program should take only a second or so to run.

c) The quantum uncertainty in the position of a particle in the nth level of a harmonic oscillator can be quantified by its root-mean-square position $\sqrt{\langle x^2 \rangle}$, where

$$\langle x^2 \rangle = \int_{-\infty}^{\infty} x^2 |\psi_n(x)|^2 \, dx.$$

Write a program that evaluates this integral using Gaussian quadrature on 100 points, then calculates the uncertainty (i.e., the root-mean-square position of the particle) for a given value of n. Use your program to calculate the uncertainty for $n = 5$. You should get an answer in the vicinity of $\sqrt{\langle x^2 \rangle} = 2.3$.

5.9 MULTIPLE INTEGRALS

Integrals over more than one variable are common in physics problems and can be tackled using generalizations of the methods we have already seen. Consider for instance the integral

$$I = \int_0^1 \int_0^1 f(x,y) \, dx \, dy. \tag{5.78}$$

We can rewrite this by defining a function $F(y)$ thus

$$F(y) = \int_0^1 f(x,y) \, dx. \tag{5.79}$$

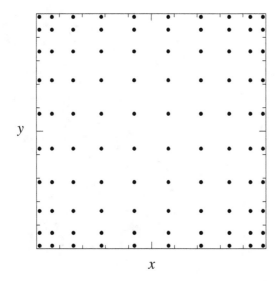

y

x

Figure 5.5: Sample points for Gaussian quadrature in two dimensions. If one applies Eq. (5.82) to integrate the function $f(x, y)$ in two dimensions, using Gaussian quadrature with $N = 10$ points along each axis, the resulting set of sample points in the two-dimensional space looks like this.

Then

$$I = \int_0^1 F(y)\, dy. \tag{5.80}$$

Thus one way to do the multiple integral numerically is first to evaluate $F(y)$ for a suitable set of y values, which means performing the integral in Eq. (5.79), then using those values of $F(y)$ to do the integral in Eq. (5.80). For instance, if we do the integrals by Gaussian quadrature with the same number N of points for both x and y integrals, we have

$$F(y) \simeq \sum_{i=1}^{N} w_i f(x_i, y) \quad \text{and} \quad I \simeq \sum_{j=1}^{N} w_j F(y_j). \tag{5.81}$$

An alternative way to look at the calculation is to substitute the first sum into the second to get the *Gauss–Legendre product formula*:

$$I \simeq \sum_{i=1}^{N} \sum_{j=1}^{N} w_i w_j f(x_i, y_j). \tag{5.82}$$

This expression has a form similar to the standard integration formula for single integrals, Eq. (5.52), with a sum over values of the function $f(x, y)$ at a set of

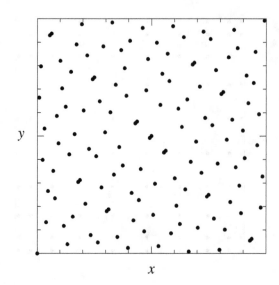

y

x

Figure 5.6: 128-point Sobol sequence. The Sobol sequence is one example of a low-discrepancy point set that gives good results for integrals in high dimensions. This figure shows a Sobol sequence of 128 points in two dimensions.

sample points, multiplied by appropriate weights. Equation (5.82) represents a kind of two-dimensional Gaussian quadrature, with weights $w_i w_j$ distributed over a two-dimensional grid of points as shown in Fig. 5.5.

Once you look at it this way, however, you realize that in principle there's no reason why the sample points have to be on a grid. They could be anywhere—we can use any set of 2D locations and suitable weights that give a good estimate of the integral. Just as Gaussian quadrature gives the best choice of points for an integral in one dimension, so we can ask what the best choice is for two dimensions, or for higher dimensions like three or four. It turns out, however, that the answer to this question is not known in general. There are some results for special cases, but no general answer. Various point sets have been proposed for use with 2D integrals that appear to give reasonable results, but there is no claim that they are the best possible choices. Typically they are selected because they have some other desirable properties, such as nesting, and not because they give the most accurate answer. One common choice of point set is the *Sobol sequence*, shown for $N = 128$ points in Fig. 5.6. Sobol sequences and similar sets of points are known as *low-discrepancy point sets* or sometimes *quasi-random point sets* (although the latter name is a poor

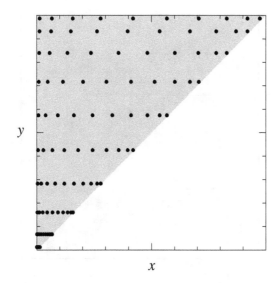

Figure 5.7: Integration over a non-rectangular domain. When the limits of multiple integrals depend on one another they can produce arbitrarily shaped domains of integration. This figure shows the triangular domain that results from the integral in Eq. (5.83). The gray region is the domain of integration. Note how the points become squashed together towards the bottom of the plot.

one because there's nothing random about them). Another common way to choose the sample points is to choose them completely randomly, which leads to the method known as Monte Carlo integration. Choosing points at random may seem like an odd idea, but as we will see it can be a useful approach for certain types of integrals, particularly integrals over very many variables. We will look at Monte Carlo integration in Section 10.2, after we study random number generators.

In the integral of Eq. (5.78) the limits of both integrals were constant, which made the domain of integration rectangular in xy space. It's not uncommon, however, for the limits of one integral to depend on the other, as here:

$$I = \int_0^1 dy \int_0^y dx\, f(x,y). \tag{5.83}$$

We can use the same approach as before to evaluate this integral. We define

$$F(y) = \int_0^y f(x,y)\, dx, \tag{5.84}$$

Figure 5.8: A complicated integration domain. Integration domains can be arbitrarily complicated in their shapes. They can even contain holes, or take on complex topologies in higher dimensions such as tori or knotted topologies.

so that

$$I = \int_0^1 F(y)\, dy, \tag{5.85}$$

and then do both integrals with any method we choose, such as Gaussian quadrature. The result, again, is a two-dimensional integration rule, but now with the sample points arranged in a triangular space as shown in Fig. 5.7.

This method will work, and will probably give reasonable answers, but it's not ideal. In particular note how the sample points are crammed together in the lower left corner of the integration domain but much farther apart at the top. This means, all other things being equal, that we'll have lower accuracy for the part of the integral at the top. It would be better if the accuracy were roughly uniform.

And things can get worse still. Suppose the domain of integration takes some more complicated shape like Fig. 5.8. We will not come across any examples this complicated in this book, but if we did there would be various techniques we could use. One is the Monte Carlo integration method mentioned above, which we study in detail in Section 10.2. Another is to set the integrand to zero everywhere outside the domain of integration and then integrate it using a standard method over some larger, regularly shaped domain, such as a rectangle, that completely encloses the irregular one. There are many more sophisticated techniques as well, but we will not need them for the moment.

Exercise 5.14: Gravitational pull of a uniform sheet

A uniform square sheet of metal is floating motionless in space:

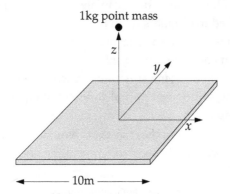

The sheet is $10\,\text{m}$ on a side and of negligible thickness, and it has a mass of 10 metric tonnes.

a) Consider the gravitational force due to the plate felt by a point mass of $1\,\text{kg}$ a distance z from the center of the square, in the direction perpendicular to the sheet, as shown above. Show that the component of the force along the z-axis is

$$F_z = G\sigma z \iint_{-L/2}^{L/2} \frac{dx\,dy}{(x^2 + y^2 + z^2)^{3/2}},$$

where $G = 6.674 \times 10^{-11}\,\text{m}^3\,\text{kg}^{-1}\,\text{s}^{-2}$ is Newton's gravitational constant and σ is the mass per unit area of the sheet.

b) Write a program to calculate and plot the force as a function of z from $z = 0$ to $z = 10\,\text{m}$. For the double integral use (double) Gaussian quadrature, as in Eq. (5.82), with 100 sample points along each axis.

c) You should see a smooth curve, except at very small values of z, where the force should drop off suddenly to zero. This drop is not a real effect, but an artifact of the way we have done the calculation. Explain briefly where this artifact comes from and suggest a strategy to remove it, or at least to decrease its size.

This calculation can thought of as a model for the gravitational pull of a galaxy. Most of the mass in a spiral galaxy (such as our own Milky Way) lies in a thin plane or disk passing through the galactic center, and the gravitational pull exerted by that plane on bodies outside the galaxy can be calculated by just the methods we have employed here.

5.10 DERIVATIVES

The opposite of a numerical integral is a numerical derivative. You hear a lot less about numerical derivatives than integrals, however, for a number of reasons:

1. The basic techniques for numerical derivatives are quite simple, so they don't take long to explain.
2. Derivatives of known functions can always be calculated analytically, so there's less need to calculate them numerically.
3. There are some significant practical problems with numerical derivatives, which means they are used less often than numerical integrals. (There are, however, some situations in which they are important, particularly in the solution of partial differential equations, which we will look at in Chapter 9.)

For all of these reasons this is a short section—you need to know about numerical derivatives, but we won't spend too much time on them.

5.10.1 FORWARD AND BACKWARD DIFFERENCES

The standard definition of a derivative, the one you see in the calculus books, is

$$\frac{df}{dx} = \lim_{h \to 0} \frac{f(x+h) - f(x)}{h}. \tag{5.86}$$

The basic method for calculating numerical derivatives is precisely an implementation of this formula. We can't take the limit $h \to 0$ in practice, but we can make h very small and then calculate

$$\frac{df}{dx} \simeq \frac{f(x+h) - f(x)}{h}. \tag{5.87}$$

This approximation to the derivative is called the *forward difference*, because it's measured in the forward (i.e., positive) direction from the point of interest x. You can think of it in geometric terms as shown in Fig. 5.9—it's simply the slope of the curve $f(x)$ measured over a small interval of width h in the forward direction from x.

There is also the *backward difference*, which has the mirror image definition

$$\frac{df}{dx} \simeq \frac{f(x) - f(x-h)}{h}. \tag{5.88}$$

The forward and backward differences typically give about the same answer and in many cases you can use either. Most often one uses the forward difference. There are a few special cases where one is preferred over the other,

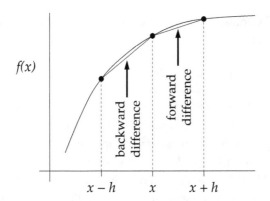

Figure 5.9: Forward and backward differences. The forward and backward differences provide two different approximations to the derivative of a function $f(x)$ at the point x in terms of the slopes of small segments measured in the forward (i.e., positive) direction from x and the backward (negative) direction, respectively.

particularly when there is a discontinuity in the derivative of the function at the point x or when the domain of the function is bounded and you want the value of the derivative on the boundary, in which case only one or other of the two difference formulas will work. The rest of the time, however, there is little to choose between them.

Before using either the forward or backward difference we must choose a value for h. To work out what the best value is we need to look at the errors and inaccuracies involved in calculating numerical derivatives.

5.10.2 ERRORS

Calculations of derivatives using forward and backward differences are not perfectly accurate. There are two sources of error. The first is rounding error of the type discussed in Section 4.2. The second is the approximation error that arises because we cannot take the limit $h \to 0$, so our differences are not really true derivatives. By contrast with numerical integrals, where, as we have seen, rounding error is usually negligible, it turns out that both sources of error are important when we calculate a derivative.

To understand why this is let us focus on the forward difference and consider the Taylor expansion of $f(x)$ about x:

$$f(x+h) = f(x) + hf'(x) + \tfrac{1}{2}h^2 f''(x) + \ldots \qquad (5.89)$$

where f' and f'' denote the first and second derivatives of f. Rearranging this expression, we get

$$f'(x) = \frac{f(x+h) - f(x)}{h} - \tfrac{1}{2}hf''(x) + \dots \tag{5.90}$$

When we calculate the forward difference we calculate only the first part on the right-hand side, and neglect the term in $f''(x)$ and all higher terms. The size of these neglected terms measures the approximation error on the forward difference. Thus, to leading order in h, the absolute magnitude of the approximation error is $\tfrac{1}{2}h\,|f''(x)|$, which is linear in h so that, as we would expect, we should get more accurate answers if we use smaller values of h.

But now here is the problem: as we saw in Section 4.2, subtracting numbers from one another on a computer can give rise to big rounding errors (in fractional terms) if the numbers are close to one another. And that's exactly what happens here—the numbers $f(x+h)$ and $f(x)$ that we are subtracting will be very close to one another if we make h small. Thus if we make h too small, we will get a large rounding error in our result. This puts us in a difficult situation: we want to make h small to make the forward difference approximation as accurate as possible, but if we make it too small we will get a large rounding error. To get the best possible answer, we are going to have to find a compromise.

In Section 4.2 we saw that the computer can typically calculate a number such as $f(x)$ to an accuracy of $Cf(x)$, where the value of the error constant C can vary but is typically about $C = 10^{-16}$ in Python. Since $f(x+h)$ is normally close in value to $f(x)$, the accuracy of our value for $f(x+h)$ will also be about the same, and the absolute magnitude of the total rounding error on $f(x+h) - f(x)$ will, in the worst case, be about $2C\,|f(x)|$—it might be better than this if the two errors go in opposite directions and happen to cancel out, but we cannot assume that this will be the case. Then the worst-case rounding error on the complete forward difference, Eq. (5.87), will be $2C|f(x)|/h$.

Meanwhile, the approximation error is, as we have said, about $\tfrac{1}{2}h\,|f''(x)|$ from Eq. (5.90), which means that the total error ϵ on our derivative, in the worst case, is

$$\epsilon = \frac{2C\,|f(x)|}{h} + \tfrac{1}{2}h\,|f''(x)|. \tag{5.91}$$

We want to find the value of h that minimizes this error, so we differentiate with respect to h and set the result equal to zero, which gives

$$-\frac{2C\,|f(x)|}{h^2} + \tfrac{1}{2}\,|f''(x)| = 0, \tag{5.92}$$

or equivalently

$$h = \sqrt{4C \left| \frac{f(x)}{f''(x)} \right|}. \tag{5.93}$$

Substituting this value back into Eq. (5.91) we find that the error on our derivative is

$$\epsilon = h\,|f''(x)| = \sqrt{4C\,|f(x)f''(x)|}. \tag{5.94}$$

Thus, for instance, if $f(x)$ and $f''(x)$ are of order 1, we should choose h to be roughly of order $\sqrt{C}$, which will be typically about 10^{-8}, and the final error on our result will also be about $\sqrt{C}$ or 10^{-8}. A similar analysis can be applied to the backward difference, and gives the same end result.

In other words, we can get about half of the usual numerical precision on our derivatives but not better. If the precision is, as here, about 16 digits, then we can get 8 digits of precision on our derivatives. This is substantially poorer than most of the calculations we have seen so far in this book, and could be a significant source of error for calculations that require high accuracy.

5.10.3 Central differences

We have seen that forward and backward differences are not very accurate. What can we do to improve the situation? A simple improvement is to use the *central difference*:

$$\frac{df}{dx} \simeq \frac{f(x+h/2) - f(x-h/2)}{h}. \tag{5.95}$$

The central difference is similar to the forward and backward differences, approximating the derivative using the difference between two values of $f(x)$ at points a distance h apart. What's changed is that the two points are now placed symmetrically around x, one at a distance $\frac{1}{2}h$ in the forward (i.e., positive) direction and the other at a distance $\frac{1}{2}h$ in the backward (negative) direction.

To calculate the approximation error on the central difference we write two Taylor expansions:

$$f(x+h/2) = f(x) + \tfrac{1}{2}hf'(x) + \tfrac{1}{8}h^2 f''(x) + \tfrac{1}{48}h^3 f'''(x) + \ldots \tag{5.96}$$
$$f(x-h/2) = f(x) - \tfrac{1}{2}hf'(x) + \tfrac{1}{8}h^2 f''(x) - \tfrac{1}{48}h^3 f'''(x) + \ldots \tag{5.97}$$

Subtracting the second expression from the first and rearranging for $f'(x)$, we get

$$f'(x) = \frac{f(x+h/2) - f(x-h/2)}{h} - \tfrac{1}{24}h^2 f'''(x) + \ldots \tag{5.98}$$

To leading order the magnitude of the error is now $\frac{1}{24}h^2\,|f'''(x)|$, which is one order in h higher than before. There is also, as before, a rounding error; its size is unchanged from our previous calculation, having magnitude $2C\,|f(x)|/h$, so the magnitude of the total error on our estimate of the derivative is

$$\epsilon = \frac{2C\,|f(x)|}{h} + \frac{1}{24}h^2\,|f'''(x)|.\tag{5.99}$$

Differentiating to find the minimum and rearranging, we find that the optimal value of h is

$$h = \left(24C\left|\frac{f(x)}{f'''(x)}\right|\right)^{1/3},\tag{5.100}$$

and substituting this back into Eq. (5.99) we find the optimal error itself to be

$$\epsilon = \tfrac{1}{8}h^2\,|f'''(x)| = \left(\tfrac{9}{8}C^2[f(x)]^2|f'''(x)|\right)^{1/3}.\tag{5.101}$$

Thus, for instance, if $f(x)$ and $f'''(x)$ are of order 1, the ideal value of h is going to be around $h \simeq C^{1/3}$, which is typically about 10^{-5} but the error itself will be around $C^{2/3}$, or about 10^{-10}.

Thus the central difference is indeed more accurate than the forward and backward differences, by a factor of 100 or so in this case, though we get this accuracy by using a *larger* value of h. This may seem slightly surprising, but it is the correct result.

EXAMPLE 5.4: DERIVATIVE OF A SAMPLED FUNCTION

As an example application of the central difference, suppose we are given the values of a function $f(x)$ measured at regularly spaced sample points a distance h apart—see Fig. 5.10. One often gets such samples from data collected in the laboratory, for example. Now suppose we want to calculate the derivative of f at one of these points (case (a) in the figure). We could use a forward or backward difference based on the sample at x and one of the adjacent ones, or we could use a central difference. However, if we use a central difference, which is based on points equally spaced on either side of x, then we must use the points at $x+h$ and $x-h$. We cannot, as in Eq. (5.95), use points at $x+h/2$ and $x-h/2$ because there are no such points—we only have the samples we are given. The formula for the central difference in this case will thus be

$$\frac{df}{dx} \simeq \frac{f(x+h) - f(x-h)}{2h}.\tag{5.102}$$

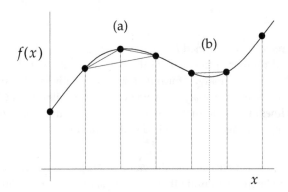

Figure 5.10: Derivative of a sampled function. (a) If we only know the function at a set of sample points spaced a distance h apart then we must chose between calculating the forward or backward difference between adjacent samples, or the central difference between samples $2h$ apart. We cannot calculate a central difference using the standard formula, Eq. (5.95), because we do not know the value of the function at $x \pm \frac{1}{2}h$. (b) We can, however, calculate the value of the derivative at a point half way between two samples (dotted line) using the standard formula.

This means that the interval between the points we use is $2h$ for the central difference, but only h for the forward and backward differences. So which will give a better answer? The central difference because it's a better approximation or the forward difference because of its smaller interval?

From Eq. (5.94) we see the error on the forward difference is $h\,|f''(x)|$ and from Eq. (5.101) the error on the central difference—with h replaced by $2h$—is $h^2\,|f'''(x)|$. Which is smaller depends on the value of h. For the central difference to give the more accurate answer, we require $h^2\,|f'''(x)| < h\,|f''(x)|$ or

$$h < \left| \frac{f''(x)}{f'''(x)} \right|. \tag{5.103}$$

If h is larger than this then the forward difference is actually the better approximation in this case.

But now suppose that instead of calculating the value of the derivative at one of the sample points itself, we want to calculate it at a point x that lies half way between two of the samples—case (b) in Fig. 5.10. Viewed from that point we do have samples at $x + h/2$ and $x - h/2$, so now we can use the original form of the central difference, Eq. (5.95), with an interval only h wide, as with the forward difference. This calculation will give a more accurate answer, but only at the expense of calculating the result at a point in between the samples.

Exercise 5.15: Create a user-defined function $f(x)$ that returns the value $1 + \frac{1}{2}\tanh 2x$, then use a central difference to calculate the derivative of the function in the range $-2 \leq x \leq 2$. Calculate an analytic formula for the derivative and make a graph with your numerical result and the analytic answer on the same plot. It may help to plot the exact answer as lines and the numerical one as dots. (Hint: In Python the tanh function is found in the math package, and it's called simply tanh.)

Exercise 5.16: Even when we can find the value of $f(x)$ for any value of x the forward difference can still be more accurate than the central difference for sufficiently large h. For what values of h will the approximation error on the forward difference of Eq. (5.87) be smaller than on the central difference of Eq. (5.95)?

5.10.4 HIGHER-ORDER APPROXIMATIONS FOR DERIVATIVES

One way to think about the numerical derivatives of the previous sections is that we are fitting a straight line through two points, such as the points $f(x)$ and $f(x+h)$, and then asking about the slope of that line at the point x. The trapezoidal rule of Section 5.1.1 does a similar thing for integrals, approximating a curve by a straight line between sample points and estimating the area under the curve using that line. We saw that we can make a higher-order—and usually better—approximation to an integral by fitting a quadratic or higher-order polynomial instead of a straight line, and this led to the Simpson and Newton–Cotes rules for integrals. We can take a similar approach with derivatives by fitting a polynomial to a set of sample points and then calculating the derivative of the polynomial at x.

Consider, for example, fitting a quadratic curve $y = ax^2 + bx + c$ to the function $f(x)$. We require three sample points to make the fit and, as with the central difference of Section 5.10.3, the best results are obtained by placing the points symmetrically about the point of interest x. Suppose, for example, that we are interested in the derivative at $x = 0$, so we place our three points at $-h$, 0, and $+h$, for some h that we choose. Requiring that our quadratic is equal to $f(x)$ at these three points gives us three equations thus:

$$ah^2 - bh + c = f(-h), \qquad c = f(0), \qquad ah^2 + bh + c = f(h), \qquad (5.104)$$

In principle, we can now solve these equations for the three parameters a, b, and c. (This is the same calculation that we did in Section 5.1.2 for Simpson's rule.) However, in this case, we don't need the whole solution, because we don't need all of the parameters. Given the quadratic fit $y = ax^2 + bx + c$, the derivative of the curve at the point $x = 0$ is

$$\frac{dy}{dx} = \left[2ax + b\right]_{x=0} = b. \tag{5.105}$$

So we need only the one parameter b, which we can get from Eq. (5.104) by subtracting the first equation from the third to give $2bh = f(h) - f(-h)$ and rearranging. Thus our approximation for the derivative at $x = 0$ is

$$\frac{df}{dx} \simeq \frac{f(h) - f(-h)}{2h}. \tag{5.106}$$

We have done this calculation for the derivative at $x = 0$, but the same result applies at any other point—we can slide the whole function up or down the x-axis, to put any point x at the origin and then calculate the derivative from the formula above. Or, equivalently, we can just write

$$\frac{df}{dx} \simeq \frac{f(x+h) - f(x-h)}{2h} \tag{5.107}$$

for general x.

This is the correct result for the quadratic approximation, but it's a disappointing result, since Eq. (5.107) is nothing other than the central difference approximation for sample points $2h$ apart, which we already saw in Eq. (5.102). In other words, the higher-order approximation has not helped us in this case.

However, going to still higher orders does help. If we use a cubic or quartic approximation, we do get improved estimates of the derivative. At higher orders there is a distinction between the odd- and even-order approximations. For the odd-order ones the sample points fall at "half-way" points, as with the central difference of Eq. (5.95). For instance, to get the four sample points required for a cubic approximation, symmetrically distributed about zero, we would choose them to fall at $x = -\frac{3}{2}h, -\frac{1}{2}h, \frac{1}{2}h$, and $\frac{3}{2}h$. For even-order approximations, on the other hand, the samples fall at "integer" points; the five points for the quartic approximation, for instance, fall at $-2h, -h, 0, h$, and $2h$.

The methodology for deriving the higher-order approximations follows the same pattern as for the quadratic case: we write down the required value of the polynomial at each of the sample points, which gives us a set of simultaneous equations in the polynomial coefficients. As before, we actually need only one of those coefficients, the coefficient of the linear term in the polynomial. Solving for this coefficient gives us our expression for the derivative. At each order the expression is a linear combination of the samples, divided by h. We

Degree	$f(-\tfrac{5}{2}h)$	$f(-2h)$	$f(-\tfrac{3}{2}h)$	$f(-h)$	$f(-\tfrac{1}{2}h)$	$f(0)$	$f(\tfrac{1}{2}h)$	$f(h)$	$f(\tfrac{3}{2}h)$	$f(2h)$	$f(\tfrac{5}{2}h)$	Error
1					-1		1					$O(h^2)$
2				$-\tfrac{1}{2}$				$\tfrac{1}{2}$				$O(h^2)$
3			$\tfrac{1}{24}$		$-\tfrac{27}{24}$		$\tfrac{27}{24}$		$-\tfrac{1}{24}$			$O(h^4)$
4		$\tfrac{1}{12}$		$-\tfrac{2}{3}$				$\tfrac{2}{3}$		$-\tfrac{1}{12}$		$O(h^4)$
5	$-\tfrac{3}{640}$		$\tfrac{25}{384}$		$-\tfrac{75}{64}$		$\tfrac{75}{64}$		$-\tfrac{25}{384}$		$\tfrac{3}{640}$	$O(h^6)$

Table 5.1: Coefficients for numerical derivatives. The coefficients for central approximations to the first derivative of $f(x)$ at $x = 0$. To derive the full expression for an approximation, multiply the samples listed in the top row of the table by the coefficients in one of the other rows, then divide by h. For instance, the cubic approximation would be $[\tfrac{1}{24}f(-\tfrac{3}{2}h) - \tfrac{27}{24}f(-\tfrac{1}{2}h) + \tfrac{27}{24}f(\tfrac{1}{2}h) - \tfrac{1}{24}f(\tfrac{3}{2}h)]/h$. For derivatives at points other than $x = 0$ the same coefficients apply—one just uses the appropriate sample points around the value x of interest. The final column of the table gives the order of the approximation error on the derivative.

will not go through the derivations in detail, but Table 5.1 gives the coefficients of the combinations for the first five approximations.

Each of the approximations given in the table is exact, apart from rounding error, if the function being differentiated is actually a polynomial of the appropriate (or lower) degree, so that the polynomial fit is a perfect one. Most of the time, however, this will not be the case and there will be an approximation error involved in calculating the derivative. One can calculate this error to leading order for each of the approximations by a method analogous to our calculations for the forward, backward, and central differences: we perform Taylor expansions about $x = 0$ to derive expressions for $f(x)$ at each of the sample points, then plug these expressions into the formula for the derivative. The order in h of the resulting error is listed in the final column of Table 5.1. As before, this approximation error must be balanced against the rounding error and a suitable value of h chosen to minimize the overall error in the derivative.

An interesting point to notice about Table 5.1 is that the coefficient for $f(0)$ in all the approximations is zero. The value of the function exactly at the point of interest never plays a role in the evaluation of the derivative. Another (not unrelated) point is that the order in h of the error given in the final column does not go up uniformly with the degree of the polynomial—it is the same for the even-degree polynomials as for the next-lower odd-degree ones. We saw a special case of this result for the quadratic: the quadratic fit just gives us an ordinary central difference and therefore necessarily has an error $O(h^2)$, the same as the central difference derived from the linear fit. In general, the odd-degree approximations give us slightly more accurate results than the even-degree ones—the error is of the same order in h but the constant of proportionality is smaller. On the other hand, the odd-degree approximations require samples at the half-way points, as we have noted, which can be inconvenient. As discussed in Example 5.4, we sometimes have samples at only the "integer" points, in which case we must use the even-degree approximations.

We will not be using quadratic or higher-order derivative approximations in the remainder of this book—the forward, backward, and central differences will be all we need. But it is worth knowing about them nonetheless; such things come in handy every once in a while.

5.10.5 SECOND DERIVATIVES

We can also derive numerical approximations for the second derivative of a function $f(x)$. The second derivative is, by definition, the derivative of the first derivative, so we can calculate it by applying our first-derivative formulas twice. For example, starting with the central difference formula, Eq. (5.95), we can write expressions for the first derivative at $x + h/2$ and $x - h/2$ thus:

$$f'(x + h/2) \simeq \frac{f(x+h) - f(x)}{h}, \qquad f'(x - h/2) \simeq \frac{f(x) - f(x-h)}{h}. \quad (5.108)$$

Then we apply the central difference again to get an expression for the second derivative:

$$\begin{aligned} f''(x) &\simeq \frac{f'(x + h/2) - f'(x - h/2)}{h} \\ &= \frac{[f(x+h) - f(x)]/h - [f(x) - f(x-h)]/h}{h} \\ &= \frac{f(x+h) - 2f(x) + f(x-h)}{h^2}. \end{aligned} \quad (5.109)$$

This is the simplest approximation for the second derivative. We will use it extensively in Chapter 9 for solving second-order differential equations. Higher-order approximations exist too, but we will not use them in this book.

We can also calculate the error on Eq. (5.109). We perform two Taylor expansions of $f(x)$ thus:

$$f(x + h) = f(x) + hf'(x) + \tfrac{1}{2}h^2 f''(x) + \tfrac{1}{6}h^3 f'''(x) + \tfrac{1}{24}f''''(x) + \dots \quad (5.110)$$
$$f(x - h) = f(x) - hf'(x) + \tfrac{1}{2}h^2 f''(x) - \tfrac{1}{6}h^3 f'''(x) + \tfrac{1}{24}f''''(x) - \dots \quad (5.111)$$

Adding them together and rearranging, we find that

$$f''(x) = \frac{f(x+h) - 2f(x) + f(x-h)}{h^2} - \tfrac{1}{12}h^2 f''''(x) + \dots \quad (5.112)$$

The first term on the right is our formula for the second derivative, Eq. (5.109), and the remainder of the terms measure the error. Thus, to leading order, the absolute error inherent in our approximation to the second derivative is $\frac{1}{12}h^2 |f''''(x)|$. As before, we also need to take rounding error into account, which contributes an error of roughly $C|f(x)|$ on each value of $f(x)$ so that, in the worst case, the total rounding error in the numerator of (5.109) is $4C|f(x)|$ and the rounding error on the whole expression is $4C|f(x)|/h^2$. Then the complete error on the derivative is

$$\epsilon = \frac{4C|f(x)|}{h^2} + \tfrac{1}{12}h^2 |f''''(x)|. \tag{5.113}$$

Differentiating with respect to h and setting the result to zero then gives an optimum value of h of

$$h = \left(48C\left|\frac{f(x)}{f''''(x)}\right|\right)^{1/4}. \tag{5.114}$$

Substituting this expression back into Eq. (5.113) gives the size of the optimal error to be

$$\epsilon = \tfrac{1}{6}h^2 |f''''(x)| = \left(\tfrac{4}{3}C|f(x)f''''(x)|\right)^{1/2}. \tag{5.115}$$

So if, for instance, $f(x)$ and $f''''(x)$ are of order 1, the error will be roughly of order $\sqrt{C}$, which is typically about 10^{-8}. This is about the same accuracy as we found for the forward and backward difference approximations to the first derivative in Section 5.10.2. Thus our expression for the second derivative is not very accurate—about as good as, but not better than, the forward difference. As mentioned above, there are higher-order approximations for the second derivative that can give more accurate answers. But for our purposes Eq. (5.109) will be good enough.

5.10.6 PARTIAL DERIVATIVES

We will come across a number of situations where we need to calculate partial derivatives—derivatives of a function of several variables with respect to only one of those variables. The calculation of such partial derivatives is a simple generalization of the calculation of ordinary derivatives. If you have a function $f(x, y)$ of two variables, for instance, then the central difference approximations to derivatives with respect to x and y are

$$\frac{\partial f}{\partial x} = \frac{f(x + h/2, y) - f(x - h/2, y)}{h}, \tag{5.116}$$

$$\frac{\partial f}{\partial y} = \frac{f(x, y + h/2) - f(x, y - h/2)}{h}. \tag{5.117}$$

By analogy with our approach for the second derivative in Section 5.10.5 we can also calculate second derivatives with respect to either variable, or a mixed second derivative with respect to both, which is given by

$$\frac{\partial^2 f}{\partial x \partial y} = \frac{f(x+h/2, y+h/2) - f(x-h/2, y+h/2) - f(x+h/2, y-h/2) + f(x-h/2, y-h/2)}{h^2}.$$

(5.118)

We leave the derivation to the avid reader.

5.10.7 DERIVATIVES OF NOISY DATA

Suppose we have some measurements of a quantity that, when plotted on a graph, look like Fig. 5.11a. Perhaps they come from an experiment in the lab, for instance. The overall shape of the curve is clear from the figure, but there is some noise in the data, so the curve is not completely smooth.

Now suppose we want to calculate the first derivative of this curve. So we write a program to calculate, say, the forward difference at each point and plot the values we get. The result is shown in Fig. 5.11b. As you can see, taking the derivative has made our noise problem much worse. Now it's almost impossible to see the shape of the curve. This is a common problem with numerical derivatives—if there's any noise in the curve you're differentiating, then it can be greatly exaggerated by taking the derivative, perhaps to the point where the results are useless.

The reason for the problem is easy to see if you zoom in on a small portion of the original data, as shown in Fig. 5.12. In this figure the solid line represents the actual data, and the dotted line is a sketch of what the underlying curve, without the noise, probably looks like. (We don't usually know the underlying curve, so this is just a guess.) When viewed close-up like this, we can see that, because of the noise, the slope of the noisy line is very steep in some places, and completely different from the slope of the underlying curve. Although the noisy curve follows the underlying one reasonably closely, its derivative does not. So now, when we calculate the derivative, we generate spurious large values where there should be none.

Unfortunately, this kind of issue is common with physics data, and this is one of the reasons why numerical derivatives are used less than numerical integrals. There are, however, some things we can do to mitigate the problem, although they all also decrease the accuracy of our results:

1. The simplest thing we can do is increase the value of h. We can treat

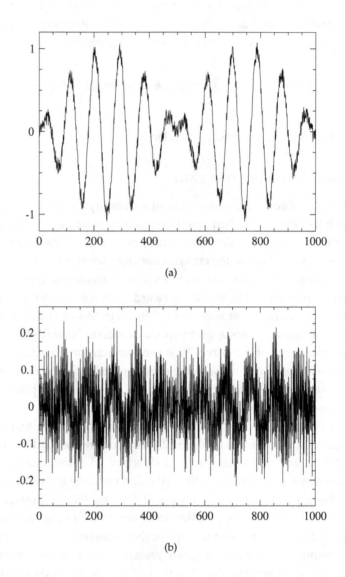

(a)

(b)

Figure 5.11: Derivative of noisy data. (a) An example of a noisy data set. The data plotted in this graph have a clear underlying form, but contain some random noise or experimental error as well. (b) The derivative of the same data calculated using a forward difference. The action of taking the derivative amplifies the noise and makes the underlying form of the result difficult to discern.

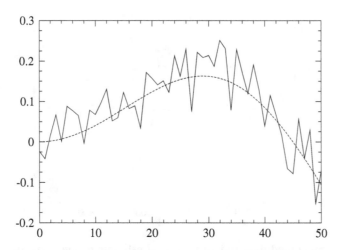

Figure 5.12: An expanded view of the noisy data. The jagged line in this plot is an enlargement of the first portion of the data from Fig. 5.11a, while the dotted line is a guess about the form of the underlying curve, without the noise.

the noise in the same way that we treat rounding error and calculate an optimum value for h that balances the error from the noise against the error in our approximation of the derivative. The end result is a formula similar to Eq. (5.93) for the forward difference or Eq. (5.100) for the central difference, but with the error constant C replaced by the fractional error introduced into the data by the noise (which is the inverse of the so-called *signal-to-noise ratio*).

2. Another approach is to fit a curve to a portion of the data near the point where we want the derivative, then differentiate the curve. For instance, we might fit a quadratic or a cubic, then differentiate that. We do not, however, fit a quadratic to just three sample points or a cubic to just four, as we did in Section 5.10.4. Instead we do a least-squares fit to find the curve that best approximates a larger number of points, even though it will not typically pass exactly through all those points. In effect, we are trying to find an approximation to the underlying smooth curve depicted in Fig. 5.12. The derivative of this curve then gives an estimate of the true derivative of the data without noise.

3. A third approach is to smooth the data in some other fashion before differentiating, which can be done, for instance, using Fourier transforms, which we will study in Chapter 7. (See Exercise 7.4 for an example

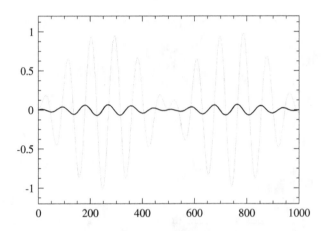

Figure 5.13: Smoothed data and an improved estimate of the derivative. The gray curve in this plot is a version of the data from Fig. 5.11a that has been smoothed to remove noise using a Fourier transform method. The black curve shows the numerical derivative of the smoothed function, which is a significant improvement over Fig. 5.11b.

of Fourier smoothing.) Figure 5.13 shows a version of the data from Fig. 5.11 that has been smoothed in this way, and the corresponding derivative, which is much cleaner now.

5.11 INTERPOLATION

We will tackle one more topic briefly in this chapter, the topic of *interpolation*, which is not directly related to integrals and derivatives, but uses similar mathematical methods, making this a good moment to look into it.

Suppose you are given the value of a function $f(x)$ at just two points $x = a, b$ and you want to know the value of the function at another point x in between. What do you do? There are a number of possibilities, of which the simplest is *linear interpolation*, which is illustrated in Fig. 5.14. We assume our function follows a straight line from $f(a)$ to $f(b)$, which in most cases is an approximation—likely the function follows a curve between the two points, as sketched in the figure. But if we make this assumption then we can calculate $f(x)$ with some elementary geometry.

The slope of the straight-line approximation is

$$m = \frac{f(b) - f(a)}{b - a},$$

(5.119)

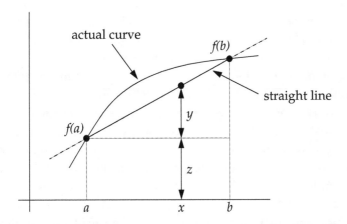

Figure 5.14: Linear interpolation. The value of $f(x)$ in between the two known points at $x = a$ and $x = b$ is estimated by assuming a straight line from $f(a)$ to $f(b)$.

and the distance marked y on the figure is given in terms of this slope by $y = m(x - a)$. The distance marked z is equal to $f(a)$, so

$$f(x) \simeq y + z = \frac{f(b) - f(a)}{b - a}(x - a) + f(a)$$
$$= \frac{(b - x)f(a) + (x - a)f(b)}{b - a}. \tag{5.120}$$

This is the fundamental formula of linear interpolation. In fact, this same formula can also be used to *extrapolate* the function to points outside the interval from a to b, although one should not extrapolate too far. The further you go, the less likely it is that the extrapolation will be accurate.

How accurate is the linear interpolation formula? The calculation of the error is similar to that for derivatives, making use of two Taylor expansions:

$$f(a) = f(x) + (a - x)f'(x) + \tfrac{1}{2}(a - x)^2 f''(x) + \dots \tag{5.121}$$
$$f(b) = f(x) + (b - x)f'(x) + \tfrac{1}{2}(b - x)^2 f''(x) + \dots \tag{5.122}$$

Substituting these into Eq. (5.120), the terms in $f'(x)$ cancel and, after rearranging a little, we find that

$$f(x) = \frac{(b - x)f(a) + (x - a)f(b)}{b - a} + (a - x)(b - x)\,f''(x) + \dots \tag{5.123}$$

The first term on the right-hand side is our linear interpolation formula; the remainder of the terms are the error. Note that the leading-order error term vanishes as x tends to either a or b, so that either $b - x$ or $a - x$ becomes small. And assuming $f''(x)$ varies slowly, the error will be largest in the middle of the interval. If we denote the width of the interval by $b - a = h$, then when we are in the middle we have $x - a = b - x = \frac{1}{2}h$ and the magnitude of the leading-order error is $\frac{1}{4}h^2 |f''(x)|$. Thus, like the central difference formula for a first derivative, the worst-case error on a linear interpolation is $O(h^2)$, and we can make the interpolation more accurate by making h smaller.

By contrast with the case of derivatives, however, we do not need to be particularly careful about rounding error when using linear interpolation. The interpolation formula, Eq. (5.120), involves the *sum* of values of $f(x)$ at two closely spaced points, not the difference, so we don't normally run into the accuracy problems that plague calculations (like calculations of derivatives) that are based on subtractions.

Can we do better than linear interpolation? Not if we know the value of the function $f(x)$ at only two points—there is no better approximation in that case. If we know the function at more than two points there are several ways to improve on linear interpolation. The most obvious is to interpolate using higher-order polynomials. If we have three points, for instance, we can fit a quadratic through them, which will usually give a better match to the underlying curve. Fitting quadratics or higher polynomials leads to a set of higher-order methods known as *Lagrange interpolation* methods.

When the number of points becomes large, however, this approach breaks down. If we have a large number N of points then you might think the best thing to do would be to fit an $(N - 1)$th order polynomial through them, but it turns out this doesn't work because very high order polynomials tend to have a lot of wiggles in them and can deviate from the fitted points badly in the intervals between points. It's better in this case to fit many lower-order polynomials such as quadratics or cubics to smaller sets of adjacent points. Unfortunately, the naive implementation of such a scheme gives rather uneven interpolations because the slope of the interpolation changes at the join-points between polynomials. A more satisfactory approach is to fit polynomials to the measured points and the derivatives at their ends, so that one gets a function that goes through the points and has a smooth slope everywhere. Such interpolations are called *splines*. The most widely used type are *cubic splines*. We won't go into these methods further, however. For our purposes, linear interpolation will be good enough.

5.17   The gamma function: A commonly occurring function in physics calculations is the gamma function $\Gamma(a)$, which is defined by the integral

$$\Gamma(a) = \int_0^\infty x^{a-1} e^{-x}\,dx.$$

There is no closed-form expression for the gamma function, but one can calculate its value for given a by performing the integral above numerically. You have to be careful how you do it, however, if you wish to get an accurate answer.

a) Write a program to make a graph of the value of the integrand $x^{a-1} e^{-x}$ as a function of x from $x = 0$ to $x = 5$, with three separate curves for $a = 2, 3,$ and 4, all on the same axes. You should find that the integrand starts at zero, rises to a maximum, and then decays again for each curve.

b) Show analytically that the maximum falls at $x = a - 1$.

c) Most of the area under the integrand falls near the maximum, so to get an accurate value of the gamma function we need to do a good job of this part of the integral. We can change the integral from 0 to ∞ to one over a finite range from 0 to 1 using the change of variables in Eq. (5.67), but this tends to squash the peak towards the edge of the $[0,1]$ range and does a poor job of evaluating the integral accurately. We can do a better job by making a different change of variables that puts the peak in the middle of the integration range, around $\frac{1}{2}$. We will use the change of variables given in Eq. (5.69), which we repeat here for convenience:

$$z = \frac{x}{c+x}.$$

For what value of x does this change of variables give $z = \frac{1}{2}$? Hence what is the appropriate choice of the parameter c that puts the peak of the integrand for the gamma function at $z = \frac{1}{2}$?

d) Before we can calculate the gamma function, there is another detail we need to attend to. The integrand $x^{a-1} e^{-x}$ can be difficult to evaluate because the factor x^{a-1} can become very large and the factor e^{-x} very small, causing numerical overflow or underflow, or both, for some values of x. Write $x^{a-1} = e^{(a-1)\ln x}$ to derive an alternative expression for the integrand that does not suffer from these problems (or at least not so much). Explain why your new expression is better than the old one.

e) Now, using the change of variables above and the value of c you have chosen, write a user-defined function `gamma(a)` to calculate the gamma function for arbitrary argument a. Use whatever integration method you feel is appropriate. Test your function by using it to calculate and print the value of $\Gamma(\frac{3}{2})$, which is known to be equal to $\frac{1}{2}\sqrt{\pi} \simeq 0.886$.

f) For integer values of a it can be shown that $\Gamma(a)$ is equal to the factorial of $a - 1$. Use your Python function to calculate $\Gamma(3)$, $\Gamma(6)$, and $\Gamma(10)$. You should get answers closely equal to $2! = 2$, $5! = 120$, and $9! = 362\,880$.

5.18 Rearranging Eq. (5.19) into a slightly more conventional form, we have:

$$\int_a^b f(x)\,dx = h\left[\tfrac{1}{2}f(a) + \tfrac{1}{2}f(b) + \sum_{k=1}^{N-1} f(a + kh)\right] + \tfrac{1}{12}h^2\left[f'(a) - f'(b)\right] + O(h^4).$$

This result gives a value for the integral on the left which has an error of order h^4—a factor of h^2 better than the error on the trapezoidal rule and as good as Simpson's rule. We can use this formula as a new rule for evaluating integrals, distinct from any of the others we have seen in this chapter. We might call it the "Euler–Maclaurin rule."

a) Write a program to calculate the value of the integral $\int_0^2 (x^4 - 2x + 1)\,dx$ using this formula. (This is the same integral that we studied in Example 5.1, whose true value is 4.4.) The order-h term in the formula is just the ordinary trapezoidal rule; the h^2 term involves the derivatives $f'(a)$ and $f'(b)$, which you should evaluate using central differences, centered on a and b respectively. Note that the size of the interval you use for calculating the central differences does not have to equal the value of h used in the trapezoidal rule part of the calculation. An interval of about 10^{-5} gives good values for the central differences.

Use your program to evaluate the integral with $N = 10$ slices and compare the accuracy of the result with that obtained from the trapezoidal rule alone with the same number of slices.

b) Good though it is, this integration method is not much used in practice. Suggest a reason why not.

5.19 Diffraction gratings: Light with wavelength λ is incident on a diffraction grating of total width w, gets diffracted, is focused with a lens of focal length f, and falls on a screen:

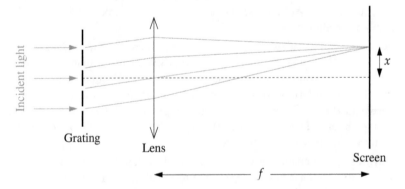

Theory tells us that the intensity of the diffraction pattern on the screen, a distance x from the central axis of the system, is given by

$$I(x) = \left| \int_{-w/2}^{w/2} \sqrt{q(u)}\, e^{i2\pi xu/\lambda f}\,du \right|^2,$$

where $q(u)$ is the intensity transmission function of the diffraction grating at a distance u from the central axis, i.e., the fraction of the incident light that the grating lets through.

a) Consider a grating with transmission function $q(u) = \sin^2 \alpha u$. What is the separation of the "slits" in this grating, expressed in terms of α?

b) Write a Python function q(u) that returns the transmission function $q(u) = \sin^2 \alpha u$ as above at position u for a grating whose slits have separation 20 μm.

c) Use your function in a program to calculate and graph the intensity of the diffraction pattern produced by such a grating having ten slits in total, if the incident light has wavelength $\lambda = 500$ nm. Assume the lens has a focal length of 1 meter and the screen is 10 cm wide. You can use whatever method you think appropriate for doing the integral. Once you've made your choice you'll also need to decide the number of sample points you'll use. What criteria play into this decision?

Notice that the integrand in the equation for $I(x)$ is complex, so you will have to use complex variables in your program. As mentioned in Section 2.2.5, there is a version of the math package for use with complex variables called cmath. In particular you may find the exp function from cmath useful because it can calculate the exponentials of complex arguments.

d) Create a visualization of how the diffraction pattern would look on the screen using a density plot (see Section 3.3). Your plot should look something like this:

e) Modify your program further to make pictures of the diffraction patterns produced by gratings with the following profiles:

i) A transmission profile that obeys $q(u) = \sin^2 \alpha u \sin^2 \beta u$, with α as before and the same total grating width w, and $\beta = \frac{1}{2}\alpha$.

ii) Two "square" slits, meaning slits with 100% transmission through the slit and 0% transmission everywhere else. Calculate the diffraction pattern for non-identical slits, one 10 μm wide and the other 20 μm wide, with a 60 μm gap between the two.

5.20   A more advanced adaptive method for the trapezoidal rule: In Section 5.3 we studied an adaptive version of the trapezoidal rule in which the number of steps is increased—and the width h of the slices correspondingly decreased—until the calculation gives a value for the integral accurate to some desired level. Although this method varies h, it still calculates the integral at any individual stage of the process using slices of equal width throughout the domain of integration. In this exercise we look at a more sophisticated form of the trapezoidal rule that uses different step sizes in different parts of the domain, which can be useful particularly for poorly behaved functions that vary rapidly in certain regions but not others. Remarkably, this method is not much more complicated to program than the ones we've already seen, if one knows the right tricks. Here's how the method works.

Suppose we wish to evaluate the integral $I = \int_a^b f(x)\,dx$ and we want an error of no more than ϵ on our answer. To put that another way, if we divide up the integral into slices of width h then we require an accuracy per slice of

$$h\,\frac{\epsilon}{b-a} = h\delta,$$

where $\delta = \epsilon/(b-a)$ is the target accuracy per unit interval.

We start by evaluating the integral using the trapezoidal rule with just a single slice of width $h_1 = b - a$. Let us call the estimate of the integral from this calculation I_1. Usually I_1 will not be very accurate, but that doesn't matter. Next we make a second estimate I_2 of the integral, again using the trapezoidal rule but now with two slices of width $h_2 = \frac{1}{2}h_1$ each. Equation (5.28) tells us that the error on this second estimate is $\frac{1}{3}(I_2 - I_1)$ to leading order. If the absolute value of this error is smaller than the required accuracy ϵ then our calculation is complete and we need go no further. I_2 is a good enough estimate of the integral.

Most likely, however, this will not be the case; the accuracy will not be good enough. If so, then we divide the integration interval into two equal parts of size $\frac{1}{2}(b - a)$ each, and we repeat the process above in each part separately, calculating estimates I_1 and I_2 using one and two slices respectively, estimating the error, and checking to see if it is less than the required accuracy, which is now $\frac{1}{2}(b - a)\delta = \frac{1}{2}\epsilon$.

We keep on repeating this process, dividing each slice in half and in half again, as many times as necessary to achieve the desired accuracy in every slice. Different slices may be divided different numbers of times, and hence we may end up with different sized slices in different parts of the integration domain. The method automatically uses whatever size and number of slices is appropriate in each region.

a) Write a program using this method to calculate the integral

$$I = \int_0^{10} \frac{\sin^2 x}{x^2}\,dx,$$

to an accuracy of $\epsilon = 10^{-4}$. Start by writing a function to calculate the integrand $f(x) = (\sin^2 x)/x^2$. Note that the limiting value of the integrand at $x = 0$ is 1. You'll probably have to include this point as a special case in your function using an if statement.

The best way to perform the integration itself is to make use of the technique of recursion, the ability of a Python function to call itself. (If you're not familiar with recursion, you may like to look at Exercise 2.13 on page 83 before doing this exercise.) Write a function step(x1,x2,f1,f2) that takes as arguments the beginning and end points x_1, x_2 of a slice and the values $f(x_1), f(x_2)$ of the integrand at those two points, and returns the value of the integral from x_1 to x_2. This function should evaluate the two estimates I_1 and I_2 of the integral from x_1 to x_2, calculated with one and two slices respectively, and the error $\frac{1}{3}(I_2 - I_1)$. If this error meets the target value, which is $(x_2 - x_1)\delta$, then the calculation is complete and the function simply returns the value I_2. If the error fails to meet the target, then the function calls itself, twice, to evaluate the integral separately on the first and second halves of the interval and returns the sum of the two results. (And then *those* functions can call themselves, and so forth, subdividing the integral as many times as necessary to reach the required accuracy.)

Hint: As icing on the cake, when the error target is met and the function returns a value for the integral in the current slice, it can, in fact, return a slightly better value than the estimate I_2. Since you will already have calculated the value of the integrand $f(x)$ at x_1, x_2, and the midpoint $x_m = \frac{1}{2}(x_1 + x_2)$ in order to evaluate I_2, you can use those results to compute the improved Simpson's rule estimate, Eq. (5.7), for this slice. You just return the value $\frac{1}{6}h[f(x_1) + 4f(x_m) + f(x_2)]$ instead of the trapezoidal rule estimate $\frac{1}{4}h[f(x_1) + 2f(x_m) + f(x_2)]$ (where $h = x_2 - x_1$). This involves very little extra work, but gives a value that is more accurate by two orders in h. (Technically, this is an example of the method of "local extrapolation," although it's perhaps not obvious what we're extrapolating in this case. We'll discuss local extrapolation again when we study adaptive methods for the solution of differential equations in Section 8.4.)

b) Why does the function step(x1,x2,f1,f2) take not only the positions x_1 and x_2 as arguments, but also the values $f(x_1)$ and $f(x_2)$? Since we know the function $f(x)$, we could just calculate these values from x_1 and x_2. Nonetheless, it is a smart move to include the values of $f(x_1)$ and $f(x_2)$ as arguments to the function. Why?

c) Modify your program to make a plot of the integrand with dots added showing where the ends of each integration slice lie. You should see larger slices in portions of the integrand that follow reasonably straight lines (because the trapezoidal rule gives an accurate value for straight-line integrands) and smaller slices in portions with more curvature.

5.21 Electric field of a charge distribution: Suppose we have a distribution of charges and we want to calculate the resulting electric field. One way to do this is to first calculate the electric potential ϕ and then take its gradient. For a point charge q at the origin, the electric potential at a distance r from the origin is $\phi = q/4\pi\epsilon_0 r$ and the electric field is $\mathbf{E} = -\nabla\phi$.

a) You have two charges, of ± 1 C, 10 cm apart. Calculate the resulting electric potential on a 1 m × 1 m square plane surrounding the charges and passing through them. Calculate the potential at 1 cm spaced points in a grid and make a visualization on the screen of the potential using a density plot.

b) Now calculate the partial derivatives of the potential with respect to x and y and hence find the electric field in the xy plane. Make a visualization of the field also. This is a little trickier than visualizing the potential, because the electric field has both magnitude and direction. One way to do it might be to make two density plots, one for the magnitude, and one for the direction, the latter using the "hsv" color scheme in `pylab`, which is a rainbow scheme that passes through all the colors but starts and ends with the same shade of red, which makes it suitable for representing things like directions or angles that go around the full circle and end up where they started. A more sophisticated visualization might use the arrow object from the `visual` package, drawing a grid of arrows with direction and length chosen to represent the field.

c) Now suppose you have a continuous distribution of charge over an $L \times L$ square. The charge density in Cm^{-2} is

$$\sigma(x, y) = q_0 \sin \frac{2\pi x}{L} \sin \frac{2\pi y}{L}.$$

Calculate and visualize the resulting electric field at 1 cm-spaced points in 1 square meter of the xy plane for the case where $L = 10\,cm$, the charge distribution is centered in the middle of the visualized area, and $q_0 = 100\,Cm^{-2}$. You will have to perform a double integral over x and y, then differentiate the potential with respect to position to get the electric field. Choose whatever integration method seems appropriate for the integrals.

5.22 Differentiating by integrating: If you are familiar with the calculus of complex variables, you may find the following technique useful and interesting.

Suppose we have a function $f(z)$ whose value we know not only on the real line but also for complex values of its argument. Then we can calculate derivatives of that function at any point z_0 by performing a contour integral, using the Cauchy derivative formula:

$$\left(\frac{d^m f}{dz^m} \right)_{z=z_0} = \frac{m!}{2\pi i} \oint \frac{f(z)}{(z - z_0)^{m+1}}\, dz,$$

where the integral is performed counterclockwise around any contour in the complex plane that surrounds the point z_0 but contains no poles in $f(z)$. Since numerical integration is significantly easier and more accurate than numerical differentiation, this formula provides us with a method for calculating derivatives—and especially multiple derivatives—accurately by turning them into integrals.

Suppose, for example, that we want to calculate derivatives of $f(z)$ at $z = 0$. Let us apply the Cauchy formula above using the trapezoidal rule to calculate the integral along a circular contour centered on the origin with radius 1. The trapezoidal rule will be slightly different from the version we are used to because the value of the interval h is now a complex number, and moreover is not constant from one slice of the integral to the next—it stays constant in modulus, but its argument changes from one slice to another.

We will divide our contour integral into N slices with sample points z_k distributed uniformly around the circular contour at the positions $z_k = e^{i2\pi k/N}$ for $k = 0 \ldots N$. Then the distance between consecutive sample points is

$$h_k = z_{k+1} - z_k = e^{i2\pi(k+1)/N} - e^{i2\pi k/N},$$

210

and, introducing the shorthand $g(z) = f(z)/z^{m+1}$ for the integrand, the trapezoidal rule approximation to the integral is

$$\oint g(z)\, dz \simeq \sum_{k=0}^{N-1} \tfrac{1}{2}[g(z_{k+1}) + g(z_k)]\left[e^{i2\pi(k+1)/N} - e^{i2\pi k/N}\right]$$

$$= \tfrac{1}{2}\left[\sum_{k=0}^{N-1} g(z_{k+1})\, e^{i2\pi(k+1)/N} - \sum_{k=0}^{N-1} g(z_k)\, e^{i2\pi k/N}\right.$$

$$\left. - \sum_{k=0}^{N-1} g(z_{k+1})\, e^{i2\pi k/N} + \sum_{k=0}^{N-1} g(z_k)\, e^{i2\pi(k+1)/N}\right].$$

Noting that $z_N = z_0$, the first two sums inside the brackets cancel each other in their entirety, and the remaining two sums are equal except for trivial phase factors, so the entire expression simplifies to

$$\oint g(z)\, dz \simeq \tfrac{1}{2}\left[e^{i2\pi/N} - e^{-i2\pi/N}\right]\sum_{k=0}^{N-1} g(z_k)\, e^{i2\pi k/N}$$

$$\simeq \frac{2\pi i}{N}\sum_{k=0}^{N-1} f(z_k)\, e^{-i2\pi km/N},$$

where we have used the definition of $g(z)$ again. Combining this result with the Cauchy formula, we then have

$$\left(\frac{d^m f}{dz^m}\right)_{z=0} \simeq \frac{m!}{N}\sum_{k=0}^{N-1} f(z_k)\, e^{-i2\pi km/N}.$$

Write a program to calculate the first twenty derivatives of $f(z) = e^{2z}$ at $z = 0$ using this formula with $N = 10000$. You will need to use the version of the exp function from the cmath package, which can handle complex arguments. You may also find the function factorial from the math package useful; it calculates factorials of integer arguments.

The correct value for the mth derivative in this case is easily shown to be 2^m, so it should be straightforward to tell if your program is working—the results should be powers of two, 2, 4, 8, 16, 32, etc. You should find that it is possible to get reasonably accurate results for all twenty derivatives rapidly using this technique. If you use standard difference formulas for the derivatives, on the other hand, you will find that you can calculate only the first three or four derivatives accurately before the numerical errors become so large that the results are useless. In this case, therefore, the Cauchy formula gives the better results.

The sum $\sum_k f(z_k)\, e^{i2\pi km/N}$ that appears in the formula above is known as the *discrete Fourier transform* of the complex samples $f(z_k)$. There exists an elegant technique for evaluating the Fourier transform for many values of m simultaneously, known as the *fast Fourier transform*, which could be useful in cases where the direct evaluation of the formula is slow. We will study the fast Fourier transform in detail in Chapter 7.

5.23 Image processing and the STM: When light strikes a surface, the amount falling per unit area depends not only on the intensity of the light, but also on the angle of incidence. If the direction the light is coming from makes an angle θ to the normal, then the light only "sees" $\cos\theta$ of area per unit of actual area on the surface:

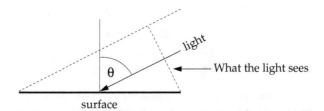

So the intensity of illumination is $a\cos\theta$, if a is the raw intensity of the light. This simple physical law is a central element of 3D computer graphics. It allows us to calculate how light falls on three-dimensional objects and hence how they will look when illuminated from various angles.

Suppose, for instance, that we are looking down on the Earth from above and we see mountains. We know the height of the mountains $w(x,y)$ as a function of position in the plane, so the equation for the Earth's surface is simply $z = w(x,y)$, or equivalently $z - w(x,y) = 0$, and the normal vector $\mathbf{v}$ to the surface is given by the gradient of $z - w(x,y)$ thus:

$$\mathbf{v} = \nabla[z - w(x,y)] = \begin{pmatrix} \partial/\partial x \\ \partial/\partial y \\ \partial/\partial z \end{pmatrix} [z - w(x,y)] = \begin{pmatrix} -\partial w/\partial x \\ -\partial w/\partial y \\ 1 \end{pmatrix}.$$

Now suppose we have incident light represented by a vector $\mathbf{a}$ that points toward the source of the light and has magnitude equal to the intensity. The dot product of the vectors $\mathbf{a}$ and $\mathbf{v}$ is

$$\mathbf{a} \cdot \mathbf{v} = |\mathbf{a}|\,|\mathbf{v}|\cos\theta,$$

where θ is the angle between the vectors. Employing the cosine rule discussed above, the intensity of illumination of the surface of the mountains is then

$$I = |\mathbf{a}|\cos\theta = \frac{\mathbf{a}\cdot\mathbf{v}}{|\mathbf{v}|} = \frac{-a_x(\partial w/\partial x) - a_y(\partial w/\partial y) + a_z}{\sqrt{(\partial w/\partial x)^2 + (\partial w/\partial y)^2 + 1}}.$$

Let's take the simple case where the light is shining horizontally with unit intensity, and the direction it's coming from makes an angle ϕ to the east-west axis, so that $\mathbf{a} = (\cos\phi, \sin\phi, 0)$. Then our intensity of illumination simplifies to

$$I = -\frac{\cos\phi\,(\partial w/\partial x) + \sin\phi\,(\partial w/\partial y)}{\sqrt{(\partial w/\partial x)^2 + (\partial w/\partial y)^2 + 1}}.$$

If we can calculate the derivatives of the height $w(x,y)$ and we know ϕ we can calculate the intensity at any point.

a) In the on-line resources you'll find a file called `altitude.txt`, which contains the altitude $w(x,y)$ in meters above sea level (or depth below sea level) of the surface of the Earth, measured on a grid of points (x,y). Write a program that reads this file and stores the data in an array. Then calculate the derivatives $\partial w/\partial x$ and $\partial w/\partial y$ at each grid point. Explain what method you used to calculate them and why. (Hint: You'll probably have to use more than one method to get every grid point, because awkward things happen at the edges of the grid.) To calculate the derivatives you'll need to know the value of h, the distance in meters between grid points, which is about 30 000 m in this case.[7]

b) Now, using your values for the derivatives, calculate the intensity for each grid point, with $\phi = 45°$, and make a density plot of the resulting values in which the brightness of each dot depends on the corresponding intensity value. If you get it working right, the plot should look like a relief map of the world—you should be able to see the continents and mountain ranges in 3D. (Common problems include a map that is upside-down or sideways, or a relief map that is "inside-out," meaning the high regions look low and *vice versa*. Work with the details of your program until you get a map that looks right to you.)

Hint: Note that the value of the intensity I from the formula above can be either positive or negative—it ranges from $+1$ to -1. What does a negative intensity mean? It means that the area in question is *in shadow*—it lies on the wrong side of the mountain to receive any light at all. You could represent this by coloring those areas of the map completely black, although in practice you will get a nicer-looking image (if arguably less true-to-life) by simply using a continuous range of grays from $+1$ to -1.

c) There is another file in the on-line resources called `stm.txt`, which contains a grid of values from scanning tunneling microscope measurements of the (111) surface of silicon. A scanning tunneling microscope (STM) is a device that measures the shape of surfaces at the atomic level by tracking a sharp tip over the surface and measuring quantum tunneling current as a function of position. The end result is a grid of values that represent the height of the surface as a function of position and the data in the file `stm.txt` contain just such a grid of values. Modify the program you just wrote to visualize the STM data and hence create a 3D picture of what the silicon surface looks like. The value of h for the derivatives in this case is around $h = 2.5$ (in arbitrary units).

---

[7]It's actually not precisely constant because we are representing the spherical Earth on a flat map, but $h = 30\,000$ m will give reasonable results.

CHAPTER 6

SOLUTION OF LINEAR AND NONLINEAR EQUATIONS

ONE of the commonest uses of computers in physics is the solution of equations or sets of equations of various kinds. In this chapter we look at methods for solving both linear and nonlinear equations. Solutions of linear equations involve techniques from linear algebra, so we will spend some time learning how linear algebra tasks such as inversion and diagonalization of matrices can be accomplished on the computer. In the second part of the chapter we will look at schemes for solving nonlinear equations.

6.1   SIMULTANEOUS LINEAR EQUATIONS

A single linear equation in one variable, such as $x - 1 = 0$, is trivial to solve. We do not need computers to do this for us. But simultaneous sets of linear equations in many variables are harder. In principle the techniques for solving them are well understood and straightforward—all of us learned them at some point in school—but they are also tedious, involving many operations, additions, subtractions, multiplications, one after another. Humans are slow and prone to error in such calculations, but computers are perfectly suited to the work and the solution of systems of simultaneous equations, particularly large systems with many variables, is a common job for computers in physics.

Let us take an example. Suppose you want to solve the following four simultaneous equations for the variables w, x, y, and z:

$$2w + x + 4y + z = -4, \tag{6.1a}$$
$$3w + 4x - y - z = 3, \tag{6.1b}$$
$$w - 4x + y + 5z = 9, \tag{6.1c}$$
$$2w - 2x + y + 3z = 7. \tag{6.1d}$$

For computational purposes the simplest way to think of these is in matrix form: they can be written as

$$
\begin{pmatrix}
2 & 1 & 4 & 1 \\
3 & 4 & -1 & -1 \\
1 & -4 & 1 & 5 \\
2 & -2 & 1 & 3
\end{pmatrix}
\begin{pmatrix}
w \\
x \\
y \\
z
\end{pmatrix}
=
\begin{pmatrix}
-4 \\
3 \\
9 \\
7
\end{pmatrix} .
\tag{6.2}
$$

Alternatively, we could write this out shorthand as

$$
\mathbf{Ax} = \mathbf{v},
\tag{6.3}
$$

where $\mathbf{x} = (w, x, y, z)$ and the matrix $\mathbf{A}$ and vector $\mathbf{v}$ take the appropriate values.

One way to solve equations of this form is to find the inverse of the matrix $\mathbf{A}$ then multiply both sides of (6.3) by it to get the solution $\mathbf{x} = \mathbf{A}^{-1}\mathbf{v}$. This sounds like a promising approach for solving equations on the computer but in practice it's not as good as you might think. Inverting the matrix $\mathbf{A}$ is a rather complicated calculation that is inefficient and cumbersome to carry out numerically. There are other ways of solving simultaneous equations that don't require us to calculate an inverse and it turns out that these are faster, simpler, and more accurate. Perhaps the most straightforward method, and the first one we will look at, is *Gaussian elimination*.

6.1.1 GAUSSIAN ELIMINATION

Suppose we wish to solve a set of simultaneous equations like Eq. (6.1). We will carry out the solution by working on the matrix form $\mathbf{Ax} = \mathbf{v}$ of the equations. As you are undoubtedly aware, the following useful rules apply:

1. We can multiply any of our simultaneous equations by a constant and it's still the same equation. For instance, we can multiply Eq. (6.1a) by 2 to get $4w + 2x + 8y + 2z = -8$ and the solution for w, x, y, z stays the same. To put that another way: *If we multiply any row of the matrix $\mathbf{A}$ by any constant, and we multiply the corresponding row of the vector $\mathbf{v}$ by the same constant, then the solution does not change.*

2. We can take any linear combination of two equations to get another correct equation. To put that another way: *If we add to or subtract from any row of $\mathbf{A}$ a multiple of any other row, and we do the same for the vector $\mathbf{v}$, then the solution does not change.*

We can use these operations to solve our equations as follows. Consider the matrix form of the equations given in Eq. (6.2) and let us perform the following steps:

1. We divide the first row by the top-left element of the matrix, which has the value 2 in this case. Recall that we must divide both the matrix itself and the corresponding element on the right-hand side of the equation, in order that the equations remain correct. Thus we get:

$$\begin{pmatrix} 1 & 0.5 & 2 & 0.5 \\ 3 & 4 & -1 & -1 \\ 1 & -4 & 1 & 5 \\ 2 & -2 & 1 & 3 \end{pmatrix} \begin{pmatrix} w \\ x \\ y \\ z \end{pmatrix} = \begin{pmatrix} -2 \\ 3 \\ 9 \\ 7 \end{pmatrix}. \tag{6.4}$$

Because we have divided both on the left and on the right, the solution to the equations is unchanged from before, but note that the top-left element of the matrix, by definition, is now equal to 1.

2. Next, note that the first element in the second row of the matrix is a 3. If we now subtract 3 times the first row from the second this element will become zero, thus:

$$\begin{pmatrix} 1 & 0.5 & 2 & 0.5 \\ 0 & 2.5 & -7 & -2.5 \\ 1 & -4 & 1 & 5 \\ 2 & -2 & 1 & 3 \end{pmatrix} \begin{pmatrix} w \\ x \\ y \\ z \end{pmatrix} = \begin{pmatrix} -2 \\ 9 \\ 9 \\ 7 \end{pmatrix}. \tag{6.5}$$

Notice again that we have performed the same subtraction on the right-hand side of the equation, to make sure the solution remains unchanged.

3. We now do a similar trick with the third and fourth rows. These have first elements equal to 1 and 2 respectively, so we subtract 1 times the first row from the third, and 2 times the first row from the fourth, which gives us the following:

$$\begin{pmatrix} 1 & 0.5 & 2 & 0.5 \\ 0 & 2.5 & -7 & -2.5 \\ 0 & -4.5 & -1 & 4.5 \\ 0 & -3 & -3 & 2 \end{pmatrix} \begin{pmatrix} w \\ x \\ y \\ z \end{pmatrix} = \begin{pmatrix} -2 \\ 9 \\ 11 \\ 11 \end{pmatrix}. \tag{6.6}$$

The end result of this series of operations is that the first column of our matrix has been reduced to the simple form $(1, 0, 0, 0)$, but the solution of the complete set of equations is unchanged.

Now we move on to the second row of the matrix and perform a similar series of operations. We divide the second row by its *second* element, to get

$$
\begin{pmatrix}
1 & 0.5 & 2 & 0.5 \\
0 & 1 & -2.8 & -1 \\
0 & -4.5 & -1 & 4.5 \\
0 & -3 & -3 & 2
\end{pmatrix}
\begin{pmatrix}
w \\ x \\ y \\ z
\end{pmatrix}
=
\begin{pmatrix}
-2 \\ 3.6 \\ 11 \\ 11
\end{pmatrix}.
\tag{6.7}
$$

Then we subtract the appropriate multiple of the second row from each of the rows below it, so as to make the second element of each of those rows zero. That is, we subtract -4.5 times the second row from the third, and -3 times the second row from the fourth, to give

$$
\begin{pmatrix}
1 & 0.5 & 2 & 0.5 \\
0 & 1 & -2.8 & -1 \\
0 & 0 & -13.6 & 0 \\
0 & 0 & -11.4 & -1
\end{pmatrix}
\begin{pmatrix}
w \\ x \\ y \\ z
\end{pmatrix}
=
\begin{pmatrix}
-2 \\ 3.6 \\ 27.2 \\ 21.8
\end{pmatrix}.
\tag{6.8}
$$

Then we move onto the third and fourth rows, and do the same thing, dividing each then subtracting from the rows below (except that the fourth row obviously doesn't have any rows below, so it only needs to be divided). The end result of the entire set of operations is the following:

$$
\begin{pmatrix}
1 & 0.5 & 2 & 0.5 \\
0 & 1 & -2.8 & -1 \\
0 & 0 & 1 & 0 \\
0 & 0 & 0 & 1
\end{pmatrix}
\begin{pmatrix}
w \\ x \\ y \\ z
\end{pmatrix}
=
\begin{pmatrix}
-2 \\ 3.6 \\ -2 \\ 1
\end{pmatrix}.
\tag{6.9}
$$

By definition, this set of equations still has the same solution for the variables w, x, y, and z as the equations we started with, but the matrix is now *upper triangular*: all the elements below the diagonal are zero. This allows us to determine the solution for the variables quite simply by the process of *backsubstitution*.

6.1.2 BACKSUBSTITUTION

Suppose we have any set of equations of the form

$$
\begin{pmatrix}
1 & a_{01} & a_{02} & a_{03} \\
0 & 1 & a_{12} & a_{13} \\
0 & 0 & 1 & a_{23} \\
0 & 0 & 0 & 1
\end{pmatrix}
\begin{pmatrix}
w \\ x \\ y \\ z
\end{pmatrix}
=
\begin{pmatrix}
v_0 \\ v_1 \\ v_2 \\ v_3
\end{pmatrix},
\tag{6.10}
$$

which is exactly the form of Eq. (6.9) generated by the Gaussian elimination procedure. We can write the equations out in full as

$$w + a_{01}x + a_{02}y + a_{03}z = v_0, \tag{6.11a}$$

$$x + a_{12}y + a_{13}z = v_1, \tag{6.11b}$$

$$y + a_{23}z = v_2, \tag{6.11c}$$

$$z = v_3. \tag{6.11d}$$

Note that we are using Python-style numbering for the elements here, starting from zero, rather than one. This isn't strictly necessary, but it will be convenient when we want to translate our calculation into computer code.

Given equations of this form, we see now that the solution for the value of z is trivial—it is given directly by Eq. (6.11d):

$$z = v_3. \tag{6.12}$$

But given this value, the solution for y is also trivial, being given by Eq. (6.11c):

$$y = v_2 - a_{23}z. \tag{6.13}$$

And we can go on. The solution for x is

$$x = v_1 - a_{12}y - a_{13}z, \tag{6.14}$$

and

$$w = v_0 - a_{01}x - a_{02}y - a_{03}z. \tag{6.15}$$

Applying these formulas to Eq. (6.9) gives

$$w = 2, \quad x = -1, \quad y = -2, \quad z = 1, \tag{6.16}$$

which is, needless to say, the correct answer.

Thus by the combination of Gaussian elimination with backsubstitution we have solved our set of simultaneous equations.

EXAMPLE 6.1: GAUSSIAN ELIMINATION WITH BACKSUBSTITUTION

We are now in a position to create a complete program for solving simultaneous equations. Here is a program to solve the equations in (6.1) using Gaussian elimination and backsubstitution:

File: gausselim.py

```
from numpy import array,empty

A = array([[ 2,  1,  4,  1 ],
           [ 3,  4, -1, -1 ],
           [ 1, -4,  1,  5 ],
           [ 2, -2,  1,  3 ]],float)
v = array([ -4, 3, 9, 7 ],float)
N = len(v)

# Gaussian elimination
for m in range(N):

    # Divide by the diagonal element
    div = A[m,m]
    A[m,:] /= div
    v[m] /= div

    # Now subtract from the lower rows
    for i in range(m+1,N):
        mult = A[i,m]
        A[i,:] -= mult*A[m,:]
        v[i] -= mult*v[m]

# Backsubstitution
x = empty(N,float)
for m in range(N-1,-1,-1):
    x[m] = v[m]
    for i in range(m+1,N):
        x[m] -= A[m,i]*x[i]

print(x)
```

There are number of features to notice about this program. We store the matrices and vectors as arrays, whose initial values are set at the start of the program. The elimination portion of the program goes through each row of the matrix, one by one, and first normalizes it by dividing by the appropriate diagonal element, then subtracts a multiple of that row from each lower row. Notice how the program uses Python's ability to perform operations on entire rows at once, which makes the calculation faster and simpler to program. The second part of the program is a straightforward version of the backsubstitution procedure. Note that the entire program is written so as to work for matrices

219

of any size: we use the variable N to represent the size, so that no matter what size of matrix we feed to the program it will perform the correct calculation.

Exercise 6.1: A circuit of resistors

Consider the following circuit of resistors:

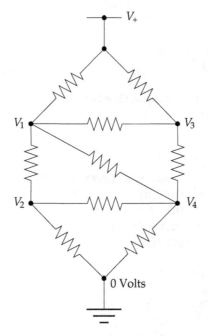

All the resistors have the same resistance R. The power rail at the top is at voltage $V_+ = 5\,\text{V}$. What are the other four voltages, V_1 to V_4?

To answer this question we use Ohm's law and the Kirchhoff current law, which says that the total net current flow out of (or into) any junction in a circuit must be zero. Thus for the junction at voltage V_1, for instance, we have

$$\frac{V_1 - V_2}{R} + \frac{V_1 - V_3}{R} + \frac{V_1 - V_4}{R} + \frac{V_1 - V_+}{R} = 0,$$

or equivalently

$$4V_1 - V_2 - V_3 - V_4 = V_+.$$

a) Write similar equations for the other three junctions with unknown voltages.

b) Write a program to solve the four resulting equations using Gaussian elimination and hence find the four voltages (or you can modify a program you already have, such as the program `gausselim.py` in Example 6.1).

6.1.3 PIVOTING

Suppose the equations we want to solve are slightly different from those of the previous section, thus:

$$\begin{pmatrix} 0 & 1 & 4 & 1 \\ 3 & 4 & -1 & -1 \\ 1 & -4 & 1 & 5 \\ 2 & -2 & 1 & 3 \end{pmatrix} \begin{pmatrix} w \\ x \\ y \\ z \end{pmatrix} = \begin{pmatrix} -4 \\ 3 \\ 9 \\ 7 \end{pmatrix}. \tag{6.17}$$

Just one thing has changed from the old equations in (6.2): the first element of the first row of the matrix is zero, where previously it was nonzero. But this makes all the difference in the world, because the first step of our Gaussian elimination procedure requires us to divide the first row of the matrix by its first element, which we can no longer do, because we would have to divide by zero. In cases like these, Gaussian elimination no longer works. So what are we to do?

The standard solution is to use *pivoting*, which means simply interchanging the rows of the matrix to get rid of the problem. Clearly we are allowed to interchange the order in which we write our simultaneous equations—it will not affect their solution—so we could, if we like, swap the first and second equations, which in matrix notation is equivalent to writing

$$\begin{pmatrix} 3 & 4 & -1 & -1 \\ 0 & 1 & 4 & 1 \\ 1 & -4 & 1 & 5 \\ 2 & -2 & 1 & 3 \end{pmatrix} \begin{pmatrix} w \\ x \\ y \\ z \end{pmatrix} = \begin{pmatrix} 3 \\ -4 \\ 9 \\ 7 \end{pmatrix}. \tag{6.18}$$

In other words, we have swapped the first and second rows of the matrix and also swapped the first and second elements of the vector on the right-hand side. Now the first element of the matrix is no longer zero, and Gaussian elimination will work just fine.

Pivoting has to be done with care. We must make sure, for instance, that in swapping equations to get rid of a problem we don't introduce another problem somewhere else. Moreover, the elements of the matrix change as the Gaussian elimination procedure progresses, so it's not always obvious in advance where problems are going to arise. A number of different rules or schemes for pivoting have been developed to guide the order in which the equations should be swapped. A good, general scheme that works well in most cases is the so-called *partial pivoting* method, which is as follows.

As we have seen the Gaussian elimination procedure works down the rows of the matrix one by one, dividing each by the appropriate diagonal element, before performing subtractions. With partial pivoting we consider rearranging the rows at each stage. When we get to the mth row, we compare it to all lower rows, looking at the value each row has in its mth element and finding the one such value that is farthest from zero—either positive or negative. If the row containing this winning value is not currently the mth row then we move it up to mth place by swapping it with the current mth row. This has the result of ensuring that the element we divide by in our Gaussian elimination is always as far from zero as possible.

If we look at Eq. (6.18), we see that in fact we inadvertently did exactly the right thing when we swapped the first and second rows of our matrix, since we moved the row with the largest first element to the top of the matrix. Now we would perform the first step of the Gaussian elimination procedure on the resulting matrix, move on to the second row and pivot again.

In practice, one should always use pivoting when applying Gaussian elimination, since you rarely know in advance when the equations you're trying to solve will present a problem. Exercise 6.2 invites you to extend the Gaussian elimination program from Example 6.1 to incorporate partial pivoting.

Exercise 6.2:

    a) Modify the program `gausselim.py` in Example 6.1 to incorporate partial pivoting (or you can write your own program from scratch if you prefer). Run your program and demonstrate that it gives the same answers as the original program when applied to Eq. (6.1).

    b) Modify the program to solve the equations in (6.17) and show that it can find the solution to these as well, even though Gaussian elimination without pivoting fails.

6.1.4 LU DECOMPOSITION

The Gaussian elimination method, combined with partial pivoting, is a reliable method for solving simultaneous equations and is widely used. For the types of calculations that crop up in computational physics, however, it is commonly used in a slightly different form from the one we have seen so far. In physics calculations it often happens that we want to solve many different sets of equations $\mathbf{A}\mathbf{x} = \mathbf{v}$ with the same matrix $\mathbf{A}$ but different right-hand sides $\mathbf{v}$. In such cases it would be wasteful to perform a full Gaussian elimination for every set of equations. The effect on the matrix **A** will be the same every time—it will always be transformed into the same upper triangular matrix—and only the vector **v** will change. It would be more efficient if we could do the full Gaussian elimination just once and somehow record or memorize the divisions, multiplications, and subtractions we perform on each side of the equation, so that we can later perform them on as many vectors **v** as we like without having to perform them again on the matrix **A**. There is a variant of Gaussian elimination that does exactly this, called *LU decomposition* (pronounced as two separate letters "L-U").

Suppose we have a matrix **A** of the form

$$\mathbf{A} = \begin{pmatrix} a_{00} & a_{01} & a_{02} & a_{03} \\ a_{10} & a_{11} & a_{12} & a_{13} \\ a_{20} & a_{21} & a_{22} & a_{23} \\ a_{30} & a_{31} & a_{32} & a_{33} \end{pmatrix}, \tag{6.19}$$

where we have numbered the elements in Python style, starting from zero. We use a 4×4 matrix here, but the generalization of the calculation to matrices of other sizes is straightforward.

Let us perform Gaussian elimination on this matrix to reduce it to upper-triangular form, but we will write the elimination process in a manner slightly different from before, using matrix notation. Recall that in the first step of the elimination process we divide the first row of the matrix by its first element a_{00}, then we subtract the first row times a_{10} from the second, times a_{20} from the third, and times a_{30} from the fourth. These operations can be neatly written as a single matrix multiplication, thus:

$$\frac{1}{a_{00}} \begin{pmatrix} 1 & 0 & 0 & 0 \\ -a_{10} & a_{00} & 0 & 0 \\ -a_{20} & 0 & a_{00} & 0 \\ -a_{30} & 0 & 0 & a_{00} \end{pmatrix} \begin{pmatrix} a_{00} & a_{01} & a_{02} & a_{03} \\ a_{10} & a_{11} & a_{12} & a_{13} \\ a_{20} & a_{21} & a_{22} & a_{23} \\ a_{30} & a_{31} & a_{32} & a_{33} \end{pmatrix} = \begin{pmatrix} 1 & b_{01} & b_{02} & b_{03} \\ 0 & b_{11} & b_{12} & b_{13} \\ 0 & b_{21} & b_{22} & b_{23} \\ 0 & b_{31} & b_{32} & b_{33} \end{pmatrix}. \tag{6.20}$$

(It's worth doing the multiplication for yourself, to see how it works out.) The matrix we're multiplying by on the left we will call $\mathbf{L}_0$. Note that this matrix is *lower triangular*—all the elements above the diagonal are zero:

$$\mathbf{L}_0 = \frac{1}{a_{00}} \begin{pmatrix} 1 & 0 & 0 & 0 \\ -a_{10} & a_{00} & 0 & 0 \\ -a_{20} & 0 & a_{00} & 0 \\ -a_{30} & 0 & 0 & a_{00} \end{pmatrix}. \tag{6.21}$$

The next step of the Gaussian elimination process involves dividing the second row of the matrix by its second element b_{11} and then subtracting appropriate multiples from lower rows. This too can be represented as a matrix multiplication:

$$\frac{1}{b_{11}} \begin{pmatrix} b_{11} & 0 & 0 & 0 \\ 0 & 1 & 0 & 0 \\ 0 & -b_{21} & b_{11} & 0 \\ 0 & -b_{31} & 0 & b_{11} \end{pmatrix} \begin{pmatrix} 1 & b_{01} & b_{02} & b_{03} \\ 0 & b_{11} & b_{12} & b_{13} \\ 0 & b_{21} & b_{22} & b_{23} \\ 0 & b_{31} & b_{32} & b_{33} \end{pmatrix} = \begin{pmatrix} 1 & c_{01} & c_{02} & c_{03} \\ 0 & 1 & c_{12} & c_{13} \\ 0 & 0 & c_{22} & c_{23} \\ 0 & 0 & c_{32} & c_{33} \end{pmatrix}. \tag{6.22}$$

Again, note that the matrix we are multiplying by—call it L_1—is lower triangular:

$$L_1 = \frac{1}{b_{11}} \begin{pmatrix} b_{11} & 0 & 0 & 0 \\ 0 & 1 & 0 & 0 \\ 0 & -b_{21} & b_{11} & 0 \\ 0 & -b_{31} & 0 & b_{11} \end{pmatrix}. \tag{6.23}$$

For the 4×4 example considered here, there are two more steps to the Gaussian elimination, which correspond to multiplying by the lower-triangular matrices

$$L_2 = \frac{1}{c_{22}} \begin{pmatrix} c_{22} & 0 & 0 & 0 \\ 0 & c_{22} & 0 & 0 \\ 0 & 0 & 1 & 0 \\ 0 & 0 & -c_{32} & c_{22} \end{pmatrix}, \quad L_3 = \frac{1}{d_{33}} \begin{pmatrix} d_{33} & 0 & 0 & 0 \\ 0 & d_{33} & 0 & 0 \\ 0 & 0 & d_{33} & 0 \\ 0 & 0 & 0 & 1 \end{pmatrix}, \tag{6.24}$$

where d_{33} is defined in the obvious way.

Putting the whole process together, the complete Gaussian elimination on the matrix A is equivalent to multiplying the matrix in succession by L_0, L_1, L_2, and L_3. Or, to put that another way, the matrix $L_3 L_2 L_1 L_0 A$ must be upper triangular, since it is the end product of a Gaussian elimination process. Knowing this it is now easy to see how we solve our original set of equations $Ax = v$ for the unknown vector x. We multiply on both sides by $L_3 L_2 L_1 L_0$ to get

$$L_3 L_2 L_1 L_0 A x = L_3 L_2 L_1 L_0 v. \tag{6.25}$$

All the quantities on the right are known, so we can perform the multiplications and get a vector. On the left we have an upper-triangular matrix times x. So our equation is now in the form of Eq. (6.10), which can be solved easily by backsubstitution.

This approach is, mathematically speaking, entirely equivalent to the Gaussian elimination of earlier sections. Practically speaking, however, it has the significant advantage that the matrices L_0 to L_3 encapsulate the entire sequence of transformations performed during the elimination process, which allows us to efficiently solve sets of equations with the same matrix A but different right-hand sides v. Every such set involves solving Eq. (6.25) for a different vector v but the matrix $L_3L_2L_1L_0A$ on the left is the same every time, so we only need to calculate it once. Since calculating the matrix is the most time-consuming part of the Gaussian elimination process, this simplification can speed up our program enormously. Once we have the matrices L_0 to L_3, the only thing we need to do on being handed a new vector v is multiply to get $L_3L_2L_1L_0v$, then apply backsubstitution to the result.

This is, in essence the entire LU decomposition method. It's just a version of Gaussian elimination expressed in terms of matrices. However, in practice the calculations are usually performed in a slightly different way that is a little simpler and more convenient than the way described here. We define the matrices

$$L = L_0^{-1}L_1^{-1}L_2^{-1}L_3^{-1}, \qquad U = L_3L_2L_1L_0A. \tag{6.26}$$

Note that U is just the matrix on the left-hand side of Eq. (6.25), which is upper triangular (hence the name U). Multiplying these two matrices together we get

$$LU = A, \tag{6.27}$$

and hence the set of equations $Ax = v$ can also be written as

$$LUx = v. \tag{6.28}$$

The nice thing about this form is that not only is U upper triangular, but L is also lower triangular. In fact L has a very simple form. Consider the matrix L_0:

$$L_0 = \frac{1}{a_{00}} \begin{pmatrix} 1 & 0 & 0 & 0 \\ -a_{10} & a_{00} & 0 & 0 \\ -a_{20} & 0 & a_{00} & 0 \\ -a_{30} & 0 & 0 & a_{00} \end{pmatrix}. \tag{6.29}$$

It takes only a moment to confirm that the inverse of this matrix is

$$L_0^{-1} = \begin{pmatrix} a_{00} & 0 & 0 & 0 \\ a_{10} & 1 & 0 & 0 \\ a_{20} & 0 & 1 & 0 \\ a_{30} & 0 & 0 & 1 \end{pmatrix}, \tag{6.30}$$

which is also lower triangular. Similarly the inverses of the other matrices are

$$
\mathbf{L}_1^{-1} = \begin{pmatrix} 1 & 0 & 0 & 0 \\ 0 & b_{11} & 0 & 0 \\ 0 & b_{21} & 1 & 0 \\ 0 & b_{31} & 0 & 1 \end{pmatrix}, \quad
\mathbf{L}_2^{-1} = \begin{pmatrix} 1 & 0 & 0 & 0 \\ 0 & 1 & 0 & 0 \\ 0 & 0 & c_{22} & 0 \\ 0 & 0 & c_{32} & 1 \end{pmatrix}, \quad
\mathbf{L}_3^{-1} = \begin{pmatrix} 1 & 0 & 0 & 0 \\ 0 & 1 & 0 & 0 \\ 0 & 0 & 1 & 0 \\ 0 & 0 & 0 & d_{33} \end{pmatrix}.
$$

(6.31)

Multiplying them all together, we find that

$$
\mathbf{L} = \mathbf{L}_0^{-1} \mathbf{L}_1^{-1} \mathbf{L}_2^{-1} \mathbf{L}_3^{-1} = \begin{pmatrix} a_{00} & 0 & 0 & 0 \\ a_{10} & b_{11} & 0 & 0 \\ a_{20} & b_{21} & c_{22} & 0 \\ a_{30} & b_{31} & c_{32} & d_{33} \end{pmatrix}.
$$

(6.32)

Thus, not only is $\mathbf{L}$ lower triangular, but its value is easy to calculate from numbers we already know.

Equation (6.27) is called the LU decomposition of the matrix $\mathbf{A}$. It tells us that $\mathbf{A}$ is equal to a product of two matrices, the first being lower triangular and the second being upper triangular.

Once we have the LU decomposition of $\mathbf{A}$ we can use it to solve $\mathbf{A}\mathbf{x} = \mathbf{v}$ directly. Here's how the process goes for a 3×3 example. The LU decomposition of $\mathbf{A}$ looks like this in general:

$$
\mathbf{A} = \begin{pmatrix} a_{00} & a_{01} & a_{02} \\ a_{10} & a_{11} & a_{12} \\ a_{20} & a_{21} & a_{22} \end{pmatrix} = \begin{pmatrix} l_{00} & 0 & 0 \\ l_{10} & l_{11} & 0 \\ l_{20} & l_{21} & l_{22} \end{pmatrix} \begin{pmatrix} u_{00} & u_{01} & u_{02} \\ 0 & u_{11} & u_{12} \\ 0 & 0 & u_{22} \end{pmatrix}.
$$

(6.33)

Then our set of simultaneous equations takes the form

$$
\begin{pmatrix} l_{00} & 0 & 0 \\ l_{10} & l_{11} & 0 \\ l_{20} & l_{21} & l_{22} \end{pmatrix} \begin{pmatrix} u_{00} & u_{01} & u_{02} \\ 0 & u_{11} & u_{12} \\ 0 & 0 & u_{22} \end{pmatrix} \begin{pmatrix} x_0 \\ x_1 \\ x_2 \end{pmatrix} = \begin{pmatrix} v_0 \\ v_1 \\ v_2 \end{pmatrix}.
$$

(6.34)

Now let us define a new vector $\mathbf{y}$ according to

$$
\begin{pmatrix} u_{00} & u_{01} & u_{02} \\ 0 & u_{11} & u_{12} \\ 0 & 0 & u_{22} \end{pmatrix} \begin{pmatrix} x_0 \\ x_1 \\ x_2 \end{pmatrix} = \begin{pmatrix} y_0 \\ y_1 \\ y_2 \end{pmatrix}.
$$

(6.35)

In terms of this vector, our equation (6.34) becomes

$$
\begin{pmatrix} l_{00} & 0 & 0 \\ l_{10} & l_{11} & 0 \\ l_{20} & l_{21} & l_{22} \end{pmatrix} \begin{pmatrix} y_0 \\ y_1 \\ y_2 \end{pmatrix} = \begin{pmatrix} v_0 \\ v_1 \\ v_2 \end{pmatrix}.
$$

(6.36)

Thus we have broken our original problem down into two separate matrix problems, Eqs. (6.35) and (6.36). If we can solve (6.36) for **y** given **v** then we can substitute the answer into (6.35) and solve to get a solution for **x**.

But both of these two new problems involves only a triangular matrix, not a full matrix and, as we saw in Section 6.1.2, such equations are easy to solve by backsubstitution. For instance, Eq. (6.36) can be solved for **y** by noting that the first line of the equation implies $l_{00}y_0 = v_0$, or

$$y_0 = \frac{v_0}{l_{00}}. \tag{6.37}$$

Then the second line gives us $l_{10}y_0 + l_{11}y_1 = v_1$, or

$$y_1 = \frac{v_1 - l_{10}y_0}{l_{11}}, \tag{6.38}$$

and the third line gives us

$$y_2 = \frac{v_2 - l_{20}y_0 - l_{21}y_1}{l_{22}}, \tag{6.39}$$

and so we have all the elements of **y**. Now we take these values, substitute them into Eq. (6.35), and then use a similar backsubstitution procedure to solve for the elements of **x**, which gives us our final solution.

This LU decomposition method is the most commonly used form of Gaussian elimination, and indeed the most commonly used method for solving simultaneous equations altogether. To avoid problems with small or zero diagonal elements, one must in general also incorporate pivoting, as discussed in Section 6.1.3, but this is straightforward to do. The standard method for solving equations then breaks down into two separate operations, the LU decomposition of the matrix **A** with partial pivoting, followed by the double backsubstitution process above to recover the final solution. For solving multiple sets of equations with the same matrix **A** the LU decomposition need be performed only once, and only the backsubstitutions need be repeated for each individual set. Exercise 6.3 invites you to try your hand at writing a program to solve simultaneous equations using this method, and you are encouraged to give it a go.

Solving simultaneous equations, however, is such a common operation, both in physics and elsewhere, that you don't need to write your own program to do it if you don't want to. Many programs to do the calculation have already been written by other people. Professionally written programs to perform LU decomposition and solve equations exist in most computer languages, including Python, and there is no reason not to use them when they're available.

It might seem like cheating to use a program written by someone else to solve a problem, but in many cases it's actually an excellent idea. Of course, it can be a very good exercise to write your own program. You can learn a lot, both about physics and about programming, by doing so. And there is no shortage of hard physics problems where no one has written a program yet and you'll have no choice but to do it yourself. But in the cases where a program is already available you shouldn't feel bad about using it.

In Python, functions for solving simultaneous equations are provided by the module `linalg` which is found in the numpy package. In particular, the module contains a function `solve`, which solves systems of linear equations of the form $\mathbf{Ax} = \mathbf{v}$ using LU decomposition and backsubstitution. You need only import the function then use it like this:

```
from numpy.linalg import solve
x = solve(A,v)
```

In most cases, if you need to solve a large system of equations (or even a small one) in Python, this is the simplest and quickest way to get an answer. There are also functions that perform the LU decomposition step separately from the backsubstitution, so that you can efficiently solve multiple sets of equations, as described above. We will not need these separate functions in this book, but if you wish to learn about them you can find more information at www.scipy.org.

Exercise 6.3: LU decomposition

This exercise invites you to write your own program to solve simultaneous equations using the method of LU decomposition.

    a) Starting, if you wish, with the program for Gaussian elimination in Example 6.1 on page 218, write a Python function that calculates the LU decomposition of a matrix. The calculation is same as that for Gaussian elimination, except that at each step of the calculation you need to extract the appropriate elements of the matrix and assemble them to form the lower diagonal matrix $\mathbf{L}$ of Eq. (6.32). Test your function by calculating the LU decomposition of the matrix from Eq. (6.2), then multiplying the $\mathbf{L}$ and $\mathbf{U}$ you get and verifying that you recover the original matrix once more.

    b) Build on your LU decomposition function to create a complete program to solve Eq. (6.2) by performing a double backsubstitution as described in this section. Solve the same equations using the function `solve` from the numpy package and verify that you get the same answer either way.

c) If you're feeling ambitious, try your hand at LU decomposition with partial pivoting. Partial pivoting works in the same way for LU decomposition as it does for Gaussian elimination, swapping rows to get the largest diagonal element as explained in Section 6.1.3, but the extension to LU decomposition requires two additional steps. First, every time you swap two rows you also have to swap the same rows in the matrix $\mathbf{L}$. Second, when you use your LU decomposition to solve a set of equations $\mathbf{Ax} = \mathbf{v}$ you will also need to perform the same sequence of swaps on the vector $\mathbf{v}$ on the right-hand side. This means you need to record the swaps as you are doing the decomposition so that you can recreate them later. The simplest way to do this is to set up a list or array in which the value of the ith element records the row you swapped with on the ith step of the process. For instance, if you swapped the first row with the second then the second with the fourth, the first two elements of the list would be 2 and 4. Solving a set of equations for given $\mathbf{v}$ involves first performing the required sequence of swaps on the elements of $\mathbf{v}$ then performing a double backsubstitution as usual. (In ordinary Gaussian elimination with pivoting, one swaps the elements of $\mathbf{v}$ as the algorithm proceeds, rather than all at once, but the difference has no effect on the results, so it's fine to perform all the swaps at once if we wish.)

Modify the function you wrote for part (a) to perform LU decomposition with partial pivoting. The function should return the matrices $\mathbf{L}$ and $\mathbf{U}$ for the LU decomposition of the swapped matrix, plus a list of the swaps made. Then modify the rest of your program to solve equations of the form $\mathbf{Ax} = \mathbf{v}$ using LU decomposition with pivoting. Test your program on the example from Eq. (6.17), which cannot be solved without pivoting because of the zero in the first element of the matrix. Check your results against a solution of the same equations using the `solve` function from numpy.

LU decomposition with partial pivoting is the most widely used method for the solution of simultaneous equations in practice. Precisely this method is used in the function `solve` from the numpy package. There's nothing wrong with using the `solve` function—it's well written, fast, and convenient. But it does nothing you haven't already done yourself if you've solved this exercise.

Exercise 6.4: Write a program to solve the resistor network problem of Exercise 6.1 on page 220 using the function `solve` from `numpy.linalg`. If you also did Exercise 6.1, you should check that you get the same answer both times.

Exercise 6.5: Here's a more complicated circuit problem:

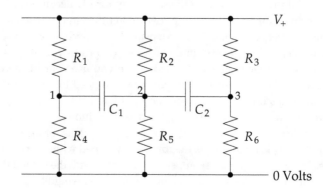

The voltage V_+ is time-varying and sinusoidal of the form $V_+ = x_+ e^{i\omega t}$ with x_+ a constant. (The voltage is actually real, of course, but things are made simpler by treating it as complex, as is common in circuit calculations.) The resistors in the circuit can be treated using Ohm's law as usual. For the capacitors the charge Q and voltage V across them are related by the capacitor law $Q = CV$, where C is the capacitance. Differentiating both sides of this expression gives the current I flowing in on one side of the capacitor and out on the other:

$$I = \frac{dQ}{dt} = C\frac{dV}{dt}.$$

a) Assuming the voltages at the points labeled 1, 2, and 3 are of the form $V_1 = x_1 e^{i\omega t}$, $V_2 = x_2 e^{i\omega t}$, and $V_3 = x_3 e^{i\omega t}$, apply Kirchhoff's law at each of the three points, along with Ohm's law and the capacitor law, to show that the constants x_1, x_2, and x_3 satisfy the equations

$$\left(\frac{1}{R_1} + \frac{1}{R_4} + i\omega C_1\right)x_1 - i\omega C_1 x_2 = \frac{x_+}{R_1},$$

$$-i\omega C_1 x_1 + \left(\frac{1}{R_2} + \frac{1}{R_5} + i\omega C_1 + i\omega C_2\right)x_2 - i\omega C_2 x_3 = \frac{x_+}{R_2},$$

$$-i\omega C_2 x_2 + \left(\frac{1}{R_3} + \frac{1}{R_6} + i\omega C_2\right)x_3 = \frac{x_+}{R_3}.$$

b) Write a program to solve for x_1, x_2, and x_3 when

$$R_1 = R_3 = R_5 = 1\,\text{k}\Omega,$$
$$R_2 = R_4 = R_6 = 2\,\text{k}\Omega,$$
$$C_1 = 1\,\mu\text{F}, \qquad C_2 = 0.5\,\mu\text{F},$$
$$x_+ = 3\,\text{V}, \qquad \omega = 1000\,\text{s}^{-1}.$$

Notice that the matrix for this problem has complex elements. You will need to define a complex array to hold it, but you can still use the solve function just as before to solve the equations—it works with either real or complex arguments. Using your solution have your program calculate and print the amplitudes of the three voltages V_1, V_2, and V_3 and their phases in degrees. (Hint: You may find the functions `polar` or `phase` in the `cmath` package useful. If z is a complex number then "`r,theta = polar(z)`" will return the modulus and phase (in radians) of z and "`theta = phase(z)`" will return the phase alone.)

6.1.5 CALCULATING THE INVERSE OF A MATRIX

As we said at the beginning of the chapter, one can solve equations of the form $\mathbf{Ax} = \mathbf{v}$ by multiplying on both sides by the inverse of $\mathbf{A}$, but this is not a very efficient way to do the calculation. Gaussian elimination or LU decomposition is faster. But what if we wanted to know the inverse of a matrix for some other reason? In fact, it's rare in physics problems that we need to do this, but just suppose for a moment that we do. How could we find the inverse?

One way to calculate a matrix inverse—the way most of us are taught in our linear algebra classes—is to compute a matrix of cofactors and divide it by the determinant. However, as with several of the problems we've examined in this book, the obvious approach is not always the best, and this method of calculating an inverse is typically quite slow and can be subject to large rounding errors. A better approach is to turn the problem of inverting the matrix into a problem of solving a set of simultaneous equations, then tackle that problem using the methods we have already seen.

Consider the form

$$\mathbf{AX} = \mathbf{V}, \qquad (6.40)$$

which is the same as before, except that $\mathbf{X}$ and $\mathbf{V}$ are $N \times N$ matrices now, not just single vectors. We can solve this equation in the same way as we did previously, using either Gaussian elimination or LU decomposition to reduce the matrix $\mathbf{A}$ to a simpler form, then solving for the elements of $\mathbf{X}$ by backsubstitution. You have to solve for each column of $\mathbf{X}$ separately, in terms of the corresponding column of $\mathbf{V}$, but in principle the calculation is just the same as the one we did before.

But now suppose we choose the value of $\mathbf{V}$ to be the identity matrix $\mathbf{I}$, so that $\mathbf{AX} = \mathbf{I}$. Then, by definition, the value of $\mathbf{X}$ is $\mathbf{A}^{-1}$, which is the inverse that we want. So when we find the solution with this choice of $\mathbf{V}$ we get the inverse of $\mathbf{A}$.

Thus, for instance, we could use the function `solve` from the `numpy.linalg` module to solve for the columns of $\mathbf{X}$ and so invert a matrix $\mathbf{A}$. This, however, is a bit cumbersome, so there is a separate function in numpy.linalg called inv that does the calculation for us in one step. To use it we simply say

```
from numpy.linalg import inv
X = inv(A)
```

which sets X equal to the value of the inverse.

As we've said, however, we don't often need the inverse in physics calculations, and indeed we will not need it for any of the calculations we do in this book, so let us move on to other things.

6.1.6 TRIDIAGONAL AND BANDED MATRICES

A special case that arises often in physics problems is the solution of $\mathbf{Ax} = \mathbf{v}$ when the matrix $\mathbf{A}$ is *tridiagonal*, thus:

$$\mathbf{A} = \begin{pmatrix} a_{00} & a_{01} & & & \\ a_{10} & a_{11} & a_{12} & & \\ & a_{21} & a_{22} & a_{23} & \\ & & a_{32} & a_{33} & a_{34} \\ & & & a_{43} & a_{44} \end{pmatrix}. \tag{6.41}$$

Here all the elements not shown are zero. That is, the matrix has nonzero elements only along the diagonal and immediately above and below it. This is a case where simple Gaussian elimination is a good choice for solving the problem—it works quickly, and pivoting is typically not used so the programming is straightforward.[1]

The point to notice about tridiagonal problems is that we don't need to go through the entire Gaussian elimination procedure to solve them. Each row only needs to be subtracted from the single row immediately below it—and not all lower rows—to make the matrix triangular. Suppose, for instance, we

---

[1]It is still possible for zeros to appear along the diagonal during the Gaussian elimination process so that pivoting would be required to find a solution as discussed in Section 6.1.3. However, the interchange of matrix rows during pivoting destroys the tridiagonal form of the matrix, meaning we cannot use the method described in this section to solve such systems and must turn to other methods, such as variants of LU decomposition. Thus we only ever use Gaussian elimination without pivoting or another method entirely, but elimination with pivoting is not used.

have a 4×4 matrix of the form

$$\mathbf{A} = \begin{pmatrix} 2 & 1 & 0 & 0 \\ 3 & 4 & -5 & 0 \\ 0 & -4 & 3 & 5 \\ 0 & 0 & 1 & 3 \end{pmatrix}. \tag{6.42}$$

The initial step of our Gaussian elimination would be to divide the first row of the matrix by 2, then subtract 3 times the result from the second row, which would give a new matrix

$$\begin{pmatrix} 1 & 0.5 & 0 & 0 \\ 0 & 2.5 & -5 & 0 \\ 0 & -4 & 3 & 5 \\ 0 & 0 & 1 & 3 \end{pmatrix}. \tag{6.43}$$

Note that the first column now has the required form $1, 0, 0, 0$. Next we would divide the second row by 2.5 and subtract -4 times the result from the third row, to get

$$\begin{pmatrix} 1 & 0.5 & 0 & 0 \\ 0 & 1 & -2 & 0 \\ 0 & 0 & -5 & 5 \\ 0 & 0 & 1 & 3 \end{pmatrix}. \tag{6.44}$$

Then we divide the third row by -5 and subtract it from the fourth to get

$$\begin{pmatrix} 1 & 0.5 & 0 & 0 \\ 0 & 1 & -2 & 0 \\ 0 & 0 & 1 & -1 \\ 0 & 0 & 0 & 4 \end{pmatrix}. \tag{6.45}$$

Finally we divide the last row by 4:

$$\begin{pmatrix} 1 & 0.5 & 0 & 0 \\ 0 & 1 & -2 & 0 \\ 0 & 0 & 1 & -1 \\ 0 & 0 & 0 & 1 \end{pmatrix}. \tag{6.46}$$

In the process we have reduced our matrix to upper-triangular form and we can now use back substitution to find the solution for $\mathbf{x}$.

The backsubstitution process is also simplified by the tridiagonal form. After the Gaussian elimination is complete we are left with an equation of the general form

$$\begin{pmatrix} 1 & a_{01} & 0 & 0 \\ 0 & 1 & a_{12} & 0 \\ 0 & 0 & 1 & a_{23} \\ 0 & 0 & 0 & 1 \end{pmatrix} \begin{pmatrix} x_0 \\ x_1 \\ x_2 \\ x_3 \end{pmatrix} = \begin{pmatrix} v_0 \\ v_1 \\ v_2 \\ v_3 \end{pmatrix},$$
(6.47)

which has the simple solution

$$x_3 = v_3,$$ (6.48a)

$$x_2 = v_2 - a_{23}x_3,$$ (6.48b)

$$x_1 = v_1 - a_{12}x_2,$$ (6.48c)

$$x_0 = v_0 - a_{01}x_1.$$ (6.48d)

When used in this way, Gaussian elimination is sometimes called the *tridiagonal matrix algorithm* or the *Thomas algorithm*. You can also solve tridiagonal problems using general solution methods like those in the function solve from numpy.linalg, but doing so will usually be slower than the Thomas algorithm, because those methods do not take take advantage of the shortcuts we have used here to simplify the calculation. Indeed for large matrices using solve can be *much* slower, so the Thomas algorithm is usually the right way to go with large tridiagonal matrices.

A related issue comes up when the matrix **A** is *banded*, meaning that it is similar to a tridiagonal matrix but can have more than one nonzero element to either side of the diagonal, like this:

$$\mathbf{A} = \begin{pmatrix} a_{00} & a_{01} & a_{02} & & & & \\ a_{10} & a_{11} & a_{12} & a_{13} & & & \\ a_{20} & a_{21} & a_{22} & a_{23} & a_{24} & & \\ & a_{31} & a_{32} & a_{33} & a_{34} & a_{35} & \\ & & a_{42} & a_{43} & a_{44} & a_{45} & a_{46} \\ & & & a_{53} & a_{54} & a_{55} & a_{56} \\ & & & & a_{64} & a_{65} & a_{66} \end{pmatrix}.$$
(6.49)

Problems containing banded matrices can be solved in the same way as for the tridiagonal case, using Gaussian elimination. If there are m nonzero elements below the diagonal then each row must now be subtracted from the m rows below it, and not just from one single row. And the equations for the backsubstitution are more complex, involving terms for each of the nonzero rows above the diagonal. These differences make the solution process slower, but it can still be much faster than using the solve function from numpy.linalg.

EXAMPLE 6.2: VIBRATION IN A ONE-DIMENSIONAL SYSTEM

Suppose we have a set of N identical masses in a row, joined by identical linear springs, thus:

For simplicity, we'll ignore gravity—the masses and springs are floating in outer space. If we jostle this system the masses will vibrate relative to one another under the action of the springs. The motions of the system could be used as a model of the vibration of atoms in a solid, which can be represented with reasonable accuracy in exactly this way. Here we examine the modes of horizontal vibration of the system.

Let us denote the displacement of the ith mass relative to its rest position by ξ_i. Then the equations of motion for the system are given by Newton's second law:

$$m\frac{d^2\xi_i}{dt^2} = k(\xi_{i+1} - \xi_i) + k(\xi_{i-1} - \xi_i) + F_i, \tag{6.50}$$

where m is the mass and k is the spring constant. The left-hand side of this equation is just mass times acceleration, while the right-hand side is the force on mass i due to the springs connecting it to the two adjacent masses, plus an extra term F_i that represents any external force imposed on mass i. The only exceptions to Eq. (6.50) are for the masses at the two ends of the line, for which there is only one spring force each, so that they obey the equations

$$m\frac{d^2\xi_1}{dt^2} = k(\xi_2 - \xi_1) + F_1, \tag{6.51}$$

$$m\frac{d^2\xi_N}{dt^2} = k(\xi_{N-1} - \xi_N) + F_N. \tag{6.52}$$

Now suppose we drive our system by applying a harmonic (i.e., sinusoidal) force to the first mass: $F_1 = Ce^{i\omega t}$, where C is a constant and we use a complex form, as one commonly does in such cases, on the understanding that we will take the real part at the end of the calculation. If we are thinking of this as a model of atoms in a solid, such a force could be created by the charge of the atoms interacting with a varying electric field, such as would be produced by an electromagnetic wave falling on the solid.

The net result of the applied force will be to make all the atoms oscillate in some fashion with angular frequency ω, so that the overall solution for the positions of the atoms will take the form

$$\xi_i(t) = x_i\,e^{i\omega t} \tag{6.53}$$

for all i. The magnitude of the quantity x_i controls the amplitude of vibration of mass i and its phase controls the phase of the vibration relative to the driving force.

Substituting Eq. (6.53) into Eqs. (6.50) to (6.52) we find that

$$-m\omega^2 x_1 = k(x_2 - x_1) + C, \tag{6.54a}$$
$$-m\omega^2 x_i = k(x_{i+1} - x_i) + k(x_{i-1} - x_i), \tag{6.54b}$$
$$-m\omega^2 x_N = k(x_{N-1} - x_N), \tag{6.54c}$$

where (6.54b) applies for all i in the range $2 \le i \le N - 1$. These equations can be rearranged to read

$$(\alpha - k)x_1 - kx_2 = C, \tag{6.55a}$$
$$\alpha x_i - kx_{i-1} - kx_{i+1} = 0, \tag{6.55b}$$
$$(\alpha - k)x_N - kx_{N-1} = 0, \tag{6.55c}$$

where $\alpha = 2k - m\omega^2$. Thus in matrix form we have:

$$\begin{pmatrix} (\alpha - k) & -k & & & & \\ -k & \alpha & -k & & & \\ & -k & \alpha & -k & & \\ & & \ddots & \ddots & \ddots & \\ & & & -k & \alpha & -k \\ & & & & -k & (\alpha - k) \end{pmatrix} \begin{pmatrix} x_1 \\ x_2 \\ x_3 \\ \vdots \\ x_{N-1} \\ x_N \end{pmatrix} = \begin{pmatrix} C \\ 0 \\ 0 \\ \vdots \\ 0 \\ 0 \end{pmatrix}. \tag{6.56}$$

This is precisely a tridiagonal set of simultaneous equations of the type we have been considering in this section, which we can solve by Gaussian elimination. Note that even though we employed a complex form for the driving force, the equations above are entirely real and hence their solution will also be real. So it is safe to write a program to solve them using real variables only.[2] (We could choose C complex, but this simply multiplies the entire solution by a constant phase factor, so it doesn't give us any new behavior.)

Here is a program to solve the problem for the case where $N = 26$, $C = 1$, $m = 1$, $k = 6$, and $\omega = 2$, so that $\alpha = 8$. The program calculates all the x_i then makes a graph of the values, which reflect the amplitudes of the oscillating masses.

---

[2]There is no problem using Gaussian elimination for complex simultaneous equations if we have to—see Exercise 6.5 on page 230 for an example. It works fine; we just don't need it in this particular case.

```
from numpy import zeros,empty                          File: springs.py
from pylab import plot,show

# Constants
N = 26
C = 1.0
m = 1.0
k = 6.0
omega = 2.0
alpha = 2*k-m*omega*omega

# Set up the initial values of the arrays
A = zeros([N,N],float)
for i in range(N-1):
    A[i,i] = alpha
    A[i,i+1] = -k
    A[i+1,i] = -k

A[0,0] = alpha - k
A[N-1,N-1] = alpha - k
v = zeros(N,float)
v[0] = C

# Perform the Gaussian elimination
for i in range(N-1):

    # Divide row i by its diagonal element
    A[i,i+1] /= A[i,i]
    v[i] /= A[i,i]

    # Now subtract it from the next row down
    A[i+1,i+1] -= A[i+1,i]*A[i,i+1]
    v[i+1] -= A[i+1,i]*v[i]

# Divide the last element of v by the last diagonal element
v[N-1] /= A[N-1,N-1]

# Backsubstitution
x = empty(N,float)
x[N-1] = v[N-1]
for i in range(N-2,-1,-1):
    x[i] = v[i] - A[i,i+1]*x[i+1]
```

```
# Make a plot using both dots and lines
plot(x)
plot(x,"ko")
show()
```

Note that in this case we did not make use of Python's array processing abilities to perform operations on entire rows of the matrix in a single step. Each row that we work with has only two nonzero elements, so it makes no sense to divide, multiply, or subtract entire rows—it's quicker to explicitly manipulate just the two nonzero elements and leave all the others alone. We also did not bother to set the diagonal elements of the matrix to one or the elements below the diagonal to zero. We know that they will end up with these values, but the values are not needed in the backsubstitution process, so we can save ourselves a little effort, and make a simpler program, by leaving these elements as they are.

Figure 6.1 shows the plot produced by the program. As we can see, the amplitude of motion of the masses varies in a wave along the length of the chain, with some masses vibrating vigorously while others remain almost stationary. This is the classic form of a standing wave in a driven spatial system, and it's not a bad model for what happens when we excite vibrations in the atoms of a solid.

The solution of tridiagonal or banded systems of equations is a relatively common operation in computational physics, so it would be a good candidate for a user-defined function, similar to the function `solve` in the module `numpy.linalg`, which solves general linear systems of equations as discussed in Section 6.1.4. One would have to write such a function only once and then one could use it whenever one wanted to solve banded systems. Such a function, called `banded`, is provided in Appendix E and can be found in the file `banded.py` in the on-line resources. This function will come in handy for a number of later problems that involve the solution of tridiagonal or banded systems. As an example, here is how we would use it to solve the problem of the masses and springs above:

File: springsb.py

```
from numpy import empty,zeros
from banded import banded
from pylab import plot,show
```

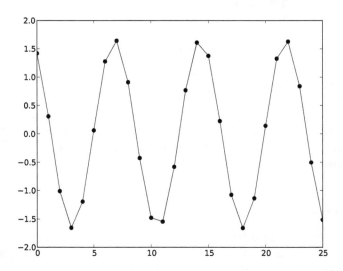

Figure 6.1: Amplitudes of vibration in a chain of identical masses coupled by springs. This figure, produced by the program springs.py given in the text, shows the amplitude with which the masses in the chain vibrate, calculated by solving the appropriate tridiagonal system of equations, Eq. (6.56), using Gaussian elimination and backsubstitution.

```
# Constants
N = 26
C = 1.0
m = 1.0
k = 6.0
omega = 2.0
alpha = 2*k-m*omega*omega

# Set up the initial values of the arrays
A = empty([3,N],float)
A[0,:] = -k
A[1,:] = alpha
A[2,:] = -k
A[1,0] = alpha - k
A[1,N-1] = alpha - k
v = zeros(N,float)
v[0] = C
```

```
# Solve the equations
x = banded(A,v,1,1)

# Make a plot using both dots and lines
plot(x)
plot(x,"ko")
show()
```

Unsurprisingly, this program produces a plot identical to Fig. 6.1. For more details of how the banded function is used, particularly the special format used for the matrix A, take a look at Appendix E.

Exercise 6.6: Starting with either the program springs.py on page 237 or the program springsb.py above, remove the code that makes a graph of the results and replace it with code that creates an animation of the masses as they vibrate back and forth, their displacements relative to their resting positions being given by the real part of Eq. (6.53). For clarity, assume that the resting positions are two units apart in a horizontal line. At a minimum your animation should show each of the individual masses, perhaps as small spheres. (Spheres of radius about 0.2 or 0.3 seem to work well.)

Exercise 6.7: A chain of resistors

Consider a long chain of resistors wired up like this:

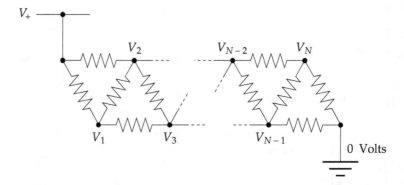

All the resistors have the same resistance R. The power rail at the top is at voltage $V_+ = 5V$. The problem is to find the voltages $V_1 \ldots V_N$ at the internal points in the circuit.

    a) Using Ohm's law and the Kirchhoff current law, which says that the total net current flow out of (or into) any junction in a circuit must be zero, show that the voltages $V_1 \ldots V_N$ satisfy the equations

$$3V_1 - V_2 - V_3 = V_+,$$
$$-V_1 + 4V_2 - V_3 - V_4 = V_+,$$
$$\vdots$$
$$-V_{i-2} - V_{i-1} + 4V_i - V_{i+1} - V_{i+2} = 0,$$
$$\vdots$$
$$-V_{N-3} - V_{N-2} + 4V_{N-1} - V_N = 0,$$
$$-V_{N-2} - V_{N-1} + 3V_N = 0.$$

Express these equations in vector form $\mathbf{Av} = \mathbf{w}$ and find the values of the matrix $\mathbf{A}$ and the vector $\mathbf{w}$.

b) Write a program to solve for the values of the V_i when there are $N = 6$ internal junctions with unknown voltages. (Hint: All the values of V_i should lie between zero and 5V. If they don't, something is wrong.)

c) Now repeat your calculation for the case where there are $N = 10\,000$ internal junctions. This part is not possible using standard tools like the solve function. You need to make use of the fact that the matrix $\mathbf{A}$ is banded and use the banded function from the file banded.py, discussed in Appendix E.

6.2   EIGENVALUES AND EIGENVECTORS

Another common matrix problem that arises in physics is the calculation of the eigenvalues and/or eigenvectors of a matrix. This problem arises in classical mechanics, quantum mechanics, electromagnetism, and other areas. Most eigenvalue problems in physics concern real symmetric matrices (or Hermitian matrices when complex numbers are involved, such as in quantum mechanics). For a symmetric matrix $\mathbf{A}$, an eigenvector $\mathbf{v}$ is a vector satisfying

$$\mathbf{Av} = \lambda\mathbf{v}, \tag{6.57}$$

where λ is the corresponding eigenvalue. For an $N \times N$ matrix there are N eigenvectors $\mathbf{v}_1 \ldots \mathbf{v}_N$ with eigenvalues $\lambda_1 \ldots \lambda_N$. The eigenvectors have the property that they are orthogonal to one another $\mathbf{v}_i \cdot \mathbf{v}_j = 0$ if $i \neq j$, and we will assume they are normalized to have unit length $\mathbf{v}_i \cdot \mathbf{v}_i = 1$ (although this is merely conventional, since Eq. (6.57) itself doesn't fix the normalization).

If we wish, we can consider the eigenvectors to be the columns of a single $N \times N$ matrix $\mathbf{V}$ and combine all the equations $\mathbf{Av}_i = \lambda_i\mathbf{v}_i$ into a single matrix equation

$$\mathbf{AV} = \mathbf{VD}, \tag{6.58}$$

where $\mathbf{D}$ is the diagonal matrix with the eigenvalues λ_i as its diagonal entries. Notice that the matrix $\mathbf{V}$ is orthogonal, meaning that its transpose $\mathbf{V}^T$ is equal to its inverse, so that $\mathbf{V}^T\mathbf{V} = \mathbf{VV}^T = \mathbf{I}$, the identity matrix.[3]

The most widely used technique for calculating eigenvalues and eigenvectors of real symmetric or Hermitian matrices on a computer is the *QR algorithm*, which we examine in this section. The QR algorithm works by calculating the matrices $\mathbf{V}$ and $\mathbf{D}$ that appear in Eq. (6.58). We'll look at the case where $\mathbf{A}$ is real and symmetric, although the Hermitian case is a simple extension.

The QR algorithm makes use of the *QR decomposition* of a matrix. In Section 6.1.4 we encountered the LU decomposition, which breaks a matrix into two parts, writing it as the product of a lower-triangular matrix $\mathbf{L}$ and an upper-triangular matrix $\mathbf{U}$. The QR decomposition is a variant on the same idea in which the matrix is written as the product $\mathbf{QR}$ of an *orthogonal* matrix $\mathbf{Q}$ and an upper-triangular matrix $\mathbf{R}$. Any square matrix can be written in this form and you'll have the opportunity shortly to write your own program to calculate QR decompositions (see Exercise 6.8), but for the moment let's just assume we have some way to calculate them.[4,5]

Suppose we have a real, square, symmetric matrix $\mathbf{A}$ and let us break it down into its QR decomposition, which we'll write as

$$\mathbf{A} = \mathbf{Q}_1\mathbf{R}_1. \tag{6.59}$$

---

[3]This follows from the fact that the ijth element of $\mathbf{V}^T\mathbf{V}$ is equal to the dot product of the ith and jth columns of $\mathbf{V}$, i.e., $\mathbf{v}_i \cdot \mathbf{v}_j$, which, as we have said is equal to 1 if $i = j$ and zero otherwise. This gives us $\mathbf{V}^T\mathbf{V} = \mathbf{I}$. That $\mathbf{VV}^T = \mathbf{I}$ as well follows from the fact that it doesn't matter in which order you multiply a matrix by its inverse—whether you multiply on the left or the right you always get the identity.

[4]The QR decomposition also provides an alternative way to solve sets of simultaneous equations. To solve a set of equations of the form $\mathbf{Ax} = \mathbf{v}$ we write $\mathbf{A}$ in its QR form to get $\mathbf{QRx} = \mathbf{v}$ then multiply on both sides by the transpose $\mathbf{Q}^T$ to get $\mathbf{Rx} = \mathbf{Q}^T\mathbf{v}$. The right-hand side is now a known vector and the matrix $\mathbf{R}$ is upper triangular, so we can solve for $\mathbf{x}$ by backsubstitution as in Section 6.1.2. This solution method is rarely used, however, because it is slower than Gaussian elimination or LU decomposition—usually about a factor of two slower in practice.

[5]There are a number of other useful matrix decompositions that crop up in physics and elsewhere. They include: the *singular value decomposition* $\mathbf{U\Sigma V}^T$, in which $\mathbf{U}$ and $\mathbf{V}$ are orthogonal matrices and $\mathbf{\Sigma}$ is a positive semi-definite diagonal matrix; the *eigendecomposition* $\mathbf{VDV}^{-1}$, where $\mathbf{V}$ is an invertible matrix and $\mathbf{D}$ is diagonal (and possibly complex); the *QL decomposition*, where $\mathbf{Q}$ is orthogonal and $\mathbf{L}$ is lower triangular; and the *polar decomposition* $\mathbf{QP}$, where $\mathbf{Q}$ is orthogonal and $\mathbf{P}$ is real, symmetric, and positive semi-definite.

Multiplying on the left by $\mathbf{Q}_1^T$, we get

$$\mathbf{Q}_1^T \mathbf{A} = \mathbf{Q}_1^T \mathbf{Q}_1 \mathbf{R}_1 = \mathbf{R}_1, \tag{6.60}$$

where we have made use of the fact that $\mathbf{Q}_1$ is orthogonal.

Now let us define a new matrix

$$\mathbf{A}_1 = \mathbf{R}_1 \mathbf{Q}_1, \tag{6.61}$$

which is just the product of the same two matrices, but in the reverse order. Combining Eqs. (6.60) and (6.61), we have

$$\mathbf{A}_1 = \mathbf{Q}_1^T \mathbf{A} \mathbf{Q}_1. \tag{6.62}$$

Now we repeat this process, forming the QR decomposition of the matrix $\mathbf{A}_1$, which we'll write as $\mathbf{A}_1 = \mathbf{Q}_2 \mathbf{R}_2$, and defining a new matrix

$$\mathbf{A}_2 = \mathbf{R}_2 \mathbf{Q}_2 = \mathbf{Q}_2^T \mathbf{A}_1 \mathbf{Q}_2 = \mathbf{Q}_2^T \mathbf{Q}_1^T \mathbf{A} \mathbf{Q}_1 \mathbf{Q}_2. \tag{6.63}$$

And then we do it again, and again, repeatedly forming the QR decomposition of the current matrix, then multiplying $\mathbf{R}$ and $\mathbf{Q}$ in the opposite order to get a new matrix. If we do a total of k steps, we generate the sequence of matrices

$$\mathbf{A}_1 = \mathbf{Q}_1^T \mathbf{A} \mathbf{Q}_1, \tag{6.64}$$

$$\mathbf{A}_2 = \mathbf{Q}_2^T \mathbf{Q}_1^T \mathbf{A} \mathbf{Q}_1 \mathbf{Q}_2, \tag{6.65}$$

$$\mathbf{A}_3 = \mathbf{Q}_3^T \mathbf{Q}_2^T \mathbf{Q}_1^T \mathbf{A} \mathbf{Q}_1 \mathbf{Q}_2 \mathbf{Q}_3, \tag{6.66}$$

$$\vdots$$

$$\mathbf{A}_k = \left(\mathbf{Q}_k^T \ldots \mathbf{Q}_1^T \right) \mathbf{A} \left(\mathbf{Q}_1 \ldots \mathbf{Q}_k \right). \tag{6.67}$$

It can be proven that if you continue this process long enough, the matrix $\mathbf{A}_k$ will eventually become diagonal. The off-diagonal entries of the matrix get smaller and smaller the more iterations of the process you do until they eventually reach zero—or as close to zero as makes no difference. Suppose we continue until we reach this point, the point where the off-diagonal elements become so small that, to whatever accuracy we choose, the matrix $\mathbf{A}_k$ approximates a diagonal matrix $\mathbf{D}$. Let us define the additional matrix

$$\mathbf{V} = \mathbf{Q}_1 \mathbf{Q}_2 \mathbf{Q}_3 \ldots \mathbf{Q}_k = \prod_{i=1}^{k} \mathbf{Q}_i, \tag{6.68}$$

which is an orthogonal matrix, since a product of orthogonal matrices is always another orthogonal matrix.[6] Then from Eq. (6.67) we have $\mathbf{D} = \mathbf{A}_k = \mathbf{V}^T \mathbf{A} \mathbf{V}$, and, multiplying on the left by $\mathbf{V}$ we find that $\mathbf{A}\mathbf{V} = \mathbf{V}\mathbf{D}$. But this is precisely the definition of the eigenvalues and eigenvectors of $\mathbf{A}$, in the matrix form of Eq. (6.58). Thus, the columns of $\mathbf{V}$ are the eigenvectors of $\mathbf{A}$ and the diagonal elements of $\mathbf{D}$ are the corresponding eigenvalues (in the same order as the eigenvectors).

This is the QR algorithm for calculating eigenvalues and eigenvectors. Although we have described it in terms of the series of matrices $\mathbf{A}_k$, only the last of these matrices is actually needed for the eigenvalues, so in practice one doesn't retain all of the matrices (which could take up a lot of memory space), instead reusing a single array to store the matrix at each step. For a given $N \times N$ starting matrix $\mathbf{A}$, the complete algorithm is as follows:

1. Create an $N \times N$ matrix $\mathbf{V}$ to hold the eigenvectors and initially set it equal to the identity matrix $\mathbf{I}$. Also choose a target accuracy ϵ for the off-diagonal elements of the eigenvalue matrix.
2. Calculate the QR decomposition $\mathbf{A} = \mathbf{Q}\mathbf{R}$.
3. Update $\mathbf{A}$ to the new value $\mathbf{A} = \mathbf{R}\mathbf{Q}$.
4. Multiply $\mathbf{V}$ on the right by $\mathbf{Q}$.
5. Check the off-diagonal elements of $\mathbf{A}$. If they are all less than ϵ, we are done. Otherwise go back to step 2.

When the algorithm ends, the diagonal elements of $\mathbf{A}$ contain the eigenvalues and the columns of $\mathbf{V}$ contain the eigenvectors.

The QR algorithm is short and simple to describe. The only tricky part is calculating the QR decomposition itself, although even this is not very complicated. It can be done in a few lines in Python. Exercise 6.8 explains how to do it and invites you to write your own program to calculate the eigenvalues and eigenvectors of a matrix. It's a good exercise to write such a program, at least once, so you can see that it's really not a complicated calculation.

Most people, however, don't write their own programs to do the QR algorithm. Like the solution of simultaneous equations in Section 6.1.4, the problem of finding eigenvalues and eigenvectors is so common, both in physics and elsewhere, that others have already taken the time to write robust general-purpose programs for its solution. In Python the module `numpy.linalg` provides functions for solving eigenvalue problems. For instance, the function

---

[6]To see this let $\mathbf{Q}_1$ and $\mathbf{Q}_2$ be orthogonal matrices, so that $\mathbf{Q}_1^T\mathbf{Q}_1 = \mathbf{Q}_2^T\mathbf{Q}_2 = \mathbf{I}$, and let $\mathbf{V} = \mathbf{Q}_1\mathbf{Q}_2$. Then $\mathbf{V}^T = \mathbf{Q}_2^T\mathbf{Q}_1^T$ and $\mathbf{V}^T\mathbf{V} = \mathbf{Q}_2^T\mathbf{Q}_1^T\mathbf{Q}_1\mathbf{Q}_2 = \mathbf{Q}_2^T\mathbf{Q}_2 = \mathbf{I}$. Hence $\mathbf{V}$ is also orthogonal.

eigh calculates the eigenvalues and eigenvectors of a real symmetric or Hermitian matrix using the QR algorithm. Here's an example:

```
from numpy import array
from numpy.linalg import eigh
A = array([[ 1, 2 ], [ 2, 1 ]], float)
x,V = eigh(A)
print(x)
print(V)
```

The function calculates the eigenvalues and eigenvectors of the matrix stored in the two-dimensional array A and returns two results, both also arrays. (We discussed functions that return more than one result in Section 2.6.) The first result returned is a one-dimensional array x containing the eigenvalues, while the second is a two-dimensional array V whose *columns* are the corresponding eigenvectors. If we run the program above, it produces the following output:

```
[-1.  3.]
[[-0.70710678  0.70710678]
 [ 0.70710678  0.70710678]]
```

Thus the program is telling us that the eigenvalues of the matrix are -1 and 3 and the corresponding eigenvectors are $(-1,1)/\sqrt{2}$ and $(1,1)/\sqrt{2}$.

Sometimes we only want the eigenvalues of a matrix and not its eigenvectors, in which case we can save ourselves some time by not calculating the matrix of eigenvectors **V**, Eq. (6.68). (This part of the calculation takes quite a lot of time because matrix multiplications involve a lot of operations, so it's a significant saving to skip this step.) The numpy.linalg module provides a separate function called eigvalsh that does this for symmetric or Hermitian matrices, calculating the eigenvalues alone, significantly faster than the function eigh. Here's an example:

```
from numpy import array
from numpy.linalg import eigvalsh
A = array([[ 1, 2 ], [ 2, 1 ]], float)
x = eigvalsh(A)
print(x)
```

The function eigvalsh returns a single, one-dimensional array containing the eigenvalues. If we run this program it prints

```
[-1.  3.]
```

telling us again that the eigenvalues are −1 and 3.

The functions eigh and eigvalsh are only for calculating the eigenvalues and eigenvectors of symmetric or Hermitian matrices. You might ask what happens if you provide an asymmetric matrix as argument to these functions? We might imagine the computer would give an error message, but in fact it does not. Instead it just ignores all elements of the matrix above the diagonal. It *assumes* that the elements above the diagonal are mirror images of the ones below and never even looks at their values. If they are not mirror images, the computer will never know, or care. It's as if the function copies the lower triangle of the matrix into the upper one before it does its calculation. You can if you wish leave the upper triangle blank—assign no values to these elements, or set them to zero. It will have no effect on the answer. In the program above we could have written

```
A = array([[ 1, 0 ], [ 2, 1 ]], float)
x = eigvalsh(A)
```

and the answer would have been exactly the same.

If the matrix is complex instead of real, eigh and eigvalsh will assume it to be Hermitian and calculate the eigenvalues and eigenvectors accordingly. Again only the lower triangle matters: the upper triangle is assumed to be the complex conjugate of the lower one (and any imaginary part of the diagonal elements is ignored).

The module linalg does also supply functions for calculating eigenvalues and eigenvectors of nonsymmetric (or non-Hermitian) matrices: the functions eig and eigvals are the equivalents of eigh and eigvalsh when the matrix is nonsymmetric. We will not have call to use these functions in this book, however, and nonsymmetric eigenvalue problems come up only rarely in physics in general.

Exercise 6.8: The QR algorithm:

In this exercise you'll write a program to calculate the eigenvalues and eigenvectors of a real symmetric matrix using the QR algorithm. The first challenge is to write a program that finds the QR decomposition of a matrix. Then we'll use that decomposition to find the eigenvalues.

As described above, the QR decomposition expresses a real square matrix $\mathbf{A}$ in the form $\mathbf{A} = \mathbf{QR}$, where $\mathbf{Q}$ is an orthogonal matrix and $\mathbf{R}$ is an upper-triangular matrix. Given an $N \times N$ matrix $\mathbf{A}$ we can compute the QR decomposition as follows.

Let us think of the matrix as a set of N column vectors $\mathbf{a}_0 \ldots \mathbf{a}_{N-1}$ thus:

$$\mathbf{A} = \begin{pmatrix} | & | & | & \cdots \\ \mathbf{a}_0 & \mathbf{a}_1 & \mathbf{a}_2 & \cdots \\ | & | & | & \cdots \end{pmatrix},$$

where we have numbered the vectors in Python fashion, starting from zero, which will be convenient when writing the program. We now define two new sets of vectors $\mathbf{u}_0 \ldots \mathbf{u}_{N-1}$ and $\mathbf{q}_0 \ldots \mathbf{q}_{N-1}$ as follows:

$$\mathbf{u}_0 = \mathbf{a}_0, \qquad\qquad \mathbf{q}_0 = \frac{\mathbf{u}_0}{|\mathbf{u}_0|},$$

$$\mathbf{u}_1 = \mathbf{a}_1 - (\mathbf{q}_0 \cdot \mathbf{a}_1)\mathbf{q}_0, \qquad\qquad \mathbf{q}_1 = \frac{\mathbf{u}_1}{|\mathbf{u}_1|},$$

$$\mathbf{u}_2 = \mathbf{a}_2 - (\mathbf{q}_0 \cdot \mathbf{a}_2)\mathbf{q}_0 - (\mathbf{q}_1 \cdot \mathbf{a}_2)\mathbf{q}_1, \qquad\qquad \mathbf{q}_2 = \frac{\mathbf{u}_2}{|\mathbf{u}_2|},$$

and so forth. The general formulas for calculating $\mathbf{u}_i$ and $\mathbf{q}_i$ are

$$\mathbf{u}_i = \mathbf{a}_i - \sum_{j=0}^{i-1}(\mathbf{q}_j \cdot \mathbf{a}_i)\mathbf{q}_j, \qquad \mathbf{q}_i = \frac{\mathbf{u}_i}{|\mathbf{u}_i|}.$$

a) Show, by induction or otherwise, that the vectors $\mathbf{q}_i$ are orthonormal, i.e., that they satisfy

$$\mathbf{q}_i \cdot \mathbf{q}_j = \begin{cases} 1 & \text{if } i = j, \\ 0 & \text{if } i \neq j. \end{cases}$$

Now, rearranging the definitions of the vectors, we have

$$\mathbf{a}_0 = |\mathbf{u}_0|\,\mathbf{q}_0,$$

$$\mathbf{a}_1 = |\mathbf{u}_1|\,\mathbf{q}_1 + (\mathbf{q}_0 \cdot \mathbf{a}_1)\mathbf{q}_0,$$

$$\mathbf{a}_2 = |\mathbf{u}_2|\,\mathbf{q}_2 + (\mathbf{q}_0 \cdot \mathbf{a}_2)\mathbf{q}_0 + (\mathbf{q}_1 \cdot \mathbf{a}_2)\mathbf{q}_1,$$

and so on. Or we can group the vectors $\mathbf{q}_i$ together as the columns of a matrix and write all of these equations as a single matrix equation

$$\mathbf{A} = \begin{pmatrix} | & | & | & \cdots \\ \mathbf{a}_0 & \mathbf{a}_1 & \mathbf{a}_2 & \cdots \\ | & | & | & \cdots \end{pmatrix} = \begin{pmatrix} | & | & | & \cdots \\ \mathbf{q}_0 & \mathbf{q}_1 & \mathbf{q}_2 & \cdots \\ | & | & | & \cdots \end{pmatrix} \begin{pmatrix} |\mathbf{u}_0| & \mathbf{q}_0 \cdot \mathbf{a}_1 & \mathbf{q}_0 \cdot \mathbf{a}_2 & \cdots \\ 0 & |\mathbf{u}_1| & \mathbf{q}_1 \cdot \mathbf{a}_2 & \cdots \\ 0 & 0 & |\mathbf{u}_2| & \cdots \end{pmatrix}.$$

(If this looks complicated it's worth multiplying out the matrices on the right to verify for yourself that you get the correct expressions for the $\mathbf{a}_i$.)

Notice now that the first matrix on the right-hand side of this equation, the matrix with columns $\mathbf{q}_i$, is orthogonal, because the vectors $\mathbf{q}_i$ are orthonormal, and the second matrix is upper triangular. In other words, we have found the QR decomposition $\mathbf{A} = \mathbf{QR}$. The matrices $\mathbf{Q}$ and $\mathbf{R}$ are

$$\mathbf{Q} = \begin{pmatrix} | & | & | & \cdots \\ \mathbf{q}_0 & \mathbf{q}_1 & \mathbf{q}_2 & \cdots \\ | & | & | & \cdots \end{pmatrix}, \qquad \mathbf{R} = \begin{pmatrix} |\mathbf{u}_0| & \mathbf{q}_0 \cdot \mathbf{a}_1 & \mathbf{q}_0 \cdot \mathbf{a}_2 & \cdots \\ 0 & |\mathbf{u}_1| & \mathbf{q}_1 \cdot \mathbf{a}_2 & \cdots \\ 0 & 0 & |\mathbf{u}_2| & \cdots \end{pmatrix}.$$

247

b) Write a Python function that takes as its argument a real square matrix $\mathbf{A}$ and returns the two matrices $\mathbf{Q}$ and $\mathbf{R}$ that form its QR decomposition. As a test case, try out your function on the matrix

$$\mathbf{A} = \begin{pmatrix} 1 & 4 & 8 & 4 \\ 4 & 2 & 3 & 7 \\ 8 & 3 & 6 & 9 \\ 4 & 7 & 9 & 2 \end{pmatrix}.$$

Check the results by multiplying $\mathbf{Q}$ and $\mathbf{R}$ together to recover the original matrix $\mathbf{A}$ again.

c) Using your function, write a complete program to calculate the eigenvalues and eigenvectors of a real symmetric matrix using the QR algorithm. Continue the calculation until the magnitude of every off-diagonal element of the matrix is smaller than 10^{-6}. Test your program on the example matrix above. You should find that the eigenvalues are 1, 21, −3, and −8.

Exercise 6.9: Asymmetric quantum well

Quantum mechanics can be formulated as a matrix problem and solved on a computer using linear algebra methods. Suppose, for example, we have a particle of mass M in a one-dimensional quantum well of width L, but not a square well like the ones you commonly find discussed in textbooks. Suppose instead that the potential $V(x)$ varies somehow inside the well:

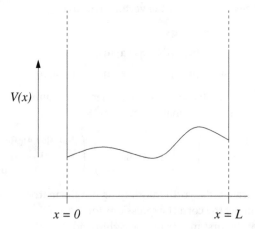

We cannot solve such problems analytically in general, but we can solve them on the computer.

In a pure state of energy E, the spatial part $\psi(x)$ of the wavefunction obeys the time-independent Schrödinger equation $\hat{H}\psi(x) = E\psi(x)$, where the Hamiltonian operator $\hat{H}$ is given by

$$\hat{H} = -\frac{\hbar^2}{2M}\frac{d^2}{dx^2} + V(x).$$

For simplicity, let's assume that the walls of the well are infinitely high, so that the wavefunction is zero outside the well, which means it must *go to* zero at $x = 0$ and $x = L$. In that case, the wavefunction can be expressed as a Fourier sine series thus:

$$\psi(x) = \sum_{n=1}^{\infty} \psi_n \sin \frac{\pi n x}{L},$$

where $\psi_1, \psi_2, \ldots$ are the Fourier coefficients.

a) Noting that, for m, n positive integers

$$\int_0^L \sin \frac{\pi m x}{L} \sin \frac{\pi n x}{L}\, dx = \begin{cases} L/2 & \text{if } m = n, \\ 0 & \text{otherwise,} \end{cases}$$

show that the Schrödinger equation $\hat{H}\psi = E\psi$ implies that

$$\sum_{n=1}^{\infty} \psi_n \int_0^L \sin \frac{\pi m x}{L} \hat{H} \sin \frac{\pi n x}{L}\, dx = \tfrac{1}{2} L E \psi_m.$$

Hence, defining a matrix $\mathbf{H}$ with elements

$$H_{mn} = \frac{2}{L} \int_0^L \sin \frac{\pi m x}{L} \hat{H} \sin \frac{\pi n x}{L}\, dx$$
$$= \frac{2}{L} \int_0^L \sin \frac{\pi m x}{L} \left[-\frac{\hbar^2}{2M} \frac{d^2}{dx^2} + V(x) \right] \sin \frac{\pi n x}{L}\, dx,$$

show that Schrödinger's equation can be written in matrix form as $\mathbf{H}\psi = E\psi$, where ψ is the vector $(\psi_1, \psi_2, \ldots)$. Thus ψ is an eigenvector of the *Hamiltonian matrix* $\mathbf{H}$ with eigenvalue E. If we can calculate the eigenvalues of this matrix, then we know the allowed energies of the particle in the well.

b) For the case $V(x) = ax/L$, evaluate the integral in H_{mn} analytically and so find a general expression for the matrix element H_{mn}. Show that the matrix is real and symmetric. You'll probably find it useful to know that

$$\int_0^L x \sin \frac{\pi m x}{L} \sin \frac{\pi n x}{L}\, dx = \begin{cases} 0 & \text{if } m \neq n \text{ and } m, n \text{ are both even or both odd,} \\ -\left(\frac{2L}{\pi}\right)^2 \frac{mn}{(m^2 - n^2)^2} & \text{if } m \neq n \text{ and one is even, one is odd,} \\ L^2/4 & \text{if } m = n. \end{cases}$$

Write a Python program to evaluate your expression for H_{mn} for arbitrary m and n when the particle in the well is an electron, the well has width 5 Å, and $a = 10\,\text{eV}$. (The mass and charge of an electron are 9.1094×10^{-31} kg and 1.6022×10^{-19} C respectively.)

c) The matrix $\mathbf{H}$ is in theory infinitely large, so we cannot calculate all its eigenvalues. But we can get a pretty accurate solution for the first few of them by cutting off the matrix after the first few elements. Modify the program you wrote for part (b) above to create a 10×10 array of the elements of **H** up to $m, n = 10$. Calculate the eigenvalues of this matrix using the appropriate function from `numpy.linalg` and hence print out, in units of electron volts, the first ten energy levels of the quantum well, within this approximation. You should find, for example, that the ground-state energy of the system is around 5.84 eV. (Hint: Bear in mind that matrix indices in Python start at zero, while the indices in standard algebraic expressions, like those above, start at one. You will need to make allowances for this in your program.)

d) Modify your program to use a 100×100 array instead and again calculate the first ten energy eigenvalues. Comparing with the values you calculated in part (c), what do you conclude about the accuracy of the calculation?

e) Now modify your program once more to calculate the wavefunction $\psi(x)$ for the ground state and the first two excited states of the well. Use your results to make a graph with three curves showing the probability density $|\psi(x)|^2$ as a function of x in each of these three states. Pay special attention to the normalization of the wavefunction—it should satisfy the condition $\int_0^L |\psi(x)|^2 \, dx = 1$. Is this true of your wavefunction?

6.3 NONLINEAR EQUATIONS

So far in this chapter we have looked at problems involving the solution of linear equations—in the all the examples we have studied the equations are simple linear combinations of the unknown variables. Many equations that arise in physics, however, are nonlinear. Nonlinear equations are, in general, significantly harder to solve than their linear counterparts. Even solving a single nonlinear equation in a single variable can be a challenging problem, and solving simultaneous equations in many variables is harder still. In this section we look at techniques for solving nonlinear equations, starting with the single-variable case.

6.3.1 THE RELAXATION METHOD

Suppose we have a single nonlinear equation that we want to solve for the value of a single variable, such as

$$x = 2 - e^{-x}. \tag{6.69}$$

There is no known analytic method for solving this equation, so we turn to computational methods. One elementary method that gives good answers in many cases is simply to iterate the equation. That is, we guess an initial value of the unknown variable x, plug it in on the right-hand side of the equation and get a new value x' on the left-hand side. For instance, we might in this case guess an initial value $x = 1$ and plug it in on the right-hand side of (6.69) to get

$$x' = 2 - e^{-1} \simeq 1.632. \qquad (6.70)$$

Then we repeat the process, taking this value and feeding it in on the right again to get

$$x'' = 2 - e^{-1.632} \simeq 1.804, \qquad (6.71)$$

and so forth. If we keep on doing this, and if we are lucky, then the value will converge to a *fixed point* of the equation, meaning it stops changing. In this particular case, that's exactly what happens. Here's a program to perform ten iterations of the calculation:

```
from math import exp
x = 1.0
for k in range(10):
    x = 2 - exp(-x)
    print(x)
```

and here's what it prints out:

```
1.63212055883
1.80448546585
1.83544089392
1.84045685534
1.84125511391
1.84138178281
1.84140187354
1.84140505985
1.84140556519
1.84140564533
```

It appears that the result is converging to a value around 1.8414, and indeed if we continue the process for a short while it converges to $x = 1.84140566044$ and stops changing. If a process like this converges to a fixed point, then the value of x you get is necessarily a solution to the original equation—you feed that value in on one side of the equation and get the same value out on the other. This is the definition of a solution.

This method is called the *relaxation method*. When it works, it's often a good way of getting a solution. It's trivial to program (the program above has only five lines) and it runs reasonably quickly (it took about fifty steps in this case to reach a solution accurate to twelve figures).

The method does have its problems. First, the equation you are solving needs to be in the simple form $x = f(x)$, where $f(x)$ is some known function, and this is not always the case. However, even when the equation you are given is not in this form, it is in most cases a straightforward task to rearrange it so that it is. If you have an equation like $\log x + x^2 - 1 = 0$, for instance, you can take the exponential of both sides and rearrange to get $x = e^{1-x^2}$.

Second, an equation may have more than one solution. Sometimes the relaxation method will converge to one solution and not to another, in which case you only learn about the one solution. You may be able to mitigate this problem somewhat by your choice of the starting value of x. Depending on the starting value, the iteration process may converge to more than one different fixed point, so you can find more than one solution to the equation. Typically if the process does converge to a given solution then you can find it by choosing a starting value near that solution. So if you have some approximate idea of the position of the solution you're looking for, you should choose a value near it for your starting point.

There are, however, some solutions to some equations that you cannot find by this method no matter what starting value you choose. Even with a starting value close to such a solution, the method will not converge to the solution. Consider, for instance, the equation discussed above:

$$x = e^{1-x^2}. \tag{6.72}$$

Again there's no general analytic method for solving this equation, but in this particular case one can see by inspection that the solution is just $x = 1$. If we try and find that solution by the relaxation method, however, starting with $x = \frac{1}{2}$ say, we get the following for the first ten values of x:

```
2.11700001661
0.03075541907
2.71571183275
0.00170346518
2.71827394058
0.00167991310
2.71827415718
0.00167991112
2.71827415720
0.00167991112
```

As you can see, instead of settling down to a fixed point, the program is oscillating back and forth between two different values. And it goes on doing this no matter how long we wait, so we never get a solution to our equation. The relaxation method has failed.

A useful trick in cases like these is to try to find an alternative way of arranging the equation to give the value of x. For instance, if we take logs of both sides of Eq. (6.72) then rearrange, we find the following alternative form for the equation:

$$x = \sqrt{1 - \log x}. \tag{6.73}$$

If we now apply the relaxation method to this form, starting with $x = \frac{1}{2}$, we get the following:

```
1.30120989105
0.85831549149
1.07367757795
0.96379990441
1.01826891043
0.99090663593
1.00455709697
0.99772403758
1.00113862994
0.99943084694
```

The method is now converging to the solution at $x = 1$, and if we go on for a few more steps we get a very good approximation to that solution.

There is a mathematical theory behind the relaxation method that explains why this rearranging trick works. Assume we have an equation of the form $x = f(x)$ that has a solution at $x = x^*$ and let us consider the behavior of the relaxation method when x is close to x^*. Performing a Taylor expansion, the value x' after an iteration of the method is given in terms of the previous value x by

$$x' = f(x) = f(x^*) + (x - x^*)f'(x^*) + \ldots \tag{6.74}$$

But by definition, x^* is a solution of the original equation, meaning that $x^* = f(x^*)$, so Eq. (6.74) can also be written as

$$x' - x^* = (x - x^*)f'(x^*), \tag{6.75}$$

where we have neglected the higher-order terms.

This equation tells us that the distance $x - x^*$ to the true solution of the equation gets multiplied on each iteration of the method by a factor of $f'(x^*)$, the derivative of the function evaluated at x^*. If the absolute value of this derivative is less than one, then the distance will get smaller on each iteration, which means we are converging to the solution. If it is greater than one, on the other hand, then we are getting farther from the solution on each step and the method will not converge. Thus the relaxation method will converge to a solution at x^* if and only if $|f'(x^*)| < 1$. This explains why the method failed for Eq. (6.72). In that case we have $f(x) = e^{1-x^2}$ and $x^* = 1$ so

$$|f'(x^*)| = \left|\left[-2xe^{1-x^2}\right]_{x=1}\right| = 2. \tag{6.76}$$

So the method will not converge.

Suppose that we find ourselves in a situation like this where the method does not converge because $|f'(x^*)| > 1$. Let us rearrange the equation $x = f(x)$ by inverting the function f to get $x = f^{-1}(x)$, where f^{-1} is the functional inverse of f. Now the method will be stable if the derivative of f^{-1} is less than one at x^*.

To calculate this derivative we define $u = f^{-1}(x)$, so that the derivative we want is du/dx. But in that case we also have $x = f(u)$, so

$$\frac{dx}{du} = f'(u). \tag{6.77}$$

But when $x = x^*$ we have $u = f^{-1}(x^*) = x^*$ and, taking reciprocals, we then find that the derivative we want is

$$\frac{du}{dx} = \frac{1}{f'(x^*)}. \tag{6.78}$$

Thus the derivative of f^{-1} at x^* is simply the reciprocal of the derivative of f. That means that if the derivative of f is greater than one then the derivative of f^{-1} must be less than one. Hence, if the relaxation method fails to converge for $x = f(x)$ it will succeed for the equivalent form $x = f^{-1}(x)$.

The bottom line is, if the method fails for a particular equation you should invert the equation and then the method will work. This is exactly what we did for Eq. (6.73).

Unfortunately, not all equations can be inverted. If your $f(x)$ is a polynomial of degree ten, for example, then inversion is not possible. In such cases it may still be possible to rearrange and get a different equation for x, but the theory above no longer applies and convergence is no longer guaranteed. You might get a convergent calculation if you're lucky, but you might not. Consider, for instance, the equation

$$x = x^2 + \sin 2x. \tag{6.79}$$

This equation has a solution at $x^* = 0$, but if we apply the relaxation method directly we will not find that solution. The method fails to converge, as we can prove by calculating $|f'(x^*)|$, which turns out to be equal to 2 in this case. Moreover, the function $f(x) = x^2 + \sin 2x$ on the right-hand side of Eq. (6.79) cannot be inverted. We can, however, still rearrange the equation thus:

$$x = \tfrac{1}{2} \sin^{-1}(x - x^2). \tag{6.80}$$

Because this is not a true inversion of the original equation, we are not guaranteed that this version will converge any better than the original one, but in fact it does—if we calculate $|f'(x^*)|$ again, we find that it is now equal to $\tfrac{1}{2}$ and hence the relaxation method will now work. This time we got lucky.

In summary, the relaxation method does not always work, but between the cases where it works first time, the ones where it can be made to work by inverting, and the ones where some other rearrangement turns out by good luck to work, the method is useful for a wide range of problems.

6.3.2   Rate of convergence of the relaxation method

An important question about the relaxation method is how fast it converges to a solution, assuming it does converge. We can answer this question by looking again at Eq. (6.75). This equation tells us that, when the method converges, the distance to the solution gets smaller by a factor of $|f'(x^*)|$ on each iteration. In other words, the distance decreases exponentially with the number of iterations we perform. Since the exponential is a rapidly decaying function this is a good thing: the relaxation method converges to a solution quickly, although there are other methods that converge faster still, such as Newton's method, which we study in Section 6.3.5.

This is a nice result as far as it goes, but it isn't very practical. In a typical application of the relaxation method what we really want to know is, "When can I stop iterating? When is the answer I have good enough?" One simple way to answer this question is just to look at the solutions you get on successive iterations of the method and observe when they stop changing. For instance, if you want an answer accurate to six figures, you continue until a few iterations have passed without any changes in the first six figures of x. If you want a quick answer to a problem and accuracy is not a big issue, this is actually not a bad approach. It's easy and it usually works.

In some cases, however, we may want to know exactly how accurate our answer is, or we may wish to stop immediately as soon as a required target accuracy is reached. If, for instance, a single iteration of the relaxation method takes a long time—minutes or hours—because the calculations involved are complex, then we may not want to perform even one more iteration than is strictly necessary. In such cases, we can take the following approach.

Let us define ϵ to be the error on our current estimate of the solution to the equation. That is, the true solution x^* is related to the current estimate x by $x^* = x + \epsilon$. Similarly let ϵ' be the error on the next estimate, so that $x^* = x' + \epsilon'$. Then Eq. (6.75) tells us that close to x^* we have

$$\epsilon' = \epsilon f'(x^*). \tag{6.81}$$

Then

$$x^* = x + \epsilon = x + \frac{\epsilon'}{f'(x^*)}, \tag{6.82}$$

and equating this with $x^* = x' + \epsilon'$ and rearranging for ϵ' we derive an expression for the error on the new estimate:

$$\epsilon' = \frac{x - x'}{1 - 1/f'(x^*)} \simeq \frac{x - x'}{1 - 1/f'(x)}, \tag{6.83}$$

where we have made use of the fact that x is close to x^*, so that $f'(x) \simeq f'(x^*)$. If we know the form of the function $f(x)$ then we can calculate its derivative and then use this formula to estimate the error ϵ' on the new value x' for the solution at each step. Then, for instance, we can simply repeat the iteration until the magnitude of this estimated error falls below some target value, ensuring that we get an answer that is as accurate as we want, without wasting any time on additional iterations.

There are some cases, however, where we don't know the full formula for $f(x)$. For instance, $f(x)$ might not be given as a mathematical formula at all, but as the output of another program that itself performs some complicated calculation. In such cases we cannot calculate the derivative $f'(x)$ directly, but we can estimate it using a numerical derivative like those we studied in Section 5.10. As we saw there, calculating a numerical derivative involves taking the difference of the values of f at two different points. In the most common application of this idea to the relaxation method we choose the points to be the values of x at successive steps of the iteration.

Suppose we have three successive estimates of x, which we denote x, x', and x''. We would like to calculate the error on the most recent estimate x'', which by Eq. (6.83) is

$$\epsilon'' = \frac{x' - x''}{1 - 1/f'(x^*)} \simeq \frac{x' - x''}{1 - 1/f'(x)}. \tag{6.84}$$

Now we approximate $f'(x)$ as

$$f'(x) \simeq \frac{f(x) - f(x')}{x - x'}. \tag{6.85}$$

But by definition $x' = f(x)$ and $x'' = f(x')$, so

$$f'(x) \simeq \frac{x' - x''}{x - x'}. \tag{6.86}$$

Substituting into Eq. (6.84), we then find that the error on the third and most recent of our estimates of the solution is given approximately by

$$\epsilon'' \simeq \frac{x' - x''}{1 - (x - x')/(x' - x'')} = \frac{(x' - x'')^2}{2x' - x - x''}. \tag{6.87}$$

Thus, if we keep track of three successive estimates of x at each stage of the calculation we can estimate the error even when we cannot calculate a derivative of $f(x)$ directly.

EXAMPLE 6.3: FERROMAGNETISM

In the mean-field theory of ferromagnetism, the strength M of magnetization of a ferromagnetic material like iron depends on temperature T according to the formula

$$M = \mu \tanh \frac{JM}{k_B T}, \tag{6.88}$$

where μ is a magnetic moment, J is a coupling constant, and k_B is Boltzmann's constant. To simplify things a little, let us make the substitutions $m = M/\mu$ and $C = \mu J/k_B$ so that

$$m = \tanh \frac{Cm}{T}. \tag{6.89}$$

It's clear that this equation always has a solution at $m = 0$, which implies a material that is not magnetized at all, but what about other solutions? Are there solutions with $m \neq 0$? There is no known method of solving for such solutions exactly, but we can find them using the computer. Let's assume that $C = 1$ for simplicity and look for solutions as a function of T, accurate to within $\pm 10^{-6}$ of the true answer. In this case, since we know the full functional form of the equation we are solving, we can evaluate the derivative in Eq. (6.83) explicitly and show that the error is given by

$$\epsilon' = \frac{m - m'}{1 - T \cosh^2(m/T)}. \tag{6.90}$$

Here's a program to find the solutions and make a plot as a function of temperature:

File: `ferromag.py`

```
from math import tanh,cosh
from numpy import linspace
from pylab import plot,show,ylim,xlabel,ylabel

# Constants
Tmax = 2.0
points = 1000
accuracy = 1e-6

# Set up lists for plotting
y = []
temp = linspace(0.01,Tmax,points)

# Temperature loop
for T in temp:
    m1 = 1.0
    error = 1.0

    # Loop until error is small enough
    while error>accuracy:
        m1,m2 = tanh(m1/T),m1
        error = abs((m1-m2)/(1-T*cosh(m2/T)**2))
    y.append(m1)

# Make the graph
plot(temp,y)
ylim(-0.1,1.1)
xlabel("Temperature")
ylabel("Magnetization")
show()
```

For each value of the temperature we iterate Eq. (6.89) starting with $m = 1$, until the magnitude of the error, estimated from Eq. (6.90), falls below the target value. Figure 6.2 shows the end result. As we can see the program does indeed find nonzero solutions of the equation, but only for values of the temperature below $T = 1$. As we approach $T = 1$ from below the value of the magnetization falls off and above $T = 1$ the program only finds the solution $m = 0$. This is a real physical phenomenon observed in experiments on real magnets. If you take a magnetized piece of iron and heat it, then the magnetization will

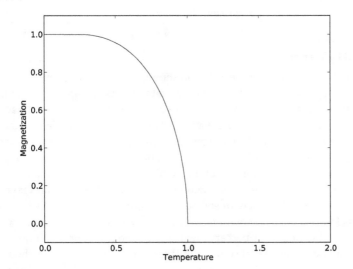

Figure 6.2: Magnetization of a model magnet as a function of temperature. This figure shows the solutions of Eq. (6.89) found by the program `ferromag.py`, as a function of temperature.

get weaker. This is perhaps not a surprise—the heat increases the motion of the atoms in the metal which disrupts the ordering of the magnetic dipoles that make up the bulk magnetization. What is perhaps more surprising is that the magnetization doesn't just decrease gradually to zero but instead passes through a *phase transition* at which it disappears suddenly and completely. In our model system this transition happens at $T = 1$, which is called the *critical temperature* of the magnet.

Exercise 6.10: Consider the equation $x = 1 - e^{-cx}$, where c is a known parameter and x is unknown. This equation arises in a variety of situations, including the physics of contact processes, mathematical models of epidemics, and the theory of random graphs.

a) Write a program to solve this equation for x using the relaxation method for the case $c = 2$. Calculate your solution to an accuracy of at least 10^{-6}.

b) Modify your program to calculate the solution for values of c from 0 to 3 in steps of 0.01 and make a plot of x as a function of c. You should see a clear transition from a regime in which $x = 0$ to a regime of nonzero x. This is another example of a phase transition. In physics this transition is known as the *percolation transition*; in epidemiology it is the *epidemic threshold*.

Exercise 6.11: Overrelaxation

If you did not already do Exercise 6.10, you should do it before this one.

The ordinary relaxation method involves iterating the equation $x' = f(x)$, starting from an initial guess, until it converges. As we have seen, this is often a fast and easy way to find solutions to nonlinear equations. However, it is possible in some cases to make the method work even faster using the technique of *overrelaxation*. Suppose our initial guess at the solution of a particular equation is, say, $x = 1$, and the final, true solution is $x = 5$. After the first step of the iterative process, we might then see a value of, say, $x = 3$. In the overrelaxation method, we observe this value and note that x is increasing, then we deliberately overshoot the calculated value, in the hope that this will get us closer to the final solution—in this case we might pass over $x = 3$ and go straight to a value of $x = 4$ perhaps, which is closer to the final solution of $x = 5$ and hence should get us to that solution quicker. The overrelaxation method provides a formula for performing this kind of overshooting in a controlled fashion and often, though not always, it does get us to our solution faster. In detail, it works as follows.

We can rewrite the equation $x' = f(x)$ in the form $x' = x + \Delta x$, where

$$\Delta x = x' - x = f(x) - x.$$

The overrelaxation method involves iteration of the modified equation

$$x' = x + (1 + \omega)\,\Delta x,$$

(keeping the definition of Δx the same). If the parameter ω is zero, then this is the same as the ordinary relaxation method, but for $\omega > 0$ the method takes the amount Δx by which the value of x would have been changed and changes by a little more. Using $\Delta x = f(x) - x$, we can also write x' as

$$x' = x + (1 + \omega)\big[f(x) - x\big] = (1 + \omega)f(x) - \omega x,$$

which is the form in which it is usually written.

For the method to work the value of ω must be chosen correctly, although there is some wiggle room—there is an optimal value, but other values close to it will typically also give good results. Unfortunately, there is no general theory that tells us what the optimal value is. Usually it is found by trial and error.

a) Derive an equivalent of Eq. (6.81) for the overrelaxation method and hence show that the error on x', the equivalent of Eq. (6.83), is given by

$$\epsilon' \simeq \frac{x - x'}{1 - 1/[(1 + \omega)f'(x) - \omega]}.$$

b) Consider again the equation $x = 1 - e^{-cx}$ that we solved in Exercise 6.10. Take the program you wrote for part (a) of that exercise, which solved the equation for the case $c = 2$, and modify it to print out the number of iterations it takes to converge to a solution accurate to 10^{-6}.

c) Now write a new program (or modify the previous one) to solve the same equation $x = 1 - e^{-cx}$ for $c = 2$, again to an accuracy of 10^{-6}, but this time using overrelaxation. Have your program print out the answers it finds along with the number of iterations it took to find them. Experiment with different values of ω to see how fast you can get the method to converge. A value of $\omega = 0.5$ is a reasonable starting point. With some trial and error you should be able to get the calculation to converge at least twice as fast as the simple relaxation method, i.e., in about half as many iterations.

d) Are there any circumstances under which using a value $\omega < 0$ would help us find a solution faster than we can with the ordinary relaxation method? (Hint: The answer is yes, but why?)

6.3.3 RELAXATION METHOD FOR TWO OR MORE VARIABLES

A further nice feature of the relaxation method is that it extends easily to the solution of simultaneous equations in two or more variables. Suppose for example that we are given two equations to solve for the two variables x and y. First we rearrange those two equations into the form

$$x = f(x, y), \qquad y = g(x, y), \tag{6.91}$$

where f and g are known functions of x and y. Then we guess initial starting values for both variables, substitute them into these equations to get new values, and repeat. If we converge to a fixed point, meaning unchanging values for both variables, then those values must be a solution to the two simultaneous equations.

More generally, if we have any number N of simultaneous equations in N unknown variables $x_1, x_2, \ldots$ we can write them in the form

$$x_1 = f_1(x_1, \ldots, x_N),$$

$$\vdots \tag{6.92}$$

$$x_N = f_N(x_1, \ldots, x_N).$$

Then we can choose starting values for all the variables and apply the equations repeatedly to find a solution. As with the relaxation method for a single variable the method is not guaranteed to converge to a solution. Depending on

CHAPTER 6 | SOLUTION OF LINEAR AND NONLINEAR EQUATIONS

the exact form of the equations it will sometimes converge and sometimes not. If it does not converge then it may be possible to rearrange the equations into a different form that will converge, as described for the one-variable case in Section 6.3.1—a little experimentation is sometimes needed to make the method work.[7]

Exercise 6.12: The biochemical process of *glycolysis*, the breakdown of glucose in the body to release energy, can be modeled by the equations

$$\frac{dx}{dt} = -x + ay + x^2y, \qquad \frac{dy}{dt} = b - ay - x^2y.$$

Here x and y represent concentrations of two chemicals, ADP and F6P, and a and b are positive constants. One of the important features of nonlinear equations like these is their *stationary points*, meaning values of x and y at which the derivatives of both variables become zero simultaneously, so that the variables stop changing and become constant in time. Setting the derivatives to zero above, the stationary points of our glycolysis equations are solutions of

$$-x + ay + x^2y = 0, \qquad b - ay - x^2y = 0.$$

a) Demonstrate analytically that the solution of these equations is

$$x = b, \qquad y = \frac{b}{a + b^2}.$$

b) Show that the equations can be rearranged to read

$$x = y(a + x^2), \qquad y = \frac{b}{a + x^2}$$

and write a program to solve these for the stationary point using the relaxation method with $a = 1$ and $b = 2$. You should find that the method fails to converge to a solution in this case.

c) Find a different way to rearrange the equations such that when you apply the relaxation method again it now converges to a fixed point and gives a solution. Verify that the solution you get agrees with part (a).

---

[7] In Section 6.3.1 we derived the condition $|f'(x^*)| < 1$ for the convergence of the single-variable relaxation method to a fixed point at $x = x^*$. The corresponding condition for the many-variable case is that $|\lambda_i| < 1$ for all $i = 1 \ldots N$, where λ_i are the eigenvalues of the *Jacobian matrix*, the $N \times N$ matrix with elements equal to the partial derivatives $\partial f_i / \partial x_j$, evaluated at the fixed point. In the jargon of linear algebra one says that the *spectral radius* of the Jacobian matrix at the fixed point must be less than 1.

262

6.3.4 BINARY SEARCH

The relaxation method of the previous section is simple and reasonably fast, and when it works it is a useful method for solving nonlinear equations. As we have seen, however, it does not always work, since it may not converge to a fixed point.

Binary search, also known as the *bisection method*, is an alternative and more robust method for solving nonlinear equations for a single variable x. In this method one specifies an interval in which one wishes to search for a solution to an equation. If a single solution exists in that interval then the binary search method will always find it. In this sense, binary search is significantly superior to the relaxation method, although it is also a little more complicated to program.

A nonlinear equation for a single variable x can always be rearranged to put all the terms on one side of the equals sign, giving an equation of form $f(x) = 0$. Thus finding the solution or solutions to such an equation is equivalent to finding the zeros, or roots, of $f(x)$. Binary search works by finding roots.

Suppose we want to find a root of $f(x)$ in the interval between x_1 and x_2, if such a root exists. We begin by calculating the values $f(x_1)$ and $f(x_2)$ of the function at the two points. Perhaps, as shown in Fig. 6.3, we discover that one of those values is positive while the other is negative. Then, so long as the function $f(x)$ is continuous, there must be at least one point (and possibly more than one) somewhere between x_1 and x_2 where the function equals zero, and hence there must exist at least one root of the function in the interval. We say that the points x_1 and x_2 *bracket* a root of the function. We can also make the inverse statement that if there is exactly one root in the interval from x_1 to x_2 then $f(x_1)$ and $f(x_2)$ must have different signs. Thus if a single root exists in the interval, we will know it. (If there's more than one root, things are more complicated, but let's stick to the simple case for the moment.)

Now suppose we take a new point $x' = \frac{1}{2}(x_1 + x_2)$ half way between x_1 and x_2 and evaluate $f(x')$. If we are incredibly lucky, it's possible x' might fall exactly at the root of the function and $f(x') = 0$, in which case we have found our root and our job is done. Much more likely, however, is that $f(x')$ is nonzero, in which case it must be either positive or negative, meaning it must have the same sign as either $f(x_1)$ or $f(x_2)$, and the opposite sign to the other. Suppose, as in the figure above, it has the same sign as $f(x_2)$ and the opposite sign to $f(x_1)$. Then, by the same argument as before, the points x_1 and x' must bracket a root of the function. But notice that x_1 and x' are closer together, by

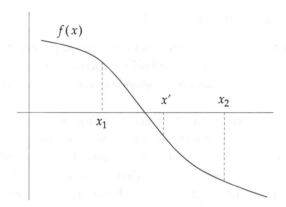

Figure 6.3: The binary search method. If $f(x_1)$ and $f(x_2)$ have opposite signs, one positive and one negative, and if $f(x)$ is continuous from x_1 to x_2, then there must be at least one root of $f(x)$ between x_1 and x_2. By looking at the sign of the function at the midpoint between x_1 and x_2, denoted x' here, we can determine whether that root lies in the left or right half of the interval.

a factor of two, than the two points we started with, x_1 and x_2. So we have narrowed the range in which our root lies. Similarly if $f(x')$ and $f(x_2)$ have opposite signs then x' and x_2 bracket the root of the function and are closer together than the original two points. Either way, we can by this process find two points that bracket our root and are closer together, by a factor of two, than the points we started with.

Now we repeat this process, finding the midpoint of our two new points, and so forth. Every time around we narrow the range in which our root lies by a factor of two, and we repeat the process until we have narrowed the range to whatever accuracy we desire. For instance, if we want to find a root to an accuracy of 10^{-6} then we repeat until the distance between our points is 10^{-6} or less. Then we take the midpoint of the final two points as our answer.

Here is an outline of the whole procedure:

1. Given an initial pair of points x_1, x_2, check that $f(x_1)$ and $f(x_2)$ have opposite signs. Also choose a target accuracy for the answer you want.

2. Calculate the midpoint $x' = \frac{1}{2}(x_1 + x_2)$ and evaluate $f(x')$.

3. If $f(x')$ has the same sign as $f(x_1)$ then set $x_1 = x'$. Otherwise set $x_2 = x'$.

4. If $|x_1 - x_2|$ is greater than the target accuracy, repeat from step 2. Otherwise, calculate $\frac{1}{2}(x_1 + x_2)$ once more and this is the final estimate of the position of the root.

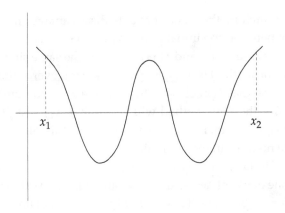

Figure 6.4: An even number of roots bracketed between two points. If the points x_1 and x_2 bracket an even number of roots—four in this case—then $f(x_1)$ and $f(x_2)$ have the same sign and the binary search method fails.

As with the relaxation method, the accuracy of the binary search method improves exponentially as the method progresses, since the error decreases by a factor of two on each step of the calculation. This means again that the method is relatively fast, because exponentials decay towards zero quickly. Indeed we can quite easily estimate how many steps the method will take to find a solution of a desired accuracy. Suppose our initial points x_1 and x_2 are a distance Δ apart. That distance is halved at each step of the binary search, so that after N steps the distance is $\Delta/2^N$. When this distance reaches our desired target accuracy—let's call it ϵ—the calculation ends. So the number of steps we have to take is given by $\Delta/2^N = \epsilon$, or

$$N = \log_2 \frac{\Delta}{\epsilon}. \tag{6.93}$$

The logarithm is a very slowly growing function of its argument, so even if Δ is much larger than ϵ, as it usually will be, the number of steps N is still quite small. For instance, suppose our two starting points are a distance $\Delta = 10^{10}$ apart and we want an accuracy of 10^{-10}, which is pretty generous. Then we have

$$N = \log_2 \frac{10^{10}}{10^{-10}} = \log_2 10^{20} \simeq 66.43\ldots \tag{6.94}$$

So it will take 67 steps to find the root. A computer can comfortably execute millions of instructions per second, so sixty or seventy is nothing. This method will work very fast.

The binary search method does have its disadvantages however. First, if there is no root between the initial points x_1 and x_2 then obviously it will not find a root. In this case, $f(x_1)$ and $f(x_2)$ will have the same sign, which tells us in advance that the method is not going to work. However, if there are an even number of roots, so that $f(x)$ crosses the zero line an even number of times as shown in Fig. 6.4, then the method will also fail: $f(x_1)$ and $f(x_2)$ will again have the same sign and there is no easy way to tell this situation from one in which there are no roots in the interval.

In fact, the way the binary search method is most often used, one does not necessarily have an initial interval in mind to search for a root. Instead one is simply interested in finding a root wherever it may be lurking. In that case any initial pair of points x_1, x_2 such that $f(x_1)$ and $f(x_2)$ have different signs is good enough. But finding such a pair is not always easy, and there is no universal method that will always work. Here are a couple of rules of thumb that are useful in practice:

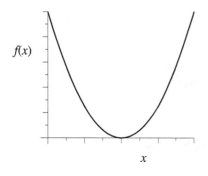

$f(x)$

x

Figure 6.5: A function with a double root. A function such as $(1 - x)^2$ that has two roots in the same place (or any even number of roots) only touches the horizontal axis but does not cross it.

1. If you know some facts about your function that can help you bracket a root then this is usually the best place to start. For instance, if you know of specific special values where the function must be positive or negative, perhaps for physical reasons, make use of those values.

2. If you are looking for a root near a particular value of x, start with values x_1, x_2 closely spaced around that value and check to see if they bracket the root. If they do not, double their distance from x in either direction and check again. Keep doubling until you find values that bracket the root.

These techniques, however, can fail on occasions. Initial bracketing of a root is one of those problems that can require a creative solution.

A more subtle, though rarer, problem with the binary search method is that it cannot find even-order polynomial roots of functions, i.e., double roots, quadruple roots, and so forth that fall in the same place, like the roots of $(1 - x)^2$ or $(2 - 3x)^4$. Functions with such roots only touch the horizontal axis at the position of the root but don't cross it (see Fig. 6.5) so it is not possible to bracket the root and the method fails. (Such roots can, however, be located by finding a maximum or minimum of the function. We look at methods for finding maxima and minima in Section 6.4.)

Finally, a significant drawback of the binary search method is that it does not extend to the solution of multiple, simultaneous equations. It works only for the solution of single equations in one variable, unlike, for instance, the relaxation method. In the next section we study a third method, Newton's method, which is significantly faster than either the relaxation method or binary search and works for simultaneous equations.

Exercise 6.13: Wien's displacement constant

Planck's radiation law tells us that the intensity of radiation per unit area and per unit wavelength λ from a black body at temperature T is

$$I(\lambda) = \frac{2\pi hc^2 \lambda^{-5}}{e^{hc/\lambda k_B T} - 1},$$

where h is Planck's constant, c is the speed of light, and k_B is Boltzmann's constant.

a) Show by differentiating that the wavelength λ at which the emitted radiation is strongest is the solution of the equation

$$5e^{-hc/\lambda k_B T} + \frac{hc}{\lambda k_B T} - 5 = 0.$$

Make the substitution $x = hc/\lambda k_B T$ and hence show that the wavelength of maximum radiation obeys the *Wien displacement law*:

$$\lambda = \frac{b}{T},$$

where the so-called *Wien displacement constant* is $b = hc/k_B x$, and x is the solution to the nonlinear equation

$$5e^{-x} + x - 5 = 0.$$

b) Write a program to solve this equation to an accuracy of $\epsilon = 10^{-6}$ using the binary search method, and hence find a value for the displacement constant.

c) The displacement law is the basis for the method of *optical pyrometry*, a method for measuring the temperatures of objects by observing the color of the thermal radiation they emit. The method is commonly used to estimate the surface temperatures of astronomical bodies, such as the Sun. The wavelength peak in the Sun's emitted radiation falls at $\lambda = 502\,\text{nm}$. From the equations above and your value of the displacement constant, estimate the surface temperature of the Sun.

Exercise 6.14: Consider a square potential well of width w, with walls of height V:

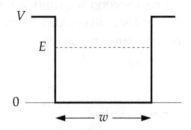

Using Schrödinger's equation, it can be shown that the allowed energies E of a single quantum particle of mass m trapped in the well are solutions of

$$\tan \sqrt{w^2 mE/2\hbar^2} = \begin{cases} \sqrt{(V - E)/E} & \text{for the even numbered states,} \\ -\sqrt{E/(V - E)} & \text{for the odd numbered states,} \end{cases}$$

where the states are numbered starting from 0, with the ground state being state 0, the first excited state being state 1, and so forth.

a) For an electron (mass 9.1094×10^{-31} kg) in a well with $V = 20$ eV and $w = 1$ nm, write a Python program to plot the three quantities

$$y_1 = \tan \sqrt{w^2 mE/2\hbar^2}, \qquad y_2 = \sqrt{\frac{V - E}{E}}, \qquad y_3 = -\sqrt{\frac{E}{V - E}},$$

on the same graph, as a function of E from $E = 0$ to $E = 20$ eV. From your plot make approximate estimates of the energies of the first six energy levels of the particle.

b) Write a second program to calculate the values of the first six energy levels in electron volts to an accuracy of 0.001 eV using binary search.

6.3.5 NEWTON'S METHOD

The relaxation and binary search methods of previous sections are both simple, fast, and useful methods for solving nonlinear equations. But, as we have seen, both have their disadvantages. The relaxation method does not always converge to a solution, while binary search requires an initial pair of points x_1, x_2 that bracket a solution, which may be hard to find, and it only works for the solution of single equations in one variable. Furthermore, although both methods are reasonably fast, they are not the fastest available, which can be an issue in cases where speed is important.

In this section we examine a third common method for solving nonlinear equations, *Newton's method*, sometimes also called the *Newton–Raphson method*,

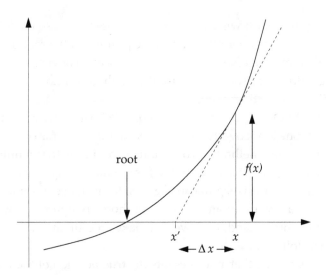

Figure 6.6: Newton's method. Newton's method takes a single estimate x of the root of a function and uses the slope of the function at that point to extrapolate a better estimate x'.

which addresses some of these problems, although (as you'll probably not be surprised to hear) it has its own set of problems too.

You may already be familiar with Newton's method. Figure 6.6 shows a graphical representation of the method for a single variable. As with the binary search method of Section 6.3.4, we convert the problem of solving an equation into one of finding the root of a function $f(x)$, represented by the curve in the figure. We start with a single guess x and then use the slope at that position (dotted line) to extrapolate and make another guess x', which will usually be better than the first guess, although it is possible, in unlucky cases, for it to be worse.

From the figure we see that the slope at x is

$$f'(x) = \frac{f(x)}{\Delta x},$$
(6.95)

and the formula for our new guess x' for the position of the root is thus

$$x' = x - \Delta x = x - \frac{f(x)}{f'(x)}.$$
(6.96)

The catch with this approach is that it requires us to know the derivative $f'(x)$, but if we do know the derivative then putting the method to work on the com-

269

puter is straightforward: we just start with a guess x and use Eq. (6.96) repeatedly to get better and better estimates of the position of the root. For functions with more than one root Newton's method typically converges to a root near the starting value of x, so one can use the method to find different roots by choosing different starting points.

How long should we go on iterating Eq. (6.96) before we are satisfied with our answer? If one is not overly concerned with running time or accuracy then, as with the relaxation method, it's often adequate simply to run until the value of x stops changing, or at least until the first so many digits stop changing. If we want an answer to six significant figures, for instance, we just run until a few iterations pass without any changes in those six figures. If we are more exacting, on the other hand, then we can make an estimate of the accuracy of the method as follows.

Suppose, as before, that x^* represents the true position of the root of interest. Then, performing a Taylor expansion about our estimate x of the root, we have

$$f(x^*) = f(x) + (x^* - x)f'(x) + \tfrac{1}{2}(x^* - x)^2 f''(x) + \dots \qquad (6.97)$$

But since x^* is a root of $f(x)$ we have $f(x^*) = 0$ by definition, so the left-hand side of the equation vanishes. Then, dividing throughout by $f'(x)$ and rearranging, we get

$$x^* = \left[x - \frac{f(x)}{f'(x)} \right] - \tfrac{1}{2}(x^* - x)^2 \frac{f''(x)}{f'(x)} + \dots \qquad (6.98)$$

Comparing with Eq. (6.96), we see that the quantity in square brackets $[\dots]$ on the right-hand side of this expression is exactly equal to our new estimate x' of the root, so

$$x^* = x' - \tfrac{1}{2}(x^* - x)^2 \frac{f''(x)}{f'(x)} + \dots \qquad (6.99)$$

In other words, the position of the root x^* is equal to our estimate x' plus an error term that goes as $(x^* - x)^2$ (plus higher order terms that we are ignoring).

Let us define the error ϵ on our first estimate x of the root in the standard fashion by $x^* = x + \epsilon$, and the error on our next estimate x' similarly by $x^* = x' + \epsilon'$. Then Eq. (6.99) tells us that

$$\epsilon' = \left[\frac{-f''(x)}{2f'(x)} \right] \epsilon^2. \qquad (6.100)$$

In other words, if the size of the error is ϵ on one step of the method it will be around ϵ^2 on the next. We say that Newton's method has *quadratic convergence*.

This implies that Newton's method converges *extremely* quickly. The relaxation and binary search methods are not bad—both have an error that shrinks exponentially with passing iterations of the calculation. But Newton's method does significantly better. Assuming that the quantity $c = -f''(x)/2f'(x)$ in Eq. (6.100) is roughly constant in the vicinity of the root, the error ϵ after N iterations will be approximately $\epsilon \simeq (c\epsilon_0)^{2^N}/c$, where ϵ_0 is the error on the initial guess. In other words the error varies with N as an exponential of an exponential, which is very fast indeed.

In practice, if we want to estimate the error on our current value of the root, we can write $x^* = x + \epsilon$ and $x^* = x' + \epsilon' = x' + c\epsilon^2$ for some constant c, and equate the values of x^* to get

$$x' - x = \epsilon - c\epsilon^2 = \epsilon(1 - c\epsilon) \simeq \epsilon, \qquad (6.101)$$

if ϵ is small. Thus a simple estimate of the error ϵ on the value of x is just the change $x' - x$ from one iteration to the next. Another way of saying this is that Newton's method converges so rapidly that to all intents and purposes the value x' on the next iteration is a good enough approximation to the true root that $x - x' \simeq x - x^* = \epsilon$. Note that this gives you the error ϵ on the old estimate x, not the error on the new estimate x'. If you want the error on x' you need to perform one more iteration.

So a good (and simple) rule of thumb for implementing Newton's method is to go on iterating Eq. (6.96) until the difference between two successive values of x is comparable with or smaller than the desired accuracy of the solution, then stop.

Newton's method has two main disadvantages. The first is that it requires us to know the derivative of $f(x)$, which we sometimes don't. As described in the following section, however, we can get around this problem by using a numerical derivative instead. A more serious problem is that, like the relaxation method, Newton's method doesn't always converge. In particular, if the value of $f'(x)$ is very small then Eq. (6.100) can give an error that actually gets larger, rather than smaller upon iteration of the method, which implies we are moving further from the root we want and not closer. Alternatively, the function $f(x)$ can simply slope in the wrong direction, as shown in Fig. 6.7.

Situations like these are usually fairly easy to spot in

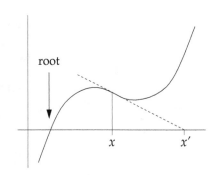

Figure 6.7: Failure of Newton's method. Newton's method can fail to converge towards a root if the shape of the function is unfavorable. In this example the initial guess x is an unlucky one and the next estimate x' actually falls further from the root.

practice—if your Newton's method calculation seems to be giving strange answers, you may be having convergence problems of this kind. In the end, the important lesson is that no root finding method is perfect, each has its good and bad points, and each is better than the others in some situations. Newton's method is certainly not perfect for every problem, although its speed and flexibility make it a good choice for many. One of the keys to doing good computational physics is to have a range of methods at your disposal, so that if one works poorly for a particular problem you have others up your sleeve.

EXAMPLE 6.4: INVERSE HYPERBOLIC TANGENT

Let us use Newton's method to calculate the inverse (or arc) hyperbolic tangent of a number u. By definition, arctanh u is the number x such that $u = \tanh x$. To put that another way, x is a root of the equation $\tanh x - u = 0$. Recalling that the derivative of $\tanh x$ is $1/\cosh^2 x$, Eq. (6.96) for this problem becomes

$$x' = x - (\tanh x - u)\cosh^2 x. \qquad (6.102)$$

Here is a Python function that calculates the inverse hyperbolic tangent using Newton's method starting from an initial guess $x = 0$:

File: atanh.py

```
from math import tanh,cosh

accuracy = 1e-12

def arctanh(u):
    x = 0.0
    delta = 1.0
    while abs(delta)>accuracy:
        delta = (tanh(x)-u)*cosh(x)**2
        x -= delta
    return x
```

We could use this function, for example, to make a plot of the arctanh function thus:

File: atanh.py

```
from numpy import linspace
from pylab import plot,show

upoints = linspace(-0.99,0.99,100)
xpoints = []
```

```
for u in upoints:
    xpoints.append(arctanh(u))
plot(upoints,xpoints)
show()
```

Exercise 6.15: The roots of a polynomial

Consider the degree-six polynomial

$$P(x) = 924x^6 - 2772x^5 + 3150x^4 - 1680x^3 + 420x^2 - 42x + 1.$$

There is no general formula for the roots of a polynomial of degree six, but one can find them easily enough using a computer.

a) Make a plot of $P(x)$ from $x = 0$ to $x = 1$ and by inspecting it find rough values for the six roots of the polynomial—the points at which the function is zero.

b) Write a Python program to solve for the positions of all six roots to at least ten decimal places of accuracy, using Newton's method.

Note that the polynomial in this example is just the sixth Legendre polynomial (mapped onto the interval from zero to one), so the calculation performed here is the same as finding the integration points for 6-point Gaussian quadrature (see Section 5.6.2), and indeed Newton's method is the method of choice for calculating Gaussian quadrature points.

6.3.6 The Secant Method

Newton's method requires us to know the derivative of $f(x)$ in order to use Eq. (6.96). If we don't have an analytic formula for the derivative then we cannot use Newton's method directly as in the previous section, but we can instead calculate the derivative numerically using the techniques we looked at in Section 5.10. As we saw there, calculating a numerical derivative involves taking the difference of the values of f at two closely spaced points. In the most common application of this idea to Newton's method we choose the points to be precisely the points x and x' at two successive steps of the method, which saves us the trouble of evaluating $f(x)$ at additional values of x, since we have evaluated it already. (We used a similar trick with the relaxation method in Section 6.3.1.) This version of Newton's method is called the *secant method*. In detail it works as follows.

We now start with two points—let us call them x_1 and x_2. (This is reminiscent of the two points used in the binary search method, but unlike binary search the points need not bracket the root. If they are on the same side of the root, that's fine.) We now calculate an approximation to the derivative of f at x_2 from the formula

$$f'(x_2) \simeq \frac{f(x_2) - f(x_1)}{x_2 - x_1}.$$

(6.103)

Then we substitute this value into Eq. (6.96) to get a new (and hopefully better) guess x_3 for the position of the root:

$$x_3 = x_2 - f(x_2)\frac{x_2 - x_1}{f(x_2) - f(x_1)}.$$

(6.104)

This new guess is based on the *two* previous values in the series, unlike the original Newton's method in which each guess was based on only one previous value. It also makes use of the values $f(x_1)$ and $f(x_2)$, though we would have to calculate these values anyway in the course of carrying out the normal Newton's method, so the secant method involves little extra work. Apart from these differences the secant method is similar to the ordinary Newton's method. It has the same fast convergence, and can break down in the same ways if we are unlucky with the shape of the function. It can also be generalized to the solution of simultaneous nonlinear equations, although the generalization is quite complicated, involving the calculation of an entire matrix of numerical derivatives, rather than just a single one—see Section 6.3.7.

Exercise 6.16: The Lagrange point

There is a magical point between the Earth and the Moon, called the L_1 Lagrange point, at which a satellite will orbit the Earth in perfect synchrony with the Moon, staying always in between the two. This works because the inward pull of the Earth and the outward pull of the Moon combine to create exactly the needed centripetal force that keeps the satellite in its orbit. Here's the setup:

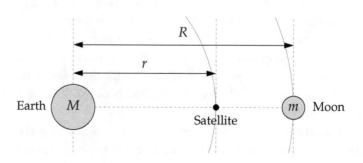

a) Assuming circular orbits, and assuming that the Earth is much more massive than either the Moon or the satellite, show that the distance r from the center of the Earth to the L_1 point satisfies

$$\frac{GM}{r^2} - \frac{Gm}{(R-r)^2} = \omega^2 r,$$

where R is the distance from the Earth to the Moon, M and m are the Earth and Moon masses, G is Newton's gravitational constant, and ω is the angular velocity of both the Moon and the satellite.

b) The equation above is a degree-five polynomial equation in r (also called a quintic equation). Such equations cannot be solved exactly in closed form, but it's straightforward to solve them numerically. Write a program that uses either Newton's method or the secant method to solve for the distance r from the Earth to the L_1 point. Compute a solution accurate to at least four significant figures.

The values of the various parameters are:

$$G = 6.674 \times 10^{-11}\, \mathrm{m^3 kg^{-1} s^{-2}},$$
$$M = 5.974 \times 10^{24}\, \mathrm{kg},$$
$$m = 7.348 \times 10^{22}\, \mathrm{kg},$$
$$R = 3.844 \times 10^{8}\, \mathrm{m},$$
$$\omega = 2.662 \times 10^{-6}\, \mathrm{s^{-1}}.$$

You will also need to choose a suitable starting value for r, or two starting values if you use the secant method.

6.3.7 NEWTON'S METHOD FOR TWO OR MORE VARIABLES

Newton's method can also be used for the solution of simultaneous nonlinear equations, meaning the solution of more than one equation for more than one variable. (We saw previously, in Section 6.3.3, that the relaxation method can also be used for simultaneous equations, but the binary search method cannot.) Any set of simultaneous equations for N variables $x_1, \ldots, x_N$ can be written in the form

$$f_1(x_1, \ldots, x_N) = 0,$$
$$\vdots \qquad\qquad\qquad (6.105)$$
$$f_N(x_1, \ldots, x_N) = 0,$$

for some set of functions $f_1, \ldots, f_N$. Note that there must be the same number of equations as there are variables if the equations are to be solvable in general.

Suppose the equations have a root at $x_1^*, \ldots, x_N^*$. Then we can write the equivalent of the Taylor expansion of Eq. (6.97) thus:

$$f_i(x_1^*, \ldots, x_N^*) = f_i(x_1, \ldots, x_N) + \sum_j (x_j^* - x_j) \frac{\partial f_i}{\partial x_j} + \ldots \tag{6.106}$$

or we could use vector notation thus:

$$\mathbf{f}(\mathbf{x}^*) = \mathbf{f}(\mathbf{x}) + \mathbf{J} \cdot (\mathbf{x}^* - \mathbf{x}) + \ldots \tag{6.107}$$

where $\mathbf{J}$ is the *Jacobian matrix*, the $N \times N$ matrix with elements $J_{ij} = \partial f_i / \partial x_j$. Since $\mathbf{x}^*$ is a root of the equations, we have $\mathbf{f}(\mathbf{x}^*) = 0$ by definition, so the left-hand side of Eq. (6.107) is zero. Neglecting higher-order terms and defining $\Delta \mathbf{x} = \mathbf{x} - \mathbf{x}^*$, we then have

$$\mathbf{J} \cdot \Delta \mathbf{x} = \mathbf{f}(\mathbf{x}). \tag{6.108}$$

This is a set of ordinary linear simultaneous equations of the form $\mathbf{Ax} = \mathbf{v}$, exactly the kind of equations that we solved in Section 6.1. Thus we could solve for $\Delta \mathbf{x}$ by using, for example, Gaussian elimination (Section 6.1.1) or we could use the Python function `solve` from the module `numpy.linalg` (Section 6.1.4). Once we have solved for $\Delta \mathbf{x}$, our new estimate $\mathbf{x}'$ of the position of the root is

$$\mathbf{x}' = \mathbf{x} - \Delta \mathbf{x}. \tag{6.109}$$

Thus, applying Newton's method for more than one variable involves evaluating the Jacobian matrix $\mathbf{J}$ of first derivatives at the point $\mathbf{x}$, then solving Eq. (6.108) for $\Delta \mathbf{x}$ and using the result to calculate the new estimate of the root. This is a more complex calculation than in the one-variable case, but all the operations are ones we already know how to do, so in principle it is straightforward. We can also calculate the derivatives numerically, as we did in Section 6.3.6, leading to a many-variable version of the secant method.

Exercise 6.17: Nonlinear circuits

Exercise 6.1 used regular simultaneous equations to solve for the behavior of circuits of resistors. Resistors are linear—current is proportional to voltage—and the resulting equations we need to solve are therefore also linear and can be solved by standard matrix methods. Real circuits, however, often include nonlinear components. To solve for the behavior of these circuits we need to solve nonlinear equations.

Consider the following simple circuit, a variation on the classic Wheatstone bridge:

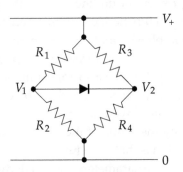

The resistors obey the normal Ohm law, but the diode obeys the diode equation:

$$I = I_0(e^{V/V_T} - 1),$$

where V is the voltage across the diode and I_0 and V_T are constants.

a) The Kirchhoff current law says that the total net current flowing into or out of every point in a circuit must be zero. Applying this law to the junction at voltage V_1 in the circuit above we get

$$\frac{V_1 - V_+}{R_1} + \frac{V_1}{R_2} + I_0 \left[e^{(V_1 - V_2)/V_T} - 1 \right] = 0.$$

Derive the corresponding equation for voltage V_2.

b) Using Newton's method, solve the two nonlinear equations for the voltages V_1 and V_2 with the conditions

$$V_+ = 5\,\text{V},$$
$$R_1 = 1\,\text{k}\Omega, \qquad R_2 = 4\,\text{k}\Omega, \qquad R_3 = 3\,\text{k}\Omega, \qquad R_4 = 2\,\text{k}\Omega,$$
$$I_0 = 3\,\text{nA}, \qquad V_T = 0.05\,\text{V}.$$

Hint: You can solve Eq. (6.108) for Δx using the function solve() from the package numpy.linalg if you want to, but in this case the matrix is only a 2×2 matrix, so it's easy to calculate the inverse directly too.

c) The electronic engineer's rule of thumb for diodes is that the voltage across a (forward biased) diode is always about 0.6 volts. Confirm that your results agree with this rule.

6.4 MAXIMA AND MINIMA OF FUNCTIONS

Closely related to the problem of finding roots is the problem of finding the maximum or minimum of a function, the point where the function is largest or smallest. Minima arise in physics in, for instance, the solution of equilibrium problems—the equilibrium point of a system is typically given by the minimum of its potential energy. Minima also appear in the variational method for solving quantum mechanics problems, which involves guessing a mathematical form for the ground-state wavefunction of a system, with a small number of parameters that specify the shape of the function, then calculating the energy of the system when it has that wavefunction. By finding the minimum of this energy with respect to the parameters one can derive an *upper bound* on the system's true ground-state energy—if you didn't guess the right wavefunction then you won't get the exact ground-state for any choice of the parameters, but you know that no wavefunction can have an energy lower than the ground state, so the lowest value found must always be greater than or equal to the ground state energy. Calculations like this, which place a limit on a physical quantity, can be enormously useful in practical situations, even though they don't give exact answers.

Minimization (or maximization) problems can have one variable or many. We can seek the minimum of a function $f(x)$ with respect to the single variable x, or the minimum of $f(x_1, x_2, x_3, \ldots)$ with respect to all of its arguments. There is, however, no analog of simultaneous equations, no problems where we simultaneously maximize or minimize many functions: in general if a value x minimizes $f(x)$, then it will not at the same time minimize $g(x)$, unless we are very lucky with f and g. So the most general minimization problem is the minimization of a single function.

Note also that a function can have more than one maximum or minimum. Conventionally, we distinguish between a *local minimum*, which is a point at which the function is lower than anywhere else in the immediate vicinity, and a *global minimum*, where the function is lower than any point anywhere—see Fig. 6.8. Most functions have only a single global minimum; only in special cases does a function have two or more points that exactly tie for lowest value. But a function may have many local minima. Both local and global minima (or maxima) are interesting from a physics point of view. The global minimum of potential energy, for example,

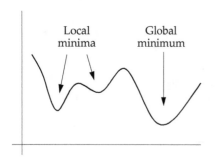

Local minima   Global minimum

Figure 6.8: Local and global minima of a function

gives the true equilibrium state of a system, but local minima correspond to *metastable equilibria*, points where the system will come to rest and remain indefinitely unless disturbed—think of water in a pond on top of a hill, or just a book sitting on a table.

Suppose then that we are given a function. One standard way to find minima or maxima, either local or global, is to differentiate and set the result equal to zero. For instance if we have a function $f(x_1, x_2, \ldots)$ then we would calculate partial derivatives with respect to each of the arguments separately and then solve the resulting set of simultaneous equations

$$\frac{\partial f}{\partial x_i} = 0 \qquad \text{for all } i. \tag{6.110}$$

We have already seen how to solve such sets of equations: if they are linear we can use standard matrix techniques (Section 6.1) and if they are nonlinear then we can use either the relaxation method (Section 6.3.1) or Newton's method (Section 6.3.5). When we use Newton's method, the combined technique is called the Gauss–Newton method, and we discuss it further in Section 6.4.2.

Unfortunately, it is often the case that we cannot calculate the derivatives of our function, because the function is not specified mathematically but is, for example, the output of another computer program. In the following sections we describe several more general methods that we use to find maxima and minima of functions whether or not we know their derivatives.

6.4.1   GOLDEN RATIO SEARCH

Consider the problem of finding a minimum or maximum of a function $f(x)$ of a single variable x. To be concrete, let us look at the minimization problem. Finding a maximum is basically the same problem—it's equivalent to finding the minimum of $-f(x)$.

A good basic method for finding the minimum of a function of a single variable is *golden ratio search*. Golden ratio search can find both local and global minima, although, like all of the methods we discuss in this chapter, it does not tell you which you have found. This makes it useful mainly for cases where you don't mind if you find only a local minimum. If you want the global minimum of a function then golden ratio search *may* find it for you, but will not tell you when it succeeds—it will tell you only that it has found *a* minimum, but not of what type. If you want the global minimum, therefore, the method may be only of limited utility. (In Section 10.4 we discuss another method,

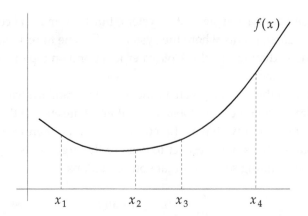

Figure 6.9: Golden ratio search. Suppose there is a minimum of the function $f(x)$ between x_1 and x_4. If the function is lower at x_2 than at x_3, as shown here, then we know the minimum must lie between x_1 and x_3; otherwise it lies between x_2 and x_4. Either way, we have narrowed the range in which it lies.

called simulated annealing, which is capable of finding the global minimum of a function, or at least a good approximation to it.)

Golden ratio search is similar in spirit to the binary search method for finding the roots of a function. In binary search we used two points to bracket a root and a third point in the middle to narrow the search. The equivalent construction for a minimum involves three points that bracket the minimum plus a fourth as shown in Fig. 6.9. Note that we have deliberately made the points unevenly spaced along the x-axis, for reasons that will become clear in a moment.

Suppose that at least one of the values $f(x_2)$ and $f(x_3)$ at the two intermediate points is less than the values $f(x_1)$ and $f(x_4)$ at the outer points. In that case we know that there must be at least one minimum of the function between x_1 and x_4, because the function goes down and up again. We can narrow down the location of this minimum by comparing the values at x_2 and x_3 to see which is smaller. If $f(x_2)$ is smaller, then the minimum must lie between x_1 and x_3 because, again, the function goes down and up again. Conversely if $f(x_3)$ is smaller, then the minimum must lie between x_2 and x_4.

By this process we can narrow down our search for the minimum to a smaller range, encompassing three of the previous four points. We then add a fourth point to those three (at a location to be specified shortly), and repeat the process again. We keep on repeating until the range in which the minimum falls becomes smaller than our required accuracy, then stop and take the middle of the range as our result for the position of the minimum.

To put this strategy into practice we must choose the distribution of the four points. For given positions of the first and last points, x_1 and x_4, the two interior points should be symmetrically distributed about the mid-point of the interval—given that we don't know where the minimum will fall it makes no sense to favor one side of the interval over the other. So we should choose

$$x_2 - x_1 = x_4 - x_3. \tag{6.111}$$

In geometric terms this says that we can move the two interior points x_2 and x_3 nearer or further from the center line, but we should always keep them symmetric about that line.

This doesn't yet fix the values of x_2 and x_3 though. We need one more equation to do that. We observe that, in order to make the method work as efficiently as possible, we want to choose the points so that the width of the interval in which the minimum falls decreases by the largest amount possible on each step of the process. If we choose x_2 and x_3 to be close to the center then it makes the current step of the process more efficient—because the size of the new interval, whichever interval it is, will be smaller. Unfortunately it also implies that the interior points on the *next* step will be far from the center of the interval which makes that step inefficient. Rather than making either step inefficient we choose instead the middle road that makes both the current step and the next one equally efficient, i.e., we choose the values of x_2 and x_3 that put the interior points equally close to the center of the interval on both steps.

Let us define z to be the ratio between the width of the bracketing interval before and after a step of the search process. If we suppose that the minimum falls in the left-hand part of the interval, between x_1 and x_3, then we have

$$z = \frac{x_4 - x_1}{x_3 - x_1} = \frac{x_2 - x_1 + x_3 - x_1}{x_3 - x_1} = \frac{x_2 - x_1}{x_3 - x_1} + 1, \tag{6.112}$$

where we have used Eq. (6.111) to eliminate x_4. (If the minimum fell in the right-hand part of the interval we would instead get $z = (x_4 - x_1)/(x_4 - x_2)$, but it is easy to show, again using Eq. (6.111), that this gives the same result.)

Now we calculate the value of z for the next step of the algorithm. Looking at Fig. 6.9 we see that it is given by

$$z = \frac{x_3 - x_1}{x_2 - x_1}, \tag{6.113}$$

(or an equivalent expression for the right-hand part of the interval). But if the algorithm is to be equally efficient on both steps then the value for z must be the same. Substituting (6.112) into (6.113), we then find that $z = 1/z + 1$, or equivalently

$$z^2 - z - 1 = 0. \tag{6.114}$$

Solving this equation for z, bearing in mind that the solution must be greater than one, gives us

$$z = \frac{1 + \sqrt{5}}{2} = 1.618\ldots \tag{6.115}$$

(And again, we can repeat the whole argument for the case where the minimum falls in the right-hand part of the interval, and we get the same result.) This value of z is called the *golden ratio*, and it is after this ratio that the golden ratio search is named. The golden ratio crops up repeatedly in many branches of physics and mathematics, as well as finding uses in art and architecture.

Knowing the value of z fixes the positions of the two interior points— Eq. (6.113) is the second equation we need to solve for their values. Alternatively, we can just observe that the point x_3 should be $1/z = 0.618$ or about 62% of the way from x_1 to x_4, and x_2 should be in the mirror-image position on the other side of the center line.

Thus the complete golden ratio search goes as follows:

1. Choose two initial outside points x_1 and x_4, then calculate the interior points x_2 and x_3 according to the golden ratio rule above. Evaluate $f(x)$ at each of the four points and check that at least one of the values at x_2 and x_3 is less than the values at both x_1 and x_4. Also choose a target accuracy for the position of the minimum.

2. If $f(x_2) < f(x_3)$ then the minimum lies between x_1 and x_3. In this case, x_3 becomes the new x_4, x_2 becomes the new x_3, and there will be a new value for x_2, chosen once again according to the golden ratio rule. Evaluate $f(x)$ at this new point.

3. Otherwise, the minimum lies between x_2 and x_4. Then x_2 becomes the new x_1, x_3 becomes the new x_2, and there will be a new value for x_3. Evaluate $f(x)$ at this new point.

4. If $x_4 - x_1$ is greater than the target accuracy, repeat from step 2. Otherwise, calculate $\frac{1}{2}(x_2 + x_3)$ and this is the final estimate of the position of the minimum.

Golden ratio search suffers from the same issue as the binary search method of Section 6.3.4, that we need to start with an interval that brackets the solution we are looking for. The same basic rules of thumb for finding such an interval apply as in that case—use whatever you know about the form of the function to guide you, or make an initial guess and keep widening it until you bracket the minimum—but, as with binary search, there is no foolproof way to find the initial interval and some trial and error may be necessary.

Also like binary search, the golden ratio method works only for functions of a single variable. There is no simple generalization of the method to functions of more than one variable. In later sections of this chapter we will look at some other methods that are suitable for minimizing or maximizing functions of more than one variable.

EXAMPLE 6.5: THE BUCKINGHAM POTENTIAL

The Buckingham potential is an approximate representation of the potential energy of interaction between atoms in a solid or gas as a function of the distance r between them:

$$V(r) = V_0\left[\left(\frac{\sigma}{r}\right)^6 - e^{-r/\sigma}\right]. \qquad (6.116)$$

A plot of the potential as a function of r/σ looks like this:

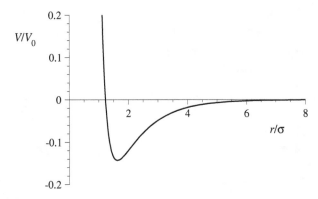

The potential contains two terms, one positive, corresponding to a short-range repulsive force between atoms, and the other negative, corresponding to a longer-range attractive force. There is an intermediate point at which these two forces balance out, corresponding to the minimum of the potential, and this point gives the resting distance between atoms. There is no known analytic expression for this resting distance, but we can find it numerically by using golden ratio search to minimize the potential. Here's a program to perform the calculation for the case where $\sigma = 1\,\text{nm}$. We do not need to know the value of the overall constant V_0, since the position of the minimum energy does not depend on it.

File: buckingham.py

```
from math import exp,sqrt

sigma = 1.0                 # Value of sigma in nm
accuracy = 1e-6             # Required accuracy in nm
z = (1+sqrt(5))/2           # Golden ratio

# Function to calculate the Buckingham potential
def f(r):
    return (sigma/r)**6 - exp(-r/sigma)

# Initial positions of the four points
x1 = sigma/10
x4 = sigma*10
x2 = x4 - (x4-x1)/z
x3 = x1 + (x4-x1)/z

# Initial values of the function at the four points
f1 = f(x1)
f2 = f(x2)
f3 = f(x3)
f4 = f(x4)

# Main loop of the search process
while x4-x1>accuracy:
    if f2<f3:
        x4,f4 = x3,f3
        x3,f3 = x2,f2
        x2 = x4 - (x4-x1)/z
        f2 = f(x2)
    else:
        x1,f1 = x2,f2
        x2,f2 = x3,f3
        x3 = x1 + (x4-x1)/z
        f3 = f(x3)

# Print the result
print("The minimum falls at",0.5*(x1+x4),"nm")
```

Note how we use a while loop to continue the search process until the interval bracketed by x_1 and x_4 falls below the desired accuracy, which is 10^{-6} nm in this case. If we run the program it prints

```
The minimum falls at 1.63051606717 nm
```

which, consulting the figure above, looks to be about the right answer.

Exercise 6.18: The temperature of a light bulb

An incandescent light bulb is a simple device—it contains a filament, usually made of tungsten, heated by the flow of electricity until it becomes hot enough to radiate thermally. Essentially all of the power consumed by such a bulb is radiated as electromagnetic energy, but some of the radiation is not in the visible wavelengths, which means it is useless for lighting purposes.

Let us define the efficiency of a light bulb to be the fraction of the radiated energy that falls in the visible band. It's a good approximation to assume that the radiation from a filament at temperature T obeys the Planck radiation law (previously encountered in Exercise 6.13 on page 267), meaning that the power radiated per unit wavelength λ obeys

$$I(\lambda) = 2\pi A h c^2 \frac{\lambda^{-5}}{e^{hc/\lambda k_B T} - 1},$$

where A is the surface area of the filament, h is Planck's constant, c is the speed of light, and k_B is Boltzmann's constant. The visible wavelengths run from $\lambda_1 = 390\,\text{nm}$ to $\lambda_2 = 750\,\text{nm}$, so the total energy radiated in the visible window is $\int_{\lambda_1}^{\lambda_2} I(\lambda)\,d\lambda$ and the total energy at all wavelengths is $\int_0^\infty I(\lambda)\,d\lambda$. Dividing one expression by the other and substituting for $I(\lambda)$ from above, we get an expression for the efficiency η of the light bulb thus:

$$\eta = \frac{\int_{\lambda_1}^{\lambda_2} \lambda^{-5}/(e^{hc/\lambda k_B T} - 1)\,d\lambda}{\int_0^\infty \lambda^{-5}/(e^{hc/\lambda k_B T} - 1)\,d\lambda},$$

where the leading constants and the area A have canceled out. Making the substitution $x = hc/\lambda k_B T$, this can also be written as

$$\eta = \frac{\int_{hc/\lambda_2 k_B T}^{hc/\lambda_1 k_B T} x^3/(e^x - 1)\,dx}{\int_0^\infty x^3/(e^x - 1)\,dx} = \frac{15}{\pi^4} \int_{hc/\lambda_2 k_B T}^{hc/\lambda_1 k_B T} \frac{x^3}{e^x - 1}\,dx,$$

where we have made use of the known exact value of the integral in the denominator.

a) Write a Python function that takes a temperature T as its argument and calculates the value of η for that temperature from the formula above. The integral in the formula cannot be done analytically, but you can do it numerically using any method of your choice. (For instance, Gaussian quadrature with 100 sample points works fine.) Use your function to make a graph of η as a function of temperature between 300 K and 10 000 K. You should see that there is an intermediate temperature where the efficiency is a maximum.

b) Calculate the temperature of maximum efficiency of the light bulb to within 1 K using golden ratio search. (Hint: An accuracy of 1 K is the equivalent of a few parts in ten thousand in this case. To get this kind of accuracy in your calculation you'll need to use values for the fundamental constants that are suitably accurate, i.e., you will need values accurate to several significant figures.)

c) Is it practical to run a tungsten-filament light bulb at the temperature you found? If not, why not?

6.4.2 THE GAUSS–NEWTON METHOD AND GRADIENT DESCENT

The golden ratio method of the previous section is a robust, reliable way to find minima or maxima of a function of a single variable. However, like the binary search that it closely resembles, it cannot be generalized to functions of more than one variable and it requires us to find an initial interval that brackets the minimum or maximum we are searching for, which is not always an easy thing to do. In this section we look at two other techniques for finding minima or maxima, the *Gauss–Newton method* and *gradient descent*, which suffer from neither of these drawbacks, although (perhaps unsurprisingly) both have some drawbacks of their own.

We described the Gauss–Newton method briefly in Section 6.4. As we pointed out there, we can find a maximum or minimum of a function $f(x)$ by differentiating it and setting the result equal to zero:

$$f'(x) = 0. \tag{6.117}$$

But this means that minima and maxima are nothing other than roots (i.e., zeros) of the derivative function f', and we have seen a variety of methods for finding roots of functions in Section 6.3. The most of efficient of these is Newton's method, which is described by Eq. (6.96). Replacing $f(x)$ in that equation by $f'(x)$, we derive the fundamental formula for the Gauss–Newton method:

$$x' = x - \frac{f'(x)}{f''(x)}. \tag{6.118}$$

If we know the form of $f(x)$ explicitly and can calculate its derivatives, this formula allows us to find its maxima and minima quickly; like the original Newton's method, it has quadratic (i.e., very fast) convergence. And, as with Newton's method, the Gauss–Newton method can be generalized to find maxima and minima of functions of more than one variable—see Eq. (6.108).

It is relatively rare, however, that we can calculate the derivatives of the function $f(x)$, so the Gauss–Newton method finds only occasional use. If we can calculate only the first derivative of a function, then we can still do an approximate version of the Gauss–Newton method by writing

$$x' = x - \gamma f'(x), \tag{6.119}$$

where γ is a constant value that represents a rough guess at $1/f''(x)$. The nice thing about this approach is that γ does not have to be very accurate for the method to work. Any value of roughly the right order of magnitude will allow you to find your minimum or maximum in a reasonable number of steps.

The method embodied in Eq. (6.119) is called the method of gradient descent. The method measures the gradient of the function at the point x and then subtracts a constant times that gradient from the function. For positive values of γ this means that the method will move "downhill" from x to x' and converge towards a minimum of the function. For negative γ it will converge towards a maximum. So by choosing the sign of γ we can control which we find. The magnitude of γ controls the rate of convergence. If γ is larger then we will converge more quickly in general, but if it is too large then we may overshoot and fail to find the maximum or minimum we are looking for. The trick with gradient descent is to choose γ appropriately so that we converge with reasonable speed without overshooting. In general, as Eq. (6.118) suggests, the value should be about equal to $1/f''(x)$, but if we cannot calculate the second derivative then we must guess a value for γ instead. Trial and error is often the quickest way to an answer. In some cases, even if one cannot calculate the second derivative analytically, one may be able to calculate it numerically using the numerical derivative formula given in Eq. (5.109) on page 197.

There are, however, many cases where we cannot calculate even the first derivative of a function for use in Eq. (6.119). In such cases we can estimate it using an approach similar to the secant method of Section 6.3.6. As there, we start with two points x_1 and x_2, and use them to calculate an approximation to the slope $f'(x)$ from

$$f'(x_2) \simeq \frac{f(x_2) - f(x_1)}{x_2 - x_1}. \tag{6.120}$$

Then we use this expression in Eq. (6.119) to calculate the next estimate x_3:

$$x_3 = x_2 - \gamma \frac{f(x_2) - f(x_1)}{x_2 - x_1}. \tag{6.121}$$

For suitable choices of γ, this expression boasts rates of convergence similar to Newton's method, while requiring us to calculate only the value of the function $f(x)$, and not its derivatives.

There are many other methods for finding maxima and minima of functions, with names such as Powell's method, the conjugate gradient method, and the BFGS method. The maximizing and minimizing of functions plays an important role not only in physics but in many other fields, including applied mathematics, computer science, biology, chemistry, engineering, and operations research, and has for this reason been the subject of much experimentation and research through the years. We will not go into these other methods here. The interested reader can find a good discussion in the book by Press *et al.*[8] We will, however, look at one further, and very important, method in Chapter 10, the method of simulated annealing, which uses ideas borrowed from physics to maximize or minimize functions and is a nice example of a computational method that is not only useful to physicists, but is inspired in the first place by the study of physics itself.

---

[8]Press, W. H., Teukolsky, S. A., Vetterling, W. T., and Flannery, B. P., *Numerical Recipes in C*, Cambridge University Press, Cambridge (1992).

CHAPTER 7

FOURIER TRANSFORMS

$\mathbf{T}$HE Fourier transform is one of the most useful, and most widely used, tools in traditional theoretical physics. It's also very useful in computational physics. It allows us to break down functions or signals into their component parts and analyze, smooth, or filter them, and it gives us a way to rapidly perform certain kinds of calculations and solve certain differential equations, such as the diffusion equation or the Schrödinger equation. In this section we look at how Fourier transforms are used in computational physics and at computational methods for calculating them.

7.1 FOURIER SERIES

As every physicist learns, a periodic function $f(x)$ defined on a finite interval $0 \leq x < L$ can be written as a *Fourier series*.[1] There are several kinds of Fourier series. If the function is even (i.e., symmetric) about the mid-point at $x = \frac{1}{2}L$ then we can use a cosine series thus:

$$f(x) = \sum_{k=0}^{\infty} \alpha_k \cos\left(\frac{2\pi k x}{L}\right), \tag{7.1}$$

where the α_k are a set of coefficients whose values depend on the shape of the function. If the function is odd (antisymmetric) about the midpoint then we can use a sine series:

$$f(x) = \sum_{k=1}^{\infty} \beta_k \sin\left(\frac{2\pi k x}{L}\right). \tag{7.2}$$

---

[1]There are some conditions the function must satisfy: it must be bounded and it can have at most a finite number of discontinuities and extrema. If you want to review the mathematics of Fourier series, a good introduction can be found in Boas, M. L., *Mathematical Methods in the Physical Sciences*, John Wiley, New York (2005).

(Note that the sum for the sine series starts at $k = 1$, because the $k = 0$ term vanishes.) And a general function, with no special symmetry, can be written as a sum of even and odd parts thus:

$$f(x) = \sum_{k=0}^{\infty} \alpha_k \cos\left(\frac{2\pi kx}{L}\right) + \sum_{k=1}^{\infty} \beta_k \sin\left(\frac{2\pi kx}{L}\right). \tag{7.3}$$

An alternative way to represent this general sine/cosine series is to make use of the identities $\cos\theta = \frac{1}{2}(e^{-i\theta} + e^{i\theta})$ and $\sin\theta = \frac{1}{2}i(e^{-i\theta} - e^{i\theta})$. Substituting these into Eq. (7.3) gives

$$f(x) = \frac{1}{2} \sum_{k=0}^{\infty} \alpha_k \left[\exp\left(-i\frac{2\pi kx}{L}\right) + \exp\left(i\frac{2\pi kx}{L}\right)\right]$$
$$+ \frac{i}{2} \sum_{k=1}^{\infty} \beta_k \left[\exp\left(-i\frac{2\pi kx}{L}\right) - \exp\left(i\frac{2\pi kx}{L}\right)\right]. \tag{7.4}$$

Collecting terms, this can be conveniently rewritten as

$$f(x) = \sum_{k=-\infty}^{\infty} \gamma_k \exp\left(i\frac{2\pi kx}{L}\right), \tag{7.5}$$

where the sum now runs from $-\infty$ to $+\infty$ and

$$\gamma_k = \begin{cases} \frac{1}{2}(\alpha_{-k} + i\beta_{-k}) & \text{if } k < 0, \\ \alpha_0 & \text{if } k = 0, \\ \frac{1}{2}(\alpha_k - i\beta_k) & \text{if } k > 0. \end{cases} \tag{7.6}$$

Since this complex series includes the sine and cosine series as special cases, we will use it for most of our calculations (although there are cases where the sine and cosine series are useful, as we will see later in the chapter). If we can find the coefficients γ_k for a particular function $f(x)$ then the Fourier series gives us a compact way of representing the entire function that comes in handy for all sorts of numerical calculations.

Note that Fourier series can be used only for periodic functions, meaning that the function in the interval from 0 to L is repeated over and over again all the way out to infinity in both the positive and negative directions. Most of the functions we deal with in the real world are not periodic. Does this mean that Fourier series cannot be used in such cases? No, it does not. If we are interested in a portion of a nonperiodic function over a finite interval from 0 to L, we can take that portion and just repeat it to create a periodic function, as shown in Fig. 7.1. Then the Fourier series formulas given above will give

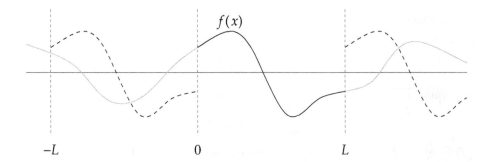

Figure 7.1: Creating a periodic function from a nonperiodic one. Most functions we deal with in physics are not periodic (solid curve). But we can calculate a Fourier series for a finite portion of such a function from $x = 0$ to $x = L$ by modifying the function as shown. We discard the parts of the function represented by the gray curves that fall outside the region of interest and replace them with an infinite set of periodic repetitions of the portion from 0 to L (dashed curves).

the correct value for the function in the interval from 0 to L (solid line in the figure). Outside of that interval they will give an incorrect answer, in the sense that they give the value of the repeated periodic function (dashed lines), not the original nonperiodic function (gray lines). But so long as we are interested only in the function in the interval from 0 to L this does not matter.

The coefficients γ_k in Eq. (7.5) in are, in general, complex. The standard way to calculate them is to evaluate the integral

$$\int_0^L f(x) \exp\left(-i\frac{2\pi k x}{L}\right) dx = \sum_{k'=-\infty}^{\infty} \gamma_{k'} \int_0^L \exp\left(i\frac{2\pi(k' - k)x}{L}\right) dx, \quad (7.7)$$

where we have substituted for $f(x)$ from Eq. (7.5). The integral on the right is straightforward. So long as $k' \neq k$ we have

$$\int_0^L \exp\left(i\frac{2\pi(k' - k)x}{L}\right) dx = \frac{L}{i2\pi(k' - k)}\left[\exp\left(i\frac{2\pi(k' - k)x}{L}\right)\right]_0^L$$

$$= \frac{L}{i2\pi(k' - k)}\left[e^{i2\pi(k' - k)} - 1\right]$$

$$= 0, \quad (7.8)$$

since $e^{i2\pi n} = 1$ for any integer n (and $k' - k$ is an integer).

The only exception to Eq. (7.8) is when $k' = k$, in which case the integral in (7.7) becomes simply $\int_0^L 1 \, dx = L$. Thus only one term in the sum in Eq. (7.7) is nonzero, the one where $k' = k$, giving

$$\int_0^L f(x) \exp\left(-i\frac{2\pi kx}{L}\right) dx = L\gamma_k, \tag{7.9}$$

or equivalently

$$\gamma_k = \frac{1}{L} \int_0^L f(x) \exp\left(-i\frac{2\pi kx}{L}\right) dx. \tag{7.10}$$

Thus given the function $f(x)$ we can find the Fourier coefficients γ_k, or given the coefficients we can find $f(x)$ from Eq. (7.5)—we can go back and forth freely between the function and the Fourier coefficients as we wish. Both, in a sense, are complete representations of the information contained in the function.

7.2 THE DISCRETE FOURIER TRANSFORM

For some functions $f(x)$ the integral in Eq. (7.10) can be performed analytically and the Fourier coefficients γ_k calculated exactly. There are, however, many cases where this is not possible: the integral may not be doable because $f(x)$ is too complicated, or $f(x)$ may not even be known in analytic form—it might be a signal measured in a laboratory experiment or the output of a computer program. In such cases we can instead calculate the Fourier coefficients numerically. We studied a number of techniques for performing integrals numerically in Chapter 5. Here we will use the trapezoidal rule of Section 5.1.1 to evaluate Eq. (7.10), with N slices of width $h = L/N$ each. Applying Eq. (5.3) we get

$$\gamma_k = \frac{1}{L}\frac{L}{N}\left[\tfrac{1}{2}f(0) + \tfrac{1}{2}f(L) + \sum_{n=1}^{N-1} f(x_n) \exp\left(-i\frac{2\pi kx_n}{L}\right)\right], \tag{7.11}$$

where the positions x_n of the sample points for the integral are

$$x_n = \frac{n}{N} L. \tag{7.12}$$

But since $f(x)$ is by hypothesis periodic we have $f(L) = f(0)$ and Eq. (7.11) simplifies to

$$\gamma_k = \frac{1}{N} \sum_{n=0}^{N-1} f(x_n) \exp\left(-i\frac{2\pi kx_n}{L}\right). \tag{7.13}$$

We can use this formula to evaluate the coefficients γ_k on the computer.

The formula has a convenient form for computational applications. It is a common occurrence that we know the value of a function $f(x)$ only at a set of equally spaced sample points x_n, exactly as in Eq. (7.13). For instance, the function might represent an audio signal—a sound wave—which would typically be sampled at regular intervals at a rate of a few thousand times a second. In that case Eq. (7.13) is perfectly suited to calculating the Fourier coefficients of the signal. A simpler way to write Eq. (7.13) in such situations is to define $y_n = f(x_n)$ to be the values of the N samples and make use of Eq. (7.12) to write

$$\gamma_k = \frac{1}{N} \sum_{n=0}^{N-1} y_n \exp\left(-i\frac{2\pi k n}{N}\right). \tag{7.14}$$

In this form, the equation doesn't require us to know the positions x_n of the sample points or even the width L of the interval in which they lie, since neither enters the formula. All we need to know are the sample values y_n and the total number of samples N.

The sum in Eq. (7.14) is a standard quantity that appears in many calculations. It is known as the *discrete Fourier transform* or DFT of the samples y_n and we will denote it c_k thus:

$$c_k = \sum_{n=0}^{N-1} y_n \exp\left(-i\frac{2\pi k n}{N}\right). \tag{7.15}$$

The quantities c_k and γ_k differ by only a factor of $1/N$ and we really don't need a separate symbol for c_k, but by convention the DFT is defined as in (7.15), without the factor $1/N$, and we will follow that convention here. A large part of this chapter is devoted to the discussion and study of the DFT, defined as in Eq. (7.15). In this discussion we will refer to the c_k (a little loosely) as "Fourier coefficients," although strictly speaking the true Fourier coefficients are $\gamma_k = c_k/N$. Clearly, however, it's trivial to calculate the γ_k once we know the c_k (and in fact, as we will see, it is hardly ever necessary).

We have derived the results above using the trapezoidal rule, which only gives an approximation to the integral of Eq. (7.10). As we now show, however, the discrete Fourier transform is, in a certain sense, *exact*. It allows us to perform exact Fourier transforms of sampled data, even though computers can't normally do integrals exactly. To demonstrate this remarkable fact, we employ the following mathematical result about exponentials. Recall the standard geometric series: $\sum_{k=0}^{N-1} a^k = (1-a^N)/(1-a)$, and put $a = e^{i2\pi m/N}$ with

m integer, to get

$$\sum_{k=0}^{N-1} e^{i2\pi km/N} = \frac{1 - e^{i2\pi m}}{1 - e^{i2\pi m/N}}. \tag{7.16}$$

Since $e^{i2\pi m} = 1$ for all integers m, the numerator vanishes and this just gives zero. The only exception is when $m = 0$ or a multiple of N, in which case the denominator also vanishes and we must be careful lest we divide by zero. In this case, however, the original sum is easy: it's just $\sum_{k=0}^{N-1} 1 = N$. Thus

$$\sum_{k=0}^{N-1} e^{i2\pi km/N} = \begin{cases} N & \text{if } m \text{ is zero or a multiple of } N, \\ 0 & \text{otherwise.} \end{cases} \tag{7.17}$$

Now let's go back to our discrete Fourier transform, Eq. (7.15), and consider the following sum:

$$\sum_{k=0}^{N-1} c_k \exp\left(i\frac{2\pi kn}{N}\right) = \sum_{k=0}^{N-1} \sum_{n'=0}^{N-1} y_{n'} \exp\left(-i\frac{2\pi kn'}{N}\right) \exp\left(i\frac{2\pi kn}{N}\right)$$
$$= \sum_{n'=0}^{N-1} y_{n'} \sum_{k=0}^{N-1} \exp\left(i\frac{2\pi k(n - n')}{N}\right), \tag{7.18}$$

where we have swapped the order of summation in the second line. Let's assume that n lies in the range $0 \leq n < N$. The final sum in Eq. (7.18) takes the form of Eq. (7.17) with $m = n - n'$, which means we can work out its value. Since n and n' are both less than N there is no way for $n - n'$ to be a nonzero multiple of N, but it could be zero if $n = n'$. Thus the sum is equal to N when $n = n'$, and zero otherwise. But that means there is only one nonzero term in the sum over n'—the one where $n' = n$—and so Eq. (7.18) simplifies to just

$$\sum_{k=0}^{N-1} c_k \exp\left(i\frac{2\pi kn}{N}\right) = Ny_n, \tag{7.19}$$

or equivalently,

$$y_n = \frac{1}{N} \sum_{k=0}^{N-1} c_k \exp\left(i\frac{2\pi kn}{N}\right). \tag{7.20}$$

This result is called the *inverse discrete Fourier transform* (or inverse DFT). It is the counterpart to the forward transform of Eq. (7.15). It tells us that, given the coefficients c_k we get from Eq. (7.15), we can recover the values of the samples y_n that they came from *exactly* (except for rounding error). This is an amazing result: even though we thought our Fourier coefficients were only approximate, it turns out that they are actually exact in the sense that we can completely recover the original samples from them.

Thus we can move freely back and forth between the samples and the coefficients c_k without losing any detail in our data—both the samples and the Fourier transform give us a complete representation of the original data. Notice that we only need the Fourier coefficients c_k up to $k = N - 1$ to recover the samples, so we only need to evaluate (7.15) for $0 \leq k < N$.

Equation (7.20) is similar, but not identical, to our original expression for the complex Fourier series, Eq. (7.5). It differs by a leading factor of $1/N$ (which compensates for the factor we removed when we defined the DFT in Eq. (7.15)), and in that the sum is over positive values of k only and runs up to $N - 1$, rather than to infinity (which is a useful feature, since it makes the sum practical on a computer where a sum to infinity would not be).

It is, however, important to appreciate that, unlike the original Fourier series, the discrete formula of Eq. (7.20) only gives us the sample values $y_n = f(x_n)$. It tells us nothing about the value of $f(x)$ between the sample points. And indeed how could it? Given that we used only the values at the sample points when we computed the c_k in Eq. (7.15), it's clear that the c_k cannot contain any information about the values in between. The original function could do anything it wanted between the sample points and we'd never know, since we didn't measure it there. To put that another way, any two functions that have the same values at the sample points will have the same DFT, no matter what they do between the points—like these two functions, for example:

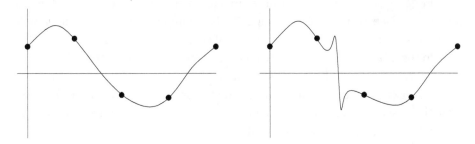

These would have the same DFT even though they are totally different between the second and third sample points.

Still, if a function is reasonably smooth, with no wild excursions between samples (like the one in the right-hand example above), then knowing the values at the sample points only is enough to get a picture of the function's general shape. And, as we have said, in many cases we are interested in a function that is represented as a set of samples in the first place and not as a continuous function, and for this kind of data the DFT is an excellent tool.

All of the results above apply whether $f(x)$ is a real function or a complex one—the DFT works equally well for either. In most practical situations, however, we're interested in real functions, in which case there are some further simplifications we can do.

Suppose all the y_n are real and consider the value of c_k for some k that is less than N but greater than $\frac{1}{2}N$, which we'll write as $k = N - r$ with $1 \le r < \frac{1}{2}N$. Then

$$c_{N-r} = \sum_{n=0}^{N-1} y_n \exp\left(-i\frac{2\pi(N-r)n}{N}\right) = \sum_{n=0}^{N-1} y_n \exp(-i2\pi n) \exp\left(i\frac{2\pi rn}{N}\right)$$

$$= \sum_{n=0}^{N-1} y_n \exp\left(i\frac{2\pi rn}{N}\right) = c_r^* \tag{7.21}$$

where c_r^* denotes the complex conjugate of c_r and we have made use of the fact that $e^{-i2\pi n} = 1$ for all integer n and that the y_n are real.

Thus for instance $c_{N-1} = c_1^*$, and $c_{N-2} = c_2^*$, and so forth. This means when we are calculating the Fourier transform of a real function we only have to calculate the coefficients c_k for $0 \le k \le \frac{1}{2}N$. The other half of the coefficients are just the complex conjugates of the first half. Thus if N is even we only have to calculate $\frac{1}{2}N + 1$ coefficients; if N is odd we have to calculate $\frac{1}{2}(N+1)$ coefficients. (It is useful to note that both these expressions can be written in Python as "N//2+1", where "//" is the integer division operator, which divides two integers one by the other and returns an integer result, rounding down if necessary—see Section 2.2.4 on page 23. For both even and odd values of N this expression gives the correct number of coefficients in the DFT.)

On the other hand, if the y_n are complex then (7.21) does not apply and we need to calculate all N Fourier coefficients.

The discrete Fourier transform is straightforward to calculate in Python. Here is a user-defined function that evaluates Eq. (7.15) for N real samples y_n for all k in the range $0 \le k \le \frac{1}{2}N$:

File: dft.py

```python
from numpy import zeros
from cmath import exp,pi

def dft(y):
    N = len(y)
    c = zeros(N//2+1,complex)
    for k in range(N//2+1):
        for n in range(N):
            c[k] += y[n]*exp(-2j*pi*k*n/N)
    return c
```

And we could write a similar function to perform the inverse Fourier transform, Eq. (7.20). Notice the use of the function exp from the cmath package—the version from cmath differs from the exp function in the math package in that it can handle complex numbers.

These lines of code represent a very direct translation of Eq. (7.15) into Python. They are not, however, the quickest way to calculate the discrete Fourier transform. There is another, indirect, way of performing the same calculation, known (prosaically) as the "fast Fourier transform," which is significantly quicker. The code above will work fine for our present purposes, but for serious, large-scale physics calculations the fast Fourier transform is much better. We will study the fast Fourier transform in Section 7.4.

7.2.1 POSITIONS OF THE SAMPLE POINTS

One thing to notice about the discrete Fourier transform, Eq. (7.13), is that we can shift the sample points along the x-axis if we want to and not much changes. Suppose that instead of using sample points $x_n = (n/N)L$ as in Eq. (7.12), we take our samples at a shifted set of points

$$x'_n = x_n + \Delta = \frac{n}{N}L + \Delta. \tag{7.22}$$

Following through the derivation leading to Eqs. (7.13) and (7.15) again, we find that the equivalent discrete Fourier transform for this set of samples is

$$
\begin{aligned}
c_k &= \sum_{n=0}^{N-1} f(x_n + \Delta) \exp\left(-i\frac{2\pi k(x_n + \Delta)}{L}\right) \\
&= \exp\left(-i\frac{2\pi k\Delta}{L}\right) \sum_{n=0}^{N-1} f(x'_n) \exp\left(-i\frac{2\pi k x_n}{L}\right) \\
&= \exp\left(-i\frac{2\pi k\Delta}{L}\right) \sum_{n=0}^{N-1} y'_n \exp\left(-i\frac{2\pi k n}{N}\right),
\end{aligned}
\tag{7.23}
$$

where $y'_n = f(x'_n)$ are the new samples. But this is just the same as the original DFT, Eq. (7.15), except that we have an extra (k-dependent) phase factor at the front. Conventionally, we absorb this phase factor into the definition of c_k and define a new coefficient $c'_k = e^{i2\pi k\Delta/L} c_k$ so that

$$c'_k = \sum_{n=0}^{N-1} y'_n \exp\left(-i\frac{2\pi k n}{N}\right), \tag{7.24}$$

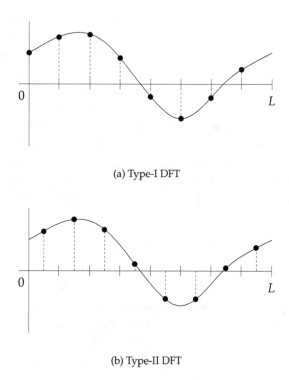

(a) Type-I DFT

(b) Type-II DFT

Figure 7.2: Sample positions for Type-I and Type-II DFTs. (a) In a Type-I discrete Fourier transform the interval from 0 to L is divided into N slices and one sample is taken at the beginning of each slice. (b) In a Type-II transform the samples are taken in the middle of the slices.

which is now in exactly the same form as before. Thus the DFT is essentially independent of where we choose to place the samples. The coefficients change by a phase factor, but that's all.

Figure 7.2a shows the choice of samples we've been using so far. The interval from 0 to L is divided into N slices and we take one sample from the beginning of each slice. There is no sample at the end of the last slice, since it would be the same as the first sample, because the function is periodic. A common alternative choice is to take samples in the middle of the slices as shown in Fig. 7.2b. Both of these schemes are widely used. The first is called a Type-I DFT; the second is called Type-II. Note that, although the formula we use to perform the DFT is the same in both cases, the coefficients we will get are different for two reasons: first, the values of the samples are different because they are measured at different points, and second the coefficients include an extra phase factor, as we've seen, in the Type-II case.

7.2.2 TWO-DIMENSIONAL FOURIER TRANSFORMS

Functions of two variables $f(x, y)$ can also be Fourier transformed, using a *two-dimensional Fourier transform*, which simply means you transform with respect to one variable then with respect to the other.

Suppose we have an $M \times N$ grid of samples y_{mn}. To carry out the two-dimensional Fourier transform, we first perform an ordinary Fourier transform on each of the M rows, following Eq. (7.15):

$$c'_{ml} = \sum_{n=0}^{N-1} y_{mn} \exp\left(-i\frac{2\pi ln}{N}\right). \tag{7.25}$$

For each row m we now have N coefficients, one for each value of l. Next we take the lth coefficient in each of the M rows and Fourier transform these M values again to get

$$c_{kl} = \sum_{m=0}^{M-1} c'_{ml} \exp\left(-i\frac{2\pi km}{M}\right). \tag{7.26}$$

Alternatively, we can substitute Eq. (7.25) into Eq. (7.26) and write a single expression for the complete Fourier transform in two dimensions:

$$c_{kl} = \sum_{m=0}^{M-1} \sum_{n=0}^{N-1} y_{mn} \exp\left[-i2\pi\left(\frac{km}{M} + \frac{ln}{N}\right)\right]. \tag{7.27}$$

The corresponding inverse transform is

$$y_{mn} = \frac{1}{MN} \sum_{k=0}^{M-1} \sum_{l=0}^{N-1} c_{kl} \exp\left[i2\pi\left(\frac{km}{M} + \frac{ln}{N}\right)\right]. \tag{7.28}$$

If the samples y_{mn} are real—as they almost always are—then there is a further point to notice. When we do the first set of Fourier transforms, Eq. (7.25), we are transforming a row of N real numbers for each value of l and hence, as discussed in Section 7.2, we end up with either $\frac{1}{2}N + 1$ independent Fourier coefficients or $\frac{1}{2}(N + 1)$, depending on whether N is even or odd—the remaining coefficients are just complex conjugates. As we have seen, however, the coefficients themselves will, in general, be complex, which means that when we perform the second set of transforms in Eq. (7.26) we are transforming M

complex numbers, not real numbers. This means that we now have to calculate all M of the Fourier coefficients—it is no longer the case that the second half are the complex conjugates of the first half. Thus the two-dimensional Fourier transform of an $M \times N$ grid of real numbers is a grid of complex numbers with $M \times (\frac{1}{2}N + 1)$ independent coefficients if N is even or $M \times \frac{1}{2}(N + 1)$ if N is odd.

Two-dimensional transforms are used, for example, in image processing, and are widely employed in astronomy to analyze photographs of the sky and reveal features that are otherwise hard to make out. They are also used in the confocal microscope and the electron microscope, two instruments that find use in many branches of science. Exercise 7.9 on page 322 invites you to try your hand at image processing using a two-dimensional Fourier transform.

7.2.3 PHYSICAL INTERPRETATION OF THE FOURIER TRANSFORM

If we were mathematicians, then the equations for the Fourier transform given in the previous sections would be all we need—they tell us exactly how the transform is defined. But for physicists it's useful to understand what the Fourier transform is telling us in physical terms.

The Fourier transform breaks a function down into a set of real or complex sinusoidal waves. Each term in a sum like Eq. (7.20) represents one wave with its own well-defined frequency. If the function $f(x)$ is a function in space then we have spatial frequencies; if it's a function in time then we have temporal frequencies, like musical notes. Saying that any function can be expressed as a Fourier transform is equivalent to saying that any function can be represented as a sum of waves of given frequencies, and the Fourier transform tells us what that sum is for any particular function: the coefficients of the transform tell us exactly how much of each frequency we have in the sum.

Thus, by looking at the output of our Fourier transform, we can get a picture of what the frequency breakdown of a signal is. Most of us are familiar with the "signal analyzers" that are built in to many home stereo systems—the animated bar charts that go up and down in time with the music. These analyzers present a graph of the frequencies present in the music and the Fourier transform conveys essentially the same information for the function $f(x)$. (Indeed, signal analyzers work precisely by performing a Fourier transform and then displaying the result.)

Thus, for example, consider the signal shown in Fig. 7.3. As we can see, the signal consists of a basic wave that goes up and down with a well-defined

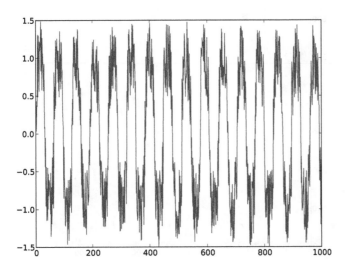

Figure 7.3: An example signal. This example of a signal has an overall wavelike shape with a well-defined frequency, but also contains some noise, or randomness.

frequency, but there is also some noise in the data as well, visible as smaller wiggles in the line. If one were to listen to this signal as sound one would hear a constant note at the frequency of the main wave, accompanied by a background hiss that comes from the noise.

Let us calculate the Fourier transform of this signal. If the signal is stored as a single column of numbers in a file called pitch.txt, we could calculate the transform using the function dft that we defined on page 296:

```
from numpy import loadtxt
from pylab import plot,xlim,show

y = loadtxt("pitch.txt",float)
c = dft(y)
plot(abs(c))
xlim(0,500)
show()
```

File: dft.py

Since the coefficients returned by the transform in the array c are in general complex, we have plotted their absolute values, which give us a measure of

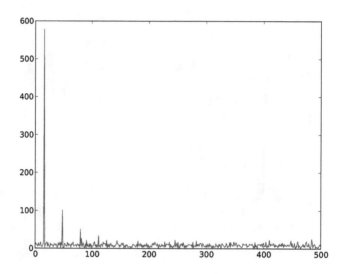

Figure 7.4: Fourier transform of Fig. 7.3. The Fourier transform of the signal shown in Fig. 7.3, as calculated by the program given in the text.

the amplitude of each of the waves in the Fourier series.

The graph produced by the program is shown in Fig. 7.4. The horizontal axis in this graph measures k, which is proportional to the frequency of the waves, and the vertical axis measures the absolute values $|c_k|$ of the corresponding Fourier coefficients. As we can see, there are a number of noticeable spikes in the plot, representing coefficients c_k with particularly large magnitudes. The first and largest of these spikes corresponds to the frequency of the main wave visible in Fig. 7.3. The remaining spikes are harmonics of the first one—multiples of its frequency whose presence tell us that the wave in the original data was not a pure sine wave. A pure sine wave could be represented fully with just a single term of the appropriate frequency in the Fourier series, but any other wave cannot and requires some additional terms to represent it.

Between the main spikes in Fig. 7.4 there are also some small, apparently random values of $|c_k|$ creating a low-level background visible as the jagged line along the bottom of the plot. These are produced by the noise in the original signal, which is "white noise," meaning it is completely random and contains on average equal amounts of all frequencies, so in a Fourier transform it appears as a uniform random background, as in the figure, with neither the high frequencies nor the low having larger Fourier coefficients.[2]

Fourier transforms have many uses in physics, but one of the most basic is as a simple tool—as here—for understanding a measurement or signal. The Fourier transform can break a signal down for us into its component parts and give us an alternative way to view it, a "spectrum analyzer" view of our data as a sum of different frequencies.

Exercise 7.1: Fourier transforms of simple functions

Write Python programs to calculate the coefficients in the discrete Fourier transforms of the following periodic functions sampled at $N = 1000$ evenly spaced points, and make plots of their amplitudes similar to the plot shown in Fig. 7.4:

a) A single cycle of a square-wave with amplitude 1
b) The sawtooth wave $y_n = n$
c) The modulated sine wave $y_n = \sin(\pi n / N) \sin(20\pi n / N)$

If you wish you can use the Fourier transform function from the file dft.py as a starting point for your program.

Exercise 7.2: Detecting periodicity

In the on-line resources there is a file called sunspots.txt, which contains the observed number of sunspots on the Sun for each month since January 1749. The file contains two columns of numbers, the first representing the month and the second being the sunspot number.

a) Write a program that reads the data in the file and makes a graph of sunspots as a function of time. You should see that the number of sunspots has fluctuated on a regular cycle for as long as observations have been recorded. Make an estimate of the length of the cycle in months.

b) Modify your program to calculate the Fourier transform of the sunspot data and then make a graph of the magnitude squared $|c_k|^2$ of the Fourier coefficients as a function of k—also called the *power spectrum* of the sunspot signal. You should see that there is a noticeable peak in the power spectrum at a nonzero value of k. The appearance of this peak tells us that there is one frequency in the Fourier series that has a higher amplitude than the others around it—meaning that there is a large sine-wave term with this frequency, which corresponds to the periodic wave you can see in the original data.

---

[2]White noise is so-called by analogy with white light, which is a mixture of all frequencies. You may occasionally see references to "pink noise" as well, which means noise with all frequencies present, so it's sort of white, but more of the low frequencies than the high, which by the same analogy would make it red. The mixture of red and white then gives pink.

c) Find the approximate value of k to which the peak corresponds. What is the period of the sine wave with this value of k? You should find that the period corresponds roughly to the length of the cycle that you estimated in part (a).

This kind of Fourier analysis is a sensitive method for detecting periodicity in signals. Even in cases where it is not clear to the eye that there is a periodic component to a signal, it may still be possible to find one using a Fourier transform.

7.3 DISCRETE COSINE AND SINE TRANSFORMS

So far we have been looking at the complex version of the Fourier series but, as mentioned in Section 7.1, one can also construct Fourier series that use sine and cosine functions in place of complex exponentials, and there are versions of the discrete Fourier transform for these sine and cosine series also. These series have their own distinct properties that make them (and particularly the cosine series) useful in certain applications, as we'll see.

Consider, for example, the cosine series of Eq. (7.1). Not every function can be represented using a series of this kind. Because of the shape of the cosine, all functions of the form (7.1) are necessarily symmetric about the midpoint of the interval at $\frac{1}{2}L$, and hence the cosine series can only be used to represent such symmetric functions. This may at first appear to be a grave disadvantage, since most functions we come across in physics are not symmetric in this way. In fact, however, it is not such a problem as it appears. Just as we can turn any function in a finite interval into a periodic function by simply repeating it endlessly, so we can turn any function into a *symmetric* periodic function by adding to it a mirror image of itself and then repeating the whole thing endlessly. The process is illustrated in Fig. 7.5. Thus cosine series can be used for essentially any function, regardless of symmetry. In practice, this is how the cosine version of the Fourier transform is always used. Given a set of samples of a function, we mirror those samples to create a symmetric function before transforming them. As a corollary, note that this implies that the total number of samples in the transform is always even, regardless of how many samples we started with, a fact that will be useful in a moment.

Once we have such a symmetric function the cosine series representing it can be found from the results we already know: it is simply the special case of Eq. (7.15) when the samples y_n are symmetric about $x = \frac{1}{2}L$. When the samples are symmetric we have $y_0 = y_N, y_1 = y_{N-1}, y_2 = y_{N-2}$, and so forth, i.e., $y_n = y_{N-n}$ for all n. Given that, as mentioned above, N is always even,

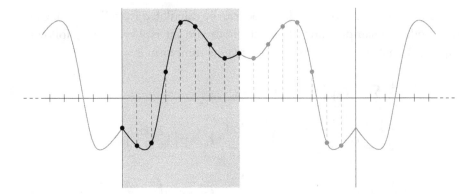

Figure 7.5: Turning a nonsymmetric function into a symmetric one. Any nonsymmetric function on a finite interval (shaded region) can be turned into a symmetric periodic function on an interval of twice the width by adding to it a mirror image of itself, then repeating the whole construction endlessly to the left and right. For functions represented as a set of samples (dots in the figure), we make a mirror-image of the samples and add them to the original set.

Eq. (7.15) then becomes

$$c_k = \sum_{n=0}^{N-1} y_n \exp\left(-i\frac{2\pi kn}{N}\right)$$

$$= \sum_{n=0}^{\frac{1}{2}N} y_n \exp\left(-i\frac{2\pi kn}{N}\right) + \sum_{n=\frac{1}{2}N+1}^{N-1} y_n \exp\left(-i\frac{2\pi kn}{N}\right)$$

$$= \sum_{n=0}^{\frac{1}{2}N} y_n \exp\left(-i\frac{2\pi kn}{N}\right) + \sum_{n=\frac{1}{2}N+1}^{N-1} y_{N-n} \exp\left(i\frac{2\pi k(N-n)}{N}\right), \qquad (7.29)$$

where we have made use of the fact that $e^{i2\pi k} = 1$ for all integer k. Now we make a change of variables $N - n \rightarrow n$ in the second sum and get

$$c_k = \sum_{n=0}^{\frac{1}{2}N} y_n \exp\left(-i\frac{2\pi kn}{N}\right) + \sum_{n=1}^{\frac{1}{2}N-1} y_n \exp\left(i\frac{2\pi kn}{N}\right)$$

$$= y_0 + y_{N/2} \cos\left(\frac{2\pi k(N/2)}{N}\right) + 2\sum_{n=1}^{\frac{1}{2}N-1} y_n \cos\left(\frac{2\pi kn}{N}\right), \qquad (7.30)$$

where we have used $\cos\theta = \frac{1}{2}(e^{-i\theta} + e^{i\theta})$. Normally the cosine transform is applied to real samples, which means that the coefficients c_k will all be real (since they are just a sum of real terms).

The inverse transform for Eq. (7.30) can be calculated from Eq. (7.20). Given that both the samples and the coefficients are all real, we have, following Eq. (7.21), that $c_{N-r} = c_r^* = c_r$, so the inverse transform is

$$
\begin{aligned}
y_n &= \frac{1}{N} \sum_{k=0}^{N-1} c_k \exp\left(i\frac{2\pi kn}{N}\right) \\
&= \frac{1}{N}\left[\sum_{k=0}^{\frac{1}{2}N} c_k \exp\left(i\frac{2\pi kn}{N}\right) + \sum_{k=\frac{1}{2}N+1}^{N-1} c_k \exp\left(i\frac{2\pi kn}{N}\right)\right] \\
&= \frac{1}{N}\left[\sum_{k=0}^{\frac{1}{2}N} c_k \exp\left(i\frac{2\pi kn}{N}\right) + \sum_{k=\frac{1}{2}N+1}^{N-1} c_{N-k} \exp\left(-i\frac{2\pi(N-k)n}{N}\right)\right] \\
&= \frac{1}{N}\left[\sum_{k=0}^{\frac{1}{2}N} c_k \exp\left(i\frac{2\pi kn}{N}\right) + \sum_{k=1}^{\frac{1}{2}N-1} c_k \exp\left(-i\frac{2\pi kn}{N}\right)\right] \\
&= \frac{1}{N}\left[c_0 + c_{N/2}\cos\left(\frac{2\pi n(N/2)}{N}\right) + 2\sum_{k=1}^{\frac{1}{2}N-1} c_k \cos\left(\frac{2\pi kn}{N}\right)\right]. \quad (7.31)
\end{aligned}
$$

Equations (7.30) and (7.31) are called the *discrete cosine transform* (or DCT) and the *inverse discrete cosine transform* (or inverse DCT) respectively. Note how the indices of both samples and Fourier coefficients run from 0 to $\frac{1}{2}N$, so that there are $\frac{1}{2}N+1$ of each in total: the DCT takes $\frac{1}{2}N+1$ real numbers and transforms them into $\frac{1}{2}N+1$ real coefficients (and the inverse DCT does the reverse). Indeed, the forward and reverse transforms in this case are actually the same mathematical expression, except for the leading factor of $1/N$ in Eq. (7.31). Thus, one sometimes says that this transform is its own inverse (except for the factor $1/N$).

There is another slightly different form of discrete cosine transform for the "Type-II" case, where the sample points are in the middle of the sample intervals rather than at their ends (see Section 7.2.1 and Fig. 7.2). In that case symmetry about the midpoint of the interval implies that $y_n = y_{N-1-n}$ and, making use once more of the fact that N is even, Eq. (7.15) becomes

$$
c_k = \sum_{n=0}^{\frac{1}{2}N-1} y_n \exp\left(-i\frac{2\pi kn}{N}\right) + \sum_{n=\frac{1}{2}N}^{N-1} y_n \exp\left(-i\frac{2\pi kn}{N}\right)
$$

$$
= \exp\left(i\frac{\pi k}{N}\right)\left[\sum_{n=0}^{\frac{1}{2}N-1} y_n \exp\left(-i\frac{2\pi k(n+\frac{1}{2})}{N}\right)\right.
$$

$$
\left. + \sum_{n=\frac{1}{2}N}^{N-1} y_{N-1-n} \exp\left(i\frac{2\pi k(N-\frac{1}{2}-n)}{N}\right)\right]
$$

$$
= \exp\left(i\frac{\pi k}{N}\right)\left[\sum_{n=0}^{\frac{1}{2}N-1} y_n \exp\left(-i\frac{2\pi k(n+\frac{1}{2})}{N}\right)\right.
$$

$$
\left. + \sum_{n=0}^{\frac{1}{2}N-1} y_n \exp\left(i\frac{2\pi k(n+\frac{1}{2})}{N}\right)\right]
$$

$$
= 2\exp\left(i\frac{\pi k}{N}\right)\sum_{n=0}^{\frac{1}{2}N-1} y_n \cos\left(\frac{2\pi k(n+\frac{1}{2})}{N}\right). \tag{7.32}
$$

Conventionally we absorb the leading phase factor into the Fourier coefficients, defining

$$
a_k = 2\sum_{n=0}^{\frac{1}{2}N-1} y_n \cos\left(\frac{2\pi k(n+\frac{1}{2})}{N}\right). \tag{7.33}
$$

Again these coefficients are purely real if the y_n are real and it's straightforward to show that the corresponding inverse transform is

$$
y_n = \frac{1}{N}\left[a_0 + 2\sum_{k=1}^{\frac{1}{2}N-1} a_k \cos\left(\frac{2\pi k(n+\frac{1}{2})}{N}\right)\right]. \tag{7.34}
$$

This Type-II DCT is arguably more elegant than the Type-I version, and it's probably the version that's most often used. Notice that it transforms $\frac{1}{2}N$ inputs y_n into $\frac{1}{2}N$ outputs a_k. For this reason, it's common to redefine $\frac{1}{2}N \to N$ and $a_k \to 2a_k$ and rewrite Eqs. (7.33) and (7.34) as

$$
a_k = \sum_{n=0}^{N-1} y_n \cos\left(\frac{\pi k(n+\frac{1}{2})}{N}\right), \qquad y_n = \frac{1}{N}\left[a_0 + 2\sum_{k=1}^{N-1} a_k \cos\left(\frac{\pi k(n+\frac{1}{2})}{N}\right)\right],
$$
$$
\tag{7.35}
$$

where N is now the actual number of inputs to the transform. This is probably the form you'll see most often in books and elsewhere. Sometimes you'll see it referred to as "the" discrete cosine transform, because it's so common.

A nice feature of the cosine transform is that, unlike the DFT, it does not assume that the samples themselves are periodic. Since the function being transformed is first mirrored, as in Fig. 7.5, the first and last of the original samples are not obliged to take the same value, as they are in the DFT. As described earlier in the chapter, any function, periodic or not, can be made periodic by repeating it endlessly, thereby making the last sample take the same value as the first, but in so doing one may create a substantial discontinuity in the function, as shown in Fig. 7.1, and such discontinuities can cause problems for DFTs. The discrete cosine transform, by contrast, does not suffer from these problems, and hence is often preferable for data that are not inherently periodic. Exercise 7.6 below invites you to investigate this point further.

One can also calculate discrete *sine* transforms, as in Eq. (7.2), although sine transforms are used less often than cosine transforms because they force the function $f(x)$ to be zero at either end of its range. Relatively few functions encountered in real-world applications do this, so the sine transform has limited applicability. It does find some use in physics, however, for representing physical functions whose boundary conditions force them to be zero at the ends of an interval. Examples are displacement of a vibrating string that is clamped at both ends or quantum wavefunctions in a closed box. We will see some examples of the use of sine series in systems like these in Chapter 9.

7.3.1 TECHNOLOGICAL APPLICATIONS OF COSINE TRANSFORMS

Discrete cosine transforms have important technological uses. For example, they form the mathematical basis for the computer image file format called *JPEG*,[3] which is used to store most of the images you see on the world wide web. Digital images are represented as regular grids of dots, or pixels, of different shades, and the shades are stored on the computer as ordinary numbers. The simplest way of storing an image is just to store all of these numbers, in order, in a computer file, but there are a lot of them and the resulting files tend to be very large. It turns out that one can store images far more economically if one makes use of discrete cosine transforms.

The JPEG format works by dividing the pixels in an image into blocks, performing DCTs on the blocks (two-dimensional Type-II DCTs to be exact), then

---

[3]The format is named after the committee that invented it, the Joint Photographic Experts Group.

looking for coefficients a_k that are small and can be discarded.[4] The remaining coefficients are stored in a file and, when you view a picture or request the relevant web page, your computer reconstitutes the picture using the inverse transform of Eq. (7.35). So there are discrete cosine transforms going on in the background pretty much every time you look at the web. The advantage of storing images this way is that in many cases *most* of the a_k are very small and can be neglected. Because only the small remaining fraction of the Fourier coefficients need to be stored, the size of the file required to store the whole picture is thus greatly reduced, and if the picture is transmitted over the Internet, for instance as part of a web page, we save on the time needed to transmit it. The disadvantage is that, because some of the Fourier data are thrown away, the picture you get back on your screen isn't quite the same picture you started with. Usually your eye can't tell the difference, but sometimes if you look carefully you can see problems in images, called *compression artifacts*, arising from the missing data.

A variant of the same technique is also used to compress moving pictures, meaning film and video, using the compression format called *MPEG*. Television broadcasts, cable and satellite TV, DVDs, and Internet video all use versions of the MPEG format to compress video images so that they can be stored and transmitted more efficiently. Again, you lose something in the compression. It's usually not very noticeable, but if you know what you're looking for you can sometimes see compression artifacts in the pictures.

A similar scheme is used to compress music, for instance in the popular file format called *MP3*. Music, or any audio signal, can be represented digitally as a set of equally spaced samples of an audio waveform. In an MP3 these samples are divided into blocks, the blocks are individually transformed using a discrete cosine transform, and then some Fourier components are discarded, saving space and allowing one to store and transmit audio recordings more efficiently. When you play an MP3, your computer, phone, or MP3 player is continually performing inverse discrete cosine transforms to reconstruct the music so you can listen to it. There is a lot of computation going on in an MP3 player.

MP3s are clever, however, in that the particular Fourier components they discard are not chosen solely on the grounds of which ones are smallest, but also with a knowledge of what the human ear can and cannot hear. For in-

---

[4]Technically the components are not thrown out altogether, but are represented with coarser dynamic range.

stance, if there are loud low-frequency sounds in a piece of music, such as bass and drums, then the ear is much less sensitive to high-frequency sounds that occur at the same time. (This is a limitation of the way the ear works, not a physical law.) So in this situation it's safe to throw out some high-frequency Fourier components and save space. Thus a piece of music stored in MP3 format is not a faithful representation of the original, but it *sounds* like one to the human ear because the aspects in which it differs from the original are aspects the human ear can't hear.

Essentially the entire digital music/audio economy—streaming music services, music downloads, Internet radio, digital TV, and all the rest of it—relies on this technology.[5] Without the discrete cosine transform it wouldn't be possible.

7.4 FAST FOURIER TRANSFORMS

The discrete Fourier transform is defined by Eq. (7.15), which is repeated here for convenience:

$$c_k = \sum_{n=0}^{N-1} y_n \exp\left(-i\frac{2\pi kn}{N}\right). \tag{7.36}$$

We gave an example of a Python function to evaluate this expression on page 296. In that function we used a for-loop to add up the terms in the sum, of which there are N, and whole the calculation is repeated for each of the $\frac{1}{2}N+1$ distinct coefficients c_k, so the total number of terms we must add together to evaluate all of the coefficients is $N(\frac{1}{2}N+1)$. Thus the computer will have to perform a little over $\frac{1}{2}N^2$ arithmetic operations to evaluate the complete DFT. In Section 4.3 we discussed a useful rule of thumb: the largest number of operations you can do in a computer program is about a billion if you want it to run in a reasonable amount of time. If we apply this rule of thumb to our Fourier transform, setting $\frac{1}{2}N^2 = 10^9$ and solving for N, we find that the largest set of samples for which we can calculate the Fourier transform in reasonable time

---

[5]The notable exception is the CD, which was introduced about ten years before the invention of MP3s. CDs use uncompressed audio—just raw sound samples—which is extremely inefficient, although in theory it also gives higher sound quality because there are no compression artifacts. Good luck hearing the difference though; these days audio compression is extremely good and it's rare that one can actually hear compression artifacts. Double-blind listening tests have been conducted where listeners are played a pristine, uncompressed recording of a piece of music, followed either by a compressed version of the same recording or by the uncompressed version again. When the best modern compression techniques are used, even expert listeners have been unable to reliably tell the difference between the two.

has about $N \simeq 45\,000$ samples.[6] For practical applications this is not a very large number. For example, $45\,000$ samples is only about enough to represent one second of music or other audio signals. Often we would like to calculate the transforms of larger data sets, sometimes much larger. Luckily it turns out that there is a clever trick for calculating the DFT much faster than is possible than by directly evaluating Eq. (7.36). This trick is called the *fast Fourier transform* or FFT. It was discovered by Carl Friedrich Gauss in 1805, when he was 28 years old.[7,8]

The fast Fourier transform is simplest when the number of samples is a power of two, so let us consider the case where $N = 2^m$ with m an integer. Consider the sum in the DFT equation, Eq. (7.36), and let us divide the terms into two equally sized groups, which we can always do when N is a power of two. Let the first group consist of the terms with n even and the second the terms with n odd. We consider the even terms first, i.e., the terms where $n = 2r$ with integer $r = 0 \dots \frac{1}{2}N - 1$. The sum of these even terms is

$$E_k = \sum_{r=0}^{\frac{1}{2}N-1} y_{2r} \exp\left(-i\frac{2\pi k(2r)}{N}\right) = \sum_{r=0}^{\frac{1}{2}N-1} y_{2r} \exp\left(-i\frac{2\pi kr}{\frac{1}{2}N}\right). \tag{7.37}$$

But this is simply another Fourier transform, just like Eq. (7.36), but with $\frac{1}{2}N$ samples instead of N. Similarly the odd terms, meaning those with $n = 2r+1$, sum to

$$\sum_{r=0}^{\frac{1}{2}N-1} y_{2r+1} \exp\left(-i\frac{2\pi k(2r+1)}{N}\right) = e^{-i2\pi k/N} \sum_{r=0}^{\frac{1}{2}N-1} y_{2r+1} \exp\left(-i\frac{2\pi kr}{\frac{1}{2}N}\right)$$
$$= e^{-i2\pi k/N} O_k, \tag{7.38}$$

---

[6]We are treating the evaluation of each term in the sum as a single operation, which is not really correct—each term involves several multiplication operations and the calculation of an exponential. However, the billion-operation rule of thumb is only an approximation anyway, so a rough estimate of the number of operations is good enough.

[7]What with Gaussian quadrature, Gaussian elimination, the Gauss–Newton method, and now the fast Fourier transform, Gauss seems to have discovered most of computational physics more than a century before the computer was invented.

[8]In much of the computational physics literature you will see the fast Fourier transform attributed not to Gauss but to the computer scientists James Cooley and John Tukey, who published a paper describing it in 1965 [Cooley, J. W. and Tukey, J. W., An algorithm for the machine calculation of complex Fourier series, *Mathematics of Computation* **19**, 297–301 (1965)]. Although Cooley and Tukey's paper was influential in popularizing the method, however, it was, unbeknownst to its authors, not the first description of the FFT. Only later did they learn that they'd been scooped by Gauss 160 years earlier.

where O_k is another Fourier transform with $\frac{1}{2}N$ samples.

The complete Fourier coefficient c_k of Eq. (7.36) is the sum of its odd and even terms:

$$c_k = E_k + e^{-i2\pi k/N} O_k. \tag{7.39}$$

In other words, the coefficient c_k in the discrete Fourier transform of $f(x)$ is just the sum of two terms E_k and O_k that are each, themselves, discrete Fourier transforms of the same function $f(x)$, but with half as many points spaced twice as far apart, plus an extra factor $e^{-i2\pi k/N}$, called a *twiddle factor*,[9] which is trivial to calculate. So if we can do the two smaller Fourier transforms, then we can calculate c_k easily.

And how do we do the smaller Fourier transforms? We just repeat the process. We split each of them into their even and odd terms and express them as the sum of the two, with a twiddle factor in between. Because we started with N a power of two, we can go on dividing the transform in half repeatedly like this, until eventually we get to the point where each transform is the transform of just a single sample. But the Fourier transform of a single sample has only a single Fourier coefficient c_0, which, putting $k = 0$, $N = 1$ in Eq. (7.36), is equal to

$$c_0 = \sum_{n=0}^{0} y_n e^0 = y_0. \tag{7.40}$$

In other words, once we get down to a single sample, the Fourier transform is trivial—it's just equal to the sample itself! So at this stage we don't need to do any more Fourier transforms; we have everything we need.

The actual calculation of the fast Fourier transform is the reverse of the process above. Starting with the individual samples, which are their own Fourier transforms, we combine them in pairs, then combine the pairs into fours, the fours into eights, and so on, creating larger and larger Fourier transforms, until we have reconstructed the full transform of the complete set of samples.

The advantage of this approach is its speed. At the first round of the calculation we have N samples. At the next round we combine these in pairs to make $\frac{1}{2}N$ transforms with two coefficients each, so we have to calculate a total of N coefficients at this round. At the round after that there are $\frac{1}{4}N$ transforms of four coefficients each—N coefficients again. Indeed it's easy to see that there are N coefficients to calculate at every level.

---

[9] I'm not making this up. That's what it's called.

And how many levels are there? After we have gone though m levels the transforms have 2^m samples each, so the total number of levels is given by $2^m = N$, or $m = \log_2 N$. Thus the number of coefficients we have to calculate in the whole calculation is $\log_2 N$ levels times N coefficients each, or $N \log_2 N$ coefficients in all. But this is *way* better than the roughly $\frac{1}{2}N^2$ terms needed to calculate the DFT the brute-force way. For instance, if we have a million samples, then the brute-force calculation would require about $\frac{1}{2}N^2 = 5 \times 10^{11}$ operations which is not a practical calculation on a typical computer. But the fast Fourier transform requires only $N \log_2 N = 2 \times 10^7$, which is entirely reasonable—you can do twenty million steps in under a second on most computers. So we've turned a calculation that was essentially impossible into one that can be done very quickly. Alternatively, employing again our rule of thumb that one should limit programs to about a billion operations, setting $N \log_2 N = 10^9$ and solving for N we get $N \simeq 40$ million samples, which is large enough for most scientific applications.

We also note another useful fact: the inverse discrete Fourier transform, Eq. (7.20), which transforms the Fourier coefficients back into samples again, has basically the same form as the forward transform—the only difference is that there's no minus sign in the exponential. This means that the inverse transform can also be performed quickly using the same tricks as for the forward transform. The resulting calculation is called, naturally, the inverse fast Fourier transform, or inverse FFT.

Finally, we have in these developments assumed that the number of samples is a power of two. In fact it is possible to make the fast Fourier transform work if the number is not a power of two, but the algebra is rather tedious, so we'll skip the details here.

7.4.1 FORMULAS FOR THE FFT

Let us look at the mathematics of the fast Fourier transform in a little more detail and consider the various stages of the decomposition process. In true Python fashion we'll number the initial stage "stage zero". At this stage we have a single Fourier transform of the entire set of samples. At the next stage, stage 1, we split the samples into two sets, at the stage after that we split again into four sets, and so forth. In general, at the mth stage of the process there will be 2^m sets consisting of $N/2^m$ samples each. The first of these sets consists of the original samples numbered 0, 2^m, $2^m \times 2$, $2^m \times 3$ and so forth. In other words it consists of sample numbers $2^m r$ with integer $r = 0 \ldots N/2^m - 1$. Similarly the second set consists of sample numbers $2^m r + 1$, and third of sample numbers $2^m r + 2$, and so on, with $r = 0 \ldots N/2^m - 1$ in each case.

The DFT of the jth set of samples is

$$\sum_{r=0}^{N/2^m-1} y_{2^m r + j} \exp\left(-i\frac{2\pi k(2^m r + j)}{N}\right) = e^{-i2\pi kj/N} \sum_{r=0}^{N/2^m-1} y_{2^m r + j} \exp\left(-i\frac{2\pi kr}{N/2^m}\right)$$

$$= e^{-i2\pi kj/N} E_k^{(m,j)} \tag{7.41}$$

for $j = 0 \ldots 2^m - 1$. Then the general version of Eq. (7.39) at the mth stage is

$$e^{-i2\pi kj/N} E_k^{(m,j)} = e^{-i2\pi kj/N} E_k^{(m+1,j)} + e^{-i2\pi k(j+2^m)/N} E_k^{(m+1,j+2^m)}, \tag{7.42}$$

or equivalently

$$E_k^{(m,j)} = E_k^{(m+1,j)} + e^{-i2\pi 2^m k/N} E_k^{(m+1,j+2^m)}. \tag{7.43}$$

Armed with this formula, the calculation of the full DFT is now relatively simple. The total number of stages—the number of times we have to split the samples in half before we get down to the level of single samples—is, as we have said, $\log_2 N$. The fast Fourier transform works by starting at the last level $m = \log_2 N$ and working backwards to $m = 0$. Initially our sets of samples all have just one sample each and one Fourier coefficient, which is equal to that one sample. Then we use Eq. (7.43) repeatedly to calculate $E_k^{(m,j)}$ for all j and k for lower and lower levels m, each level being calculated from the results for the level above it, until we get down to $m = 0$. But for $m = 0$, Eq. (7.41) gives

$$E_k^{0,0} = \sum_{r=0}^{N-1} y_r \exp\left(-i\frac{2\pi kr}{N}\right) = c_k \tag{7.44}$$

so $c_k = E_k^{0,0}$ and we have our values for the final Fourier coefficients.

There's one other detail to notice. For any value of m we need to evaluate Eq. (7.43) for $k = 0 \ldots N/2^m - 1$. But this causes problems because we evaluate the coefficients at the previous $(m+1)$th level only for the smaller range $k = 0 \ldots N/2^{m+1} - 1$. In practice, however, this is not a big deal. Consider $E_k^{(m+1,j)}$ for $k = N/2^{m+1} + s$:

$$E_{N/2^{m+1}+s}^{(m+1,j)} = \sum_{r=0}^{N/2^{m+1}-1} y_{2^{m+1}r+j} \exp\left(-i\frac{2\pi(N/2^{m+1}+s)r}{N/2^{m+1}}\right)$$

$$= \exp(-i2\pi r) \sum_{r=0}^{N/2^{m+1}-1} y_{2^{m+1}r+j} \exp\left(-i\frac{2\pi sr}{N/2^{m+1}}\right)$$

$$= E_s^{(m+1,j)}. \tag{7.45}$$

Thus the coefficients for values of k beyond $N/2^{m+1} - 1$ are simply repeats of the previous values. So long as we bear this in mind, evaluating Eq. (7.43) for all k is straightforward.

Exercise 7.7 gives you the opportunity to write your own program to perform a fast Fourier transform, and doing so is a worthwhile undertaking. You can write an FFT program in Python in only about a dozen lines, though it can be quite tricky to work out what those lines should be, given all the different indices and factors that appear in Eq. (7.43).

Most people, however, don't write their own FFT programs. As with some of the other standard calculations we've looked at, such as the solution of simultaneous equations and the evaluation of eigenvalues and eigenvectors of matrices, the FFT is such a common operation that other people have already written programs to do it for you, and Python includes functions to perform FFTs on both real and complex samples.

7.4.2 STANDARD FUNCTIONS FOR FAST FOURIER TRANSFORMS

In Python, fast Fourier transforms are provided by the module `numpy.fft`. For physics calculations the main function we use is the function `rfft`, which calculates the Fourier transform of a set of real samples in an array. (The "r" is for "real." There is another function `fft` that performs transforms of complex samples, but we will not use it in this book.)

For instance, a periodic "sawtooth" function could be represented by a set of samples like this:

```
from numpy import array
y = array([0.0,0.1,0.2,0.3,0.4,0.5,0.6,0.7,0.8,0.9],float)
```

and we could Fourier transform it like this:

```
from numpy.fft import rfft
c = rfft(y)
```

This creates a complex array c containing the coefficients of the Fourier transform of the samples in the real array y. As we have said, there are N distinct coefficients in the DFT of N samples, but if the samples are real and N is even, as here, then only the first $\frac{1}{2}N + 1$ coefficients are independent and the rest are their complex conjugates. Python knows this and the array returned by `rfft` contains only the first $\frac{1}{2}N + 1$ coefficients—Python does not bother to calculate or store the remaining ones. If you want the rest of the coefficients you're expected to calculate the complex conjugates for yourself (although in practice one rarely has cause to do this). Note also that in the example above the number of samples was ten, which is not a power of two: the `rfft` function knows how to do FFTs for any number of samples.

The module `numpy.fft` also contains a function called `irfft` for performing inverse fast Fourier transforms, meaning it evaluates the sum in Eq. (7.20) and recovers the values of the samples, but it does so quickly, using the tricks of the FFT as discussed previously. Thus if we take the output array c from the example above and do the following:

```
from numpy.fft import irfft
z = irfft(c)
print(z)
```

the program prints

```
[ 0.    0.1  0.2  0.3  0.4  0.5  0.6  0.7  0.8  0.9]
```

As you can see, we have recovered our original sample values again. Note that the `irfft` function takes as input an array that has $\frac{1}{2}N + 1$ complex elements and creates an array with N real elements. The function knows that the second half of the Fourier coefficients are the complex conjugates of the first half, so you only have to supply the first half.

The module `numpy.fft` also contains functions to calculate Fourier transforms of complex samples (`fft` and `ifft` for the forward and inverse transforms respectively) and two-dimensional transforms (`rfft2` and `irfft2` for real samples, `fft2` and `ifft2` for complex ones). The 2D Fourier transforms take a two-dimensional array as input and return a two-dimensional array. (Pretty much they just do two separate one-dimensional Fourier transforms, one for the rows and one for the columns. You could do the same thing yourself, but it saves a little effort to use the functions provided.) Note that, as discussed in Section 7.2.2, the two-dimensional Fourier transform of an $M \times N$ grid of real numbers is a grid of complex numbers with dimensions $M \times (\frac{1}{2}N + 1)$. Python knows this and returns an array of the appropriate size. Higher-dimensional transforms are also possible, in three dimensions or above, and Python has functions for these as well, but we will not use them in this book.

4

Exercise 7.3: Fourier transforms of musical instruments

In the on-line resources you will find files called `piano.txt` and `trumpet.txt`, which contain data representing the waveform of a single note, played on, respectively, a piano and a trumpet.

a) Write a program that loads a waveform from one of these files, plots it, then calculates its discrete Fourier transform and plots the magnitudes of the first 10 000 coefficients in a manner similar to Fig. 7.4. Note that you will have to use a fast Fourier transform for the calculation because there are too many samples in the files to do the transforms the slow way in any reasonable amount of time.

Apply your program to the piano and trumpet waveforms and discuss briefly what one can conclude about the sound of the piano and trumpet from the plots of Fourier coefficients.

b) Both waveforms were recorded at the industry-standard rate of 44 100 samples per second and both instruments were playing the same musical note when the recordings were made. From your Fourier transform results calculate what note they were playing. (Hint: The musical note middle C has a frequency of 261 Hz.)

Exercise 7.4: Fourier filtering and smoothing

In the on-line resources you'll find a file called `dow.txt`. It contains the daily closing value for each business day from late 2006 until the end of 2010 of the Dow Jones Industrial Average, which is a measure of average prices on the US stock market.

Write a program to do the following:

a) Read in the data from `dow.txt` and plot them on a graph.

b) Calculate the coefficients of the discrete Fourier transform of the data using the function `rfft` from `numpy.fft`, which produces an array of $\frac{1}{2}N+1$ complex numbers.

c) Now set all but the first 10% of the elements of this array to zero (i.e., set the last 90% to zero but keep the values of the first 10%).

d) Calculate the inverse Fourier transform of the resulting array, zeros and all, using the function `irfft`, and plot it on the same graph as the original data. You may need to vary the colors of the two curves to make sure they both show up on the graph. Comment on what you see. What is happening when you set the Fourier coefficients to zero?

e) Modify your program so that it sets all but the first 2% of the coefficients to zero and run it again.

Exercise 7.5: If you have not done Exercise 7.4 you should do it before this one.

The function $f(t)$ represents a square-wave with amplitude 1 and frequency 1 Hz:

$$f(t) = \begin{cases} 1 & \text{if } \lfloor 2t \rfloor \text{ is even,} \\ -1 & \text{if } \lfloor 2t \rfloor \text{ is odd,} \end{cases} \tag{7.46}$$

where $\lfloor x \rfloor$ means x rounded down to the next lowest integer. Let us attempt to smooth this function using a Fourier transform, as we did in the previous exercise. Write a program that creates an array of $N = 1000$ elements containing a thousand equally spaced samples from a single cycle of this square-wave. Calculate the discrete Fourier transform of the array. Now set all but the first ten Fourier coefficients to zero, then invert the Fourier transform again to recover the smoothed signal. Make a plot of the result and on the same axes show the original square-wave as well. You should find that the signal is not simply smoothed—there are artifacts, wiggles, in the results. Explain briefly where these come from.

Artifacts similar to these arise when Fourier coefficients are discarded in audio and visual compression schemes like those described in Section 7.3.1 and are the primary source of imperfections in digitally compressed sound and images.

7.4.3 FAST COSINE AND SINE TRANSFORMS

As we have seen, discrete cosine transforms have important technological uses and both cosine and sine transforms are used in physics calculations. The module numpy.fft does not provide functions for fast cosine and sine transforms, but it is not hard to create functions that do the job.[10] As we saw in Section 7.3, a discrete cosine transform is nothing more than an ordinary discrete Fourier transform performed on a set of samples that are symmetric around the middle of the transform interval. The most common type of discrete cosine transform, often just called "the" discrete cosine transform, is the one described by Eq. (7.35), which corresponds to samples that satisfy $y_{2N-1-n} = y_n$ with $n = 0 \ldots N - 1$. Given N samples, therefore, we can calculate the discrete cosine transform by "mirroring" those samples to create a symmetric array of twice the size, performing an ordinary discrete Fourier transform on the result, and discarding the imaginary parts of the coefficients (which should be zero anyway). Here is a Python function that does exactly that for N samples in an array y:

File: dcst.py

```
from numpy.fft import rfft
from numpy import empty,arange,exp,real,pi
```

---

[10]There does exist another Python package, called scipy, that provides functions for fast cosine and sine transforms, although we will not use it in this book. See www.scipy.org for details.

```
def dct(y):
    N = len(y)
    y2 = empty(2*N,float)
    for n in range(N):
        y2[n] = y[n]
        y2[2*N-1-n] = y[n]
    c = rfft(y2)
    phi = exp(-1j*pi*arange(N)/(2*N))
    return real(phi*c[:N])
```

Note the use of the multiplier phi, which accounts for the leading phase factor in Eq. (7.32).

Similar functions to perform the inverse cosine transform, as well as forward and inverse sine transforms, are given in Appendix E and copies can be found in the on-line resources.

Exercise 7.6: Comparison of the DFT and DCT

This exercise will be easier if you have already done Exercise 7.4.

Exercise 7.4 looked at data representing the variation of the Dow Jones Industrial Average, colloquially called "the Dow," over time. The particular time period studied in that exercise was special in one sense: the value of the Dow at the end of the period was almost the same as at the start, so the function was, roughly speaking, periodic. In the on-line resources there is another file called dow2.txt, which also contains data on the Dow but for a different time period, from 2004 until 2008. Over this period the value changed considerably from a starting level around 9000 to a final level around 14000.

a) Write a program similar to the one for Exercise 7.4, part (e), in which you read the data in the file dow2.txt and plot it on a graph. Then smooth the data by calculating its Fourier transform, setting all but the first 2% of the coefficients to zero, and inverting the transform again, plotting the result on the same graph as the original data. As in Exercise 7.4 you should see that the data are smoothed, but now there will be an additional artifact. At the beginning and end of the plot you should see large deviations away from the true smoothed function. These occur because the function is required to be periodic—its last value must be the same as its first—so it needs to deviate substantially from the correct value to make the two ends of the function meet. In some situations (including this one) this behavior is unsatisfactory. If we want to use the Fourier transform for smoothing, we would certainly prefer that it not introduce artifacts of this kind.

b) Modify your program to repeat the same analysis using discrete cosine transforms. You can use the functions from dcst.py to perform the transforms if you wish. Again discard all but the first 2% of the coefficients, invert the transform, and plot the result. You should see a significant improvement, with less distortion of the function at the ends of the interval. This occurs because, as discussed at the end of Section 7.3, the cosine transform does not force the value of the function to be the same at both ends.

It is because of the artifacts introduced by the strict periodicity of the DFT that the cosine transform is favored for many technological applications, such as audio compression. The artifacts can degrade the sound quality of compressed audio and the cosine transform generally gives better results.

The cosine transform is not wholly free of artifacts itself however. It's true it does not force the function to be periodic, but it does force the gradient to be zero at the ends of the interval (which the ordinary Fourier transform does not). You may be able to see this in your calculations for part (b) above. Look closely at the smoothed function and you should see that its slope is flat at the beginning and end of the interval. The distortion of the function introduced is less than the distortion in part (a), but it's there all the same. To reduce this effect, audio compression schemes often use overlapped cosine transforms, in which transforms are performed on overlapping blocks of samples, so that the portions at the ends of blocks, where the worst artifacts lie, need not be used.

FURTHER EXERCISES

7.7 Fast Fourier transform: Write your own program to compute the fast Fourier transform for the case where N is a power of two, based on the formulas given in Section 7.4.1. As a test of your program, use it to calculate the Fourier transform of the data in the file pitch.txt, which can be found in the on-line resources. A plot of the data is shown in Fig. 7.3. You should be able to duplicate the results for the Fourier transform shown in Fig. 7.4.

This exercise is quite tricky. You have to calculate the coefficients $E_k^{(m,j)}$ from Eq. (7.43) for all levels m, which means that first you will have to plan how the coefficients will be stored. Since, as we have seen, there are exactly N of them at every level, one way to do it would be to create a two-dimensional complex array of size $N \times (1 + \log_2 N)$, so that it has N complex numbers for each level from zero to $\log_2 N$. Then within level m you have 2^m individual transforms denoted by $j = 0 \ldots 2^m - 1$, each with $N/2^m$ coefficients indexed by k. A simple way to arrange the coefficients would be to put all the $k = 0$ coefficients in a block one after another, then all the $k = 1$ coefficients, and so forth. Then $E_k^{(m,j)}$ would be stored in the $j + 2^m k$ element of the array.

This method has the advantage of being quite simple to program, but the disadvantage of using up a lot of memory space. The array contains $N \log_2 N$ complex numbers, and a complex number typically takes sixteen bytes of memory to store. So if you had to do a large Fourier transform of, say, $N = 10^8$ numbers, it would take $16N \log_2 N \simeq 42$ gigabytes of memory, which is much more than most computers have.

An alternative approach is to notice that we do not really need to store all of the coefficients. At any one point in the calculation we only need the coefficients at the current level and the previous level (from which the current level is calculated). If one is clever one can write a program that uses only two arrays, one for the current level and one for the previous level, each consisting of N complex numbers. Then our transform of 10^8 numbers would require less than four gigabytes, which is fine on most computers.

(There is a third way of storing the coefficients that is even more efficient. If you store the coefficients in the correct order, then you can arrange things so that every time you compute a coefficient for the next level, it gets stored in the same place as the old coefficient from the previous level from which it was calculated, and which you no longer need. With this way of doing things you only need one array of N complex numbers—we say the transform is done "in place." Unfortunately, this in-place Fourier transform is much harder to work out and harder to program. If you are feeling particularly ambitious you might want to give it a try, but it's not for the faint-hearted.)

7.8 Diffraction gratings: Exercise 5.19 (page 206) looked at the physics of diffraction gratings, calculating the intensity of the diffraction patterns they produce from the equation

$$I(x) = \left| \int_{-w/2}^{w/2} \sqrt{q(u)}\, e^{i2\pi xu/\lambda f}\, du \right|^2,$$

where w is the width of the grating, λ is the wavelength of the light, f is the focal length of the lens used to focus the image, and $q(u)$ is the intensity transmission function of the diffraction grating at a distance u from the central axis, i.e., the fraction of the incident light that the grating lets through. In Exercise 5.19 we evaluated this expression directly using standard methods for performing integrals, but a more efficient way to do the calculation is to note that the integral is basically just a Fourier transform. Approximating the integral, as we did in Eq. (7.13), using the trapezoidal rule, with N points $u_n = nw/N - w/2$, we get

$$\int_{-w/2}^{w/2} \sqrt{q(u)}\, e^{i2\pi xu/\lambda f}\, du \simeq \frac{w}{N} e^{-i\pi wx/\lambda f} \sum_{n=0}^{N-1} \sqrt{q(u_n)}\, e^{i2\pi wxn/\lambda fN}$$

$$= \frac{w}{N} e^{-i\pi k} \sum_{n=0}^{N-1} y_n\, e^{i2\pi kn/N},$$

where $k = wx/\lambda f$ and $y_n = \sqrt{q(u_n)}$. Comparing with Eq. (7.15), we see that the sum in this expression is equal to the complex conjugate c_k^* of the kth coefficient of the DFT of y_n. Substituting into the expression for the intensity $I(x)$, we then have

$$I(x_k) = \frac{w^2}{N^2} |c_k|^2,$$

321

CHAPTER 7 | FOURIER TRANSFORMS

where

$$x_k = \frac{\lambda f}{w} k.$$

Thus we can calculate the intensity of the diffraction pattern at the points x_k by performing a Fourier transform.

There is a catch, however. Given that k is an integer, $k = 0 \ldots N - 1$, the points x_k at which the intensity is evaluated have spacing $\lambda f / w$ on the screen. This spacing can be large in some cases, giving us only a rather coarse picture of the diffraction pattern. For instance, in Exercise 5.19 we had $\lambda = 500\,\text{nm}$, $f = 1\,\text{m}$, and $w = 200\,\mu\text{m}$, and the screen was 10 cm wide, which means that $\lambda f / w = 2.5\,\text{mm}$ and we have only forty points on the screen. This is not enough to make a usable plot of the diffraction pattern.

One way to fix this problem is to increase the width of the grating from the given value w to a larger value $W > w$, which makes the spacing $\lambda f / W$ of the points on the screen closer. We can add the extra width on one or the other side of the grating, or both, as we prefer, but—and this is crucial—the extra portion added must be opaque, it must not transmit light, so that the physics of the system does not change. In other words, we need to "pad out" the data points y_n that measure the transmission profile of the grating with additional zeros so as to make the grating wider while keeping its transmission properties the same. For example, to increase the width to $W = 10w$, we would increase the number N of points y_n by a factor of ten, with the extra points set to zero. The extra points can be at the beginning, at the end, or split between the two—it will make no difference to the answer. Then the intensity is given by

$$I(x_k) = \frac{W^2}{N^2} |c_k|^2,$$

where

$$x_k = \frac{\lambda f}{W} k.$$

Write a Python program that uses a fast Fourier transform to calculate the diffraction pattern for a grating with transmission function $q(u) = \sin^2 \alpha u$ (the same as in Exercise 5.19), with slits of width 20 μm [meaning that $\alpha = \pi/(20\,\mu\text{m})$] and parameters as above: $w = 200\,\mu\text{m}$, $W = 10w = 2\,\text{mm}$, incident light of wavelength $\lambda = 500\,\text{nm}$, a lens with focal length of 1 meter, and a screen 10 cm wide. Choose a suitable number of points to give a good approximation to the grating transmission function and then make a graph of the diffraction intensity on the screen as a function of position x in the range $-5\,\text{cm} \le x \le 5\,\text{cm}$. If you previously did Exercise 5.19, check to make sure your answers to the two exercises agree.

7.9 Image deconvolution: You've probably seen it on TV, in one of those crime drama shows. They have a blurry photo of a crime scene and they click a few buttons on the computer and magically the photo becomes sharp and clear, so you can make out someone's face, or some lettering on a sign. Surely (like almost everything else on such TV shows) this is just science fiction? Actually, no. It's not. It's real and in this exercise you'll write a program that does it.

322

When a photo is blurred each point on the photo gets smeared out according to some "smearing distribution," which is technically called a *point spread function*. We can represent this smearing mathematically as follows. For simplicity let's assume we're working with a black and white photograph, so that the picture can be represented by a single function $a(x,y)$ which tells you the brightness at each point (x,y). And let us denote the point spread function by $f(x,y)$. This means that a single bright dot at the origin ends up appearing as $f(x,y)$ instead. If $f(x,y)$ is a broad function then the picture is badly blurred. If it is a narrow peak then the picture is relatively sharp.

In general the brightness $b(x,y)$ of the blurred photo at point (x,y) is given by

$$b(x,y) = \int_0^K \int_0^L a(x',y')f(x-x',y-y')\,dx'\,dy',$$

where $K \times L$ is the dimension of the picture. This equation is called the *convolution* of the picture with the point spread function.

Working with two-dimensional functions can get complicated, so to get the idea of how the math works, let's switch temporarily to a one-dimensional equivalent of our problem. Once we work out the details in 1D we'll return to the 2D version. The one-dimensional version of the convolution above would be

$$b(x) = \int_0^L a(x')f(x-x')\,dx'.$$

The function $b(x)$ can be represented by a Fourier series as in Eq. (7.5):

$$b(x) = \sum_{k=-\infty}^{\infty} \tilde{b}_k \exp\left(i\frac{2\pi kx}{L}\right),$$

where

$$\tilde{b}_k = \frac{1}{L}\int_0^L b(x)\exp\left(-i\frac{2\pi kx}{L}\right)dx$$

are the Fourier coefficients, Eq. (7.10). Substituting for $b(x)$ in this equation gives

$$\tilde{b}_k = \frac{1}{L}\int_0^L \int_0^L a(x')f(x-x')\exp\left(-i\frac{2\pi kx}{L}\right)dx'\,dx$$

$$= \frac{1}{L}\int_0^L \int_0^L a(x')f(x-x')\exp\left(-i\frac{2\pi k(x-x')}{L}\right)\exp\left(-i\frac{2\pi kx'}{L}\right)dx'\,dx.$$

Now let us change variables to $X = x - x'$, and we get

$$\tilde{b}_k = \frac{1}{L}\int_0^L a(x')\exp\left(-i\frac{2\pi kx'}{L}\right)\int_{-x'}^{L-x'} f(X)\exp\left(-i\frac{2\pi kX}{L}\right)dX\,dx'.$$

If we make $f(x)$ a periodic function in the standard fashion by repeating it infinitely many times to the left and right of the interval from 0 to L, then the second integral above can be written as

$$\int_{-x'}^{L-x'} f(X) \exp\left(-i\frac{2\pi kX}{L}\right) dX = \int_{-x'}^{0} f(X) \exp\left(-i\frac{2\pi kX}{L}\right) dX$$

$$+ \int_{0}^{L-x'} f(X) \exp\left(-i\frac{2\pi kX}{L}\right) dX$$

$$= \exp\left(i\frac{2\pi kL}{L}\right) \int_{L-x'}^{L} f(X) \exp\left(-i\frac{2\pi kX}{L}\right) dX + \int_{0}^{L-x'} f(X) \exp\left(-i\frac{2\pi kX}{L}\right) dX$$

$$= \int_{0}^{L} f(X) \exp\left(-i\frac{2\pi kX}{L}\right) dX,$$

which is simply L times the Fourier transform $\tilde{f}_k$ of $f(x)$. Substituting this result back into our equation for $\tilde{b}_k$ we then get

$$\tilde{b}_k = \int_{0}^{L} a(x') \exp\left(-i\frac{2\pi kx'}{L}\right) \tilde{f}_k \, dx' = L \, \tilde{a}_k \tilde{f}_k.$$

In other words, apart from the factor of L, the Fourier transform of the blurred photo is the product of the Fourier transforms of the unblurred photo and the point spread function.

Now it is clear how we deblur our picture. We take the blurred picture and Fourier transform it to get $\tilde{b}_k = L \tilde{a}_k \tilde{f}_k$. We also take the point spread function and Fourier transform it to get $\tilde{f}_k$. Then we divide one by the other:

$$\frac{\tilde{b}_k}{L\tilde{f}_k} = \tilde{a}_k$$

which gives us the Fourier transform of the *unblurred* picture. Then, finally, we do an inverse Fourier transform on $\tilde{a}_k$ to get back the unblurred picture. This process of recovering the unblurred picture from the blurred one, of reversing the convolution process, is called *deconvolution*.

Real pictures are two-dimensional, but the mathematics follows through exactly the same. For a picture of dimensions $K \times L$ we find that the two-dimensional Fourier transforms are related by

$$\tilde{b}_{kl} = KL\tilde{a}_{kl}\tilde{f}_{kl},$$

and again we just divide the blurred Fourier transform by the Fourier transform of the point spread function to get the Fourier transform of the unblurred picture.

In the digital realm of computers, pictures are not pure functions $f(x, y)$ but rather grids of samples, and our Fourier transforms are discrete transforms not continuous ones. But the math works out the same again.

The main complication with deblurring in practice is that we don't usually know the point spread function. Typically we have to experiment with different ones until we find something that works. For many cameras it's a reasonable approximation to assume the point spread function is Gaussian:

$$f(x, y) = \exp\left(-\frac{x^2 + y^2}{2\sigma^2}\right),$$

where σ is the width of the Gaussian. Even with this assumption, however, we still don't know the value of σ and we may have to experiment to find a value that works well. In the following exercise, for simplicity, we'll assume we know the value of σ.

a) On the web site you will find a file called blur.txt that contains a grid of values representing brightness on a black-and-white photo—a badly out-of-focus one that has been deliberately blurred using a Gaussian point spread function of width $\sigma = 25$. Write a program that reads the grid of values into a two-dimensional array of real numbers and then draws the values on the screen of the computer as a density plot. You should see the photo appear. If you get something wrong it might be upside-down. Work with the details of your program until you get it appearing correctly. (Hint: The picture has the sky, which is bright, at the top and the ground, which is dark, at the bottom.)

b) Write another program that creates an array, of the same size as the photo, containing a grid of samples drawn from the Gaussian $f(x,y)$ above with $\sigma = 25$. Make a density plot of these values on the screen too, so that you get a visualization of your point spread function. Remember that the point spread function is periodic (along both axes), which means that the values for negative x and y are repeated at the end of the interval. Since the Gaussian is centered on the origin, this means there should be bright patches in each of the four corners of your picture, something like this:

c) Combine your two programs and add Fourier transforms using the functions rfft2 and irfft2 from numpy.fft, to make a program that does the following:
   i) Reads in the blurred photo
   ii) Calculates the point spread function
   iii) Fourier transforms both
   iv) Divides one by the other
   v) Performs an inverse transform to get the unblurred photo
   vi) Displays the unblurred photo on the screen
When you are done, you should be able to make out the scene in the photo, although probably it will still not be perfectly sharp.

Hint: One thing you'll need to deal with is what happens when the Fourier transform of the point spread function is zero, or close to zero. In that case if you divide by it you'll get an error (because you can't divide by zero) or just a very large number (because you're dividing by something small). A workable compromise is that if a value in the Fourier transform of the point spread function is smaller than a certain amount ϵ you don't divide by it—just leave that coefficient alone. The value of ϵ is not very critical but a reasonable value seems to be 10^{-3}.

d) Bearing in mind this last point about zeros in the Fourier transform, what is it that limits our ability to deblur a photo? Why can we not perfectly unblur any photo and make it completely sharp?

We have seen this process in action here for a normal snapshot, but it is also used in many physics applications where one takes photos. For instance, it is used in astronomy to enhance photos taken by telescopes. It was famously used with images from the Hubble Space Telescope after it was realized that the telescope's main mirror had a serious manufacturing flaw and was returning blurry photos—scientists managed to partially correct the blurring using Fourier transform techniques.

CHAPTER 8

ORDINARY DIFFERENTIAL EQUATIONS

PERHAPS the most common use of computers in physics is for the solution of differential equations. In this chapter we look at techniques for solving ordinary differential equations, such as the equations of motion of rigid bodies or the equations governing the behavior of electrical circuits. In the following chapter we look at techniques for partial differential equations, such as the wave equation and the diffusion equation.

8.1 FIRST-ORDER DIFFERENTIAL EQUATIONS WITH ONE VARIABLE

We begin our study of differential equations by looking at ordinary differential equations, meaning those for which there is only one independent variable, such as time, and all dependent variables are functions solely of that one independent variable. The simplest type of ordinary differential equation is a first-order equation with one dependent variable, such as

$$\frac{\mathrm{d}x}{\mathrm{d}t} = \frac{2x}{t}. \tag{8.1}$$

This equation, however, can be solved exactly by hand by separating the variables. There's no need to use a computer in this case. But suppose instead that you had

$$\frac{\mathrm{d}x}{\mathrm{d}t} = \frac{2x}{t} + \frac{3x^2}{t^3}. \tag{8.2}$$

Now the equation is no longer separable and moreover it's nonlinear, meaning that powers or other nonlinear functions of the dependent variable x appear in the equation. Nonlinear equations can rarely be solved analytically, but they can be solved numerically. Computers don't care whether a differential equation is linear or nonlinear—the techniques used to solve it are the same either way.

The general form of a first-order one-variable ordinary differential equation is

$$\frac{dx}{dt} = f(x, t), \tag{8.3}$$

where $f(x, t)$ is some function we specify. In Eq. (8.2) we had $f(x, t) = 2x/t + 3x^2/t^3$. The independent variable is denoted t in this example, because in physics the independent variable is often time. But of course there are other possibilities. We could just as well have written our equation as

$$\frac{dy}{dx} = f(x, y). \tag{8.4}$$

In this chapter we will stick with t for the independent variable, but it's worth bearing in mind that there are plenty of examples where the independent variable is not time.

To calculate a full solution to Eq. (8.3) we also require an initial condition or boundary condition—we have to specify the value of x at one particular value of t, for instance at $t = 0$. In all the problems we'll tackle in this chapter we will assume that we're given both the equation and its initial or boundary conditions.

8.1.1 EULER'S METHOD

Suppose we are given an equation of the form (8.3) and an initial condition that fixes the value of x for some t. Then we can write the value of x a short interval h later using a Taylor expansion thus:

$$x(t + h) = x(t) + h\frac{dx}{dt} + \tfrac{1}{2}h^2\frac{d^2x}{dt^2} + \ldots$$
$$= x(t) + hf(x, t) + O(h^2), \tag{8.5}$$

where we have used Eq. (8.3) and $O(h^2)$ is a shorthand for terms that go as h^2 or higher. If h is small then h^2 is very small, so we can neglect the terms in h^2 and get

$$x(t + h) = x(t) + hf(x, t). \tag{8.6}$$

If we know the value of x at time t we can use this equation to calculate the value a short time later. Then we can just repeat the exercise to calculate x another interval h after that, and so forth, and thereby calculate x at a succession of evenly spaced points for as long as we want. We don't get $x(t)$ for all values of t from this calculation, only at a finite set of points, but if h is small enough we can get a pretty good picture of what the solution to the equation looks like. As we saw in Section 3.1, we can make a convincing plot of a curve by approximating it with a set of closely spaced points.

Thus, for instance, we might be given a differential equation for x and an initial condition at $t = a$ and asked to make a graph of $x(t)$ for values of t from a to b. To do this, we would divide the interval from a to b into steps of size h and use (8.6) repeatedly to calculate $x(t)$, then plot the results. This method for solving differential equations is called *Euler's method*, after its inventor, Leonhard Euler.

EXAMPLE 8.1: EULER'S METHOD

Let us use Euler's method to solve the differential equation

$$\frac{dx}{dt} = -x^3 + \sin t \qquad (8.7)$$

with the initial condition $x = 0$ at $t = 0$. Here is a program to do the calculation from $t = 0$ to $t = 10$ in 1000 steps and plot the result:

```
from math import sin                                    File: euler.py
from numpy import arange
from pylab import plot,xlabel,ylabel,show

def f(x,t):
    return -x**3 + sin(t)

a = 0.0          # Start of the interval
b = 10.0         # End of the interval
N = 1000         # Number of steps
h = (b-a)/N      # Size of a single step
x = 0.0          # Initial condition

tpoints = arange(a,b,h)
xpoints = []
for t in tpoints:
    xpoints.append(x)
    x += h*f(x,t)

plot(tpoints,xpoints)
xlabel("t")
ylabel("x(t)")
show()
```

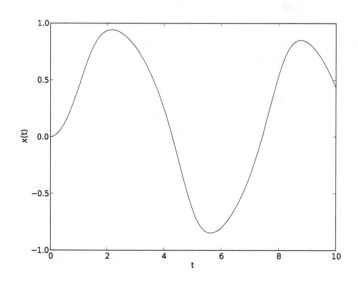

Figure 8.1: Numerical solution of an ordinary differential equation. A solution to Eq. (8.7) from $x = 0$ to $x = 10$, calculated using Euler's method.

If we run this program it produces the picture shown in Fig. 8.1, which, as we'll see, turns out to be a pretty good approximation to the shape of the true solution to the equation. In this case, Euler's method does a good job.

In general, Euler's method is not bad. It gives reasonable answers in many cases. In practice, however, we never actually use Euler's method. Why not? Because there is a better method that's very little extra work to program, much more accurate, and runs just as fast and often faster. This is the so-called Runge–Kutta method, which we'll look at in a moment. First, however, let's look a little more closely at Euler's method, to understand why it's not ideal.[1]

Euler's method only gives approximate solutions. The approximation arises because we neglected the h^2 term (and all higher-order terms) in Eq. (8.5). The

---

[1]It's not completely correct to say that we never use Euler's method. We never use it for solving *ordinary* differential equations, but in Section 9.3 we will see that Euler's method is useful for solving partial differential equations. It's true in that case also that Euler's method is not very accurate, but there are other bigger sources of inaccuracy when solving partial differential equations which mean that the inaccuracy of Euler's method is moot, and in such situations its simplicity makes it the method of choice.

size of the h^2 term is $\frac{1}{2}h^2\,\mathrm{d}^2x/\mathrm{d}t^2$, which tells us the error introduced on a single step of the method, to leading order, and this error gets smaller as h gets smaller so we can make the step more accurate by making h small.

But we don't just take a single step when we use Euler's method. We take many. If we want to calculate a solution from $t = a$ to $t = b$ using steps of size h, then the total number of steps we need is $N = (b - a)/h$. Let us denote the values of t at which the steps fall by $t_k = a + kh$. Then the total, cumulative error incurred as we solve our differential equation all the way from a to b is given by the sum of the individual errors on each step thus:

$$\sum_{k=0}^{N-1} \frac{1}{2}h^2\left(\frac{\mathrm{d}^2x}{\mathrm{d}t^2}\right)_{t=t_k} = \frac{1}{2}h\sum_{k=0}^{N-1} h\left(\frac{\mathrm{d}f}{\mathrm{d}t}\right)_{t=t_k} \simeq \frac{1}{2}h\int_a^b \frac{\mathrm{d}f}{\mathrm{d}t}\,\mathrm{d}t$$
$$= \frac{1}{2}h\big[f(x(b),b) - f(x(a),a)\big], \tag{8.8}$$

where we have approximated the sum by an integral, which is a good approximation if h is small.

Notice that the final expression for the total error is linear in h, even though the individual errors are of order h^2, meaning that the total error goes down by a factor of two when we make h half as large. In principle this allows us to make the error as small as we like, although when we make h smaller we also increase the number of steps $N = (b - a)/h$ and hence the calculation will take proportionately longer—a calculation that's twice as accurate will take twice as long.

Perhaps this doesn't sound too bad. If that's the way it had to be, we could live with it. But it doesn't have to be that way. The Runge–Kutta method does much better.

8.1.2 THE RUNGE–KUTTA METHOD

You might think that the way to improve on Euler's method would be to use the Taylor expansion of Eq. (8.5) again, but keep terms to higher order. For instance, in addition to the order h term we could keep the order h^2 term, which is equal to

$$\frac{1}{2}h^2\frac{\mathrm{d}^2x}{\mathrm{d}t^2} = \frac{1}{2}h^2\frac{\mathrm{d}f}{\mathrm{d}t}. \tag{8.9}$$

This would give us a more accurate expression for $x(t + h)$, and in some cases this approach might work, but in a lot of cases it would not. It requires us to know the derivative $\mathrm{d}f/\mathrm{d}t$, which we can calculate only if we have an explicit expression for f. Often we have no such expression because, for instance, the

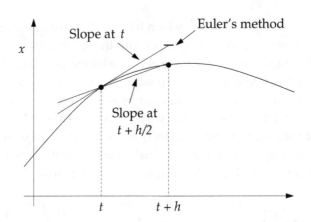

Figure 8.2: Euler's method and the second-order Runge–Kutta method. Euler's method is equivalent to taking the slope dx/dt at time t and extrapolating it into the future to time $t + h$. A better approximation is to perform the extrapolation using the slope at time $t + \frac{1}{2}h$.

function f is calculated as the output of another computer program or function and therefore doesn't have a mathematical formula. And even if f is known explicitly, a method that requires us to calculate its derivative is less convenient than the Runge–Kutta method, which gives higher accuracy and doesn't require any derivatives.

The Runge–Kutta method is really a set of methods—there are many of them of different orders, which give results of varying degrees of accuracy. In fact technically Euler's method is a Runge–Kutta method. It is the first-order Runge–Kutta method. Let us look at the next method in the series, the second-order method, also sometimes called the *midpoint method*, for reasons that will shortly become clear.

Euler's method can be represented in graphical fashion as shown in Fig. 8.2. The curve represents the true form of $x(t)$, which we are trying to calculate. The differential equation $dx/dt = f(x,t)$ tells us that the slope of the solution is equal to the function $f(x,t)$, so that, given the value of x at time t we can calculate the slope at that point, as shown in the figure. Then we extrapolate that slope to time $t + h$ and it gives us an estimate of the value of $x(t + h)$, which is labeled "Euler's method" in the figure. If the curve of $x(t)$ were in fact a straight line between t and $t + h$, then this method would give a perfect estimate of $x(t + h)$. But if it's curved, as in the picture, then the estimate is only approximate, and the error introduced is the difference between the estimate and the true value of $x(t + h)$.

Now suppose we do the same calculation but instead use the slope at the midpoint $t + \frac{1}{2}h$ to do our extrapolation, as shown in the figure. If we extrapolate using this slope we get a different estimate of $x(t + h)$ which is usually significantly better than Euler's method. This is the basis for the second-order Runge–Kutta method.

In mathematical terms the method involves performing a Taylor expansion around $t + \frac{1}{2}h$ to get the value of $x(t + h)$ thus:

$$x(t + h) = x(t + \tfrac{1}{2}h) + \tfrac{1}{2}h\left(\frac{dx}{dt}\right)_{t+\frac{1}{2}h} + \tfrac{1}{8}h^2\left(\frac{d^2x}{dt^2}\right)_{t+\frac{1}{2}h} + O(h^3). \qquad (8.10)$$

Similarly we can derive an expression for $x(t)$:

$$x(t) = x(t + \tfrac{1}{2}h) - \tfrac{1}{2}h\left(\frac{dx}{dt}\right)_{t+\frac{1}{2}h} + \tfrac{1}{8}h^2\left(\frac{d^2x}{dt^2}\right)_{t+\frac{1}{2}h} + O(h^3). \qquad (8.11)$$

Subtracting the second expression from the first and rearranging then gives

$$x(t + h) = x(t) + h\left(\frac{dx}{dt}\right)_{t+\frac{1}{2}h} + O(h^3)$$
$$= x(t) + hf\left(x(t + \tfrac{1}{2}h), t + \tfrac{1}{2}h\right) + O(h^3). \qquad (8.12)$$

Notice that the term in h^2 has completely disappeared. The error term is now $O(h^3)$, so our approximation is a whole factor of h more accurate than before. If h is small this could make a big difference to the accuracy of the calculation.

Though it looks promising, there is a problem with this approach: Eq. (8.12) requires a knowledge of $x(t + \frac{1}{2}h)$, which we don't have. We only know the value at $x(t)$. We get around this by approximating $x(t + \frac{1}{2}h)$ using Euler's method $x(t + \frac{1}{2}h) = x(t) + \frac{1}{2}hf(x, t)$ and then substituting into the equation above. The complete calculation for a single step can be written like this:

$$k_1 = hf(x, t), \qquad (8.13a)$$
$$k_2 = hf(x + \tfrac{1}{2}k_1, t + \tfrac{1}{2}h), \qquad (8.13b)$$
$$x(t + h) = x(t) + k_2. \qquad (8.13c)$$

Notice how the first equation gives us a value for k_1 which, when inserted into the second equation, gives us our estimate of $x(t + \frac{1}{2}h)$. Then the resulting value of k_2, inserted into the third equation, gives us the final Runge–Kutta estimate for $x(t + h)$.

These are the equations for the second-order Runge–Kutta method. As with the methods for performing integrals that we studied a Chapter 5, a "second-order" method, in this context, is a method *accurate* to order h^2, meaning that the *error* is of order h^3. Euler's method, by contrast, is a first-order method with an error of order h^2. Note that these designations refer to just a single step of each method. As discussed in Section 8.1.1, real calculations involve doing many steps one after another, with errors that accumulate, so that the accuracy of the final calculation is poorer (typically one order in h poorer) than the individual steps.

The second-order Runge–Kutta method is only a little more complicated to program than Euler's method, but gives much more accurate results for any given value of h. Or, alternatively, we could make h bigger—and so take fewer steps—while still getting the same level of accuracy as Euler's method, thus creating a program that achieves the same result as Euler's method but runs faster.

We are not entirely done with our derivation yet, however. Since we don't have an exact value of $x(t + \frac{1}{2}h)$ and had to approximate it using Euler's method, there is an extra source of error in Eq. (8.12), coming from this second approximation, in addition to the $O(h^3)$ error we have already acknowledged. How do we know that this second error isn't larger than $O(h^3)$ and doesn't make the accuracy of our calculation worse?

We can show that in fact this is not a problem by expanding the quantity $f(x + \frac{1}{2}k_1, t + \frac{1}{2}h)$ in Eq. (8.13b) in its first argument only, around $x(t + \frac{1}{2}h)$:

$$f(x(t) + \tfrac{1}{2}k_1, t + \tfrac{1}{2}h) = f(x(t + \tfrac{1}{2}h), t + \tfrac{1}{2}h)$$

$$+ \, [x(t) + \tfrac{1}{2}k_1 - x(t + \tfrac{1}{2}h)]\left(\frac{\partial f}{\partial x}\right)_{x(t+h/2),t+h/2} + \, O([x(t) + \tfrac{1}{2}k_1 - x(t + \tfrac{1}{2}h)]^2).$$

$$(8.14)$$

But from Eq. (8.5) we have

$$x(t + \tfrac{1}{2}h) = x(t) + \tfrac{1}{2}hf(x,t) + O(h^2) = x(t) + \tfrac{1}{2}k_1 + O(h^2), \qquad (8.15)$$

so $x(t) + \frac{1}{2}k_1 - x(t + \frac{1}{2}h) = O(h^2)$ and

$$f(x(t) + \tfrac{1}{2}k_1, t + \tfrac{1}{2}h) = f(x(t + \tfrac{1}{2}h), t + \tfrac{1}{2}h) + O(h^2). \qquad (8.16)$$

This means that Eq. (8.13b) gives $k_2 = hf(x(t + \frac{1}{2}h), t + \frac{1}{2}h) + O(h^3)$, and hence there's no problem—our Euler's method approximation for $x(t + \frac{1}{2}h)$ does introduce an additional error into the calculation, but the error goes like h^3 and hence our second-order Runge–Kutta method is still accurate to $O(h^3)$ overall.

EXAMPLE 8.2: THE SECOND-ORDER RUNGE–KUTTA METHOD

Let us use the second-order Runge–Kutta method to solve the same differential equation as we solved in Example 8.1. The program is a minor modification of our program for Euler's method:

File: rk2.py

```python
from math import sin
from numpy import arange
from pylab import plot,xlabel,ylabel,show

def f(x,t):
    return -x**3 + sin(t)

a = 0.0
b = 10.0
N = 10
h = (b-a)/N

tpoints = arange(a,b,h)
xpoints = []

x = 0.0
for t in tpoints:
    xpoints.append(x)
    k1 = h*f(x,t)
    k2 = h*f(x+0.5*k1,t+0.5*h)
    x += k2

plot(tpoints,xpoints)
xlabel("t")
ylabel("x(t)")
show()
```

If we run this program repeatedly with different values for the number of points N, starting with 10, then 20, then 50, then 100, and plot the results, we get the plot shown in Fig. 8.3. The figure reveals that the solution with 10 points is quite poor, as is the solution with 20. But the solutions for 50 and 100 points look very similar, indicating that the method has converged to a result close to the true solution, and indeed a comparison with Fig. 8.1 shows good agreement with our Euler's method solution, which used 1000 points.

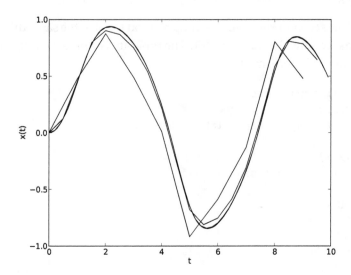

Figure 8.3: Solutions calculated with the second-order Runge–Kutta method. Solutions to Eq. (8.7) calculated using the second-order Runge–Kutta method with $N = 10$, $20, 50$, and 100 steps.

8.1.3 THE FOURTH-ORDER RUNGE–KUTTA METHOD

We can take this approach further. By performing Taylor expansions around various points and then taking the right linear combinations of them, we can arrange for terms in h^3, h^4, and so on to cancel out of our expressions, and so get more and more accurate rules for solving differential equations. The downside is that the equations become more complicated as we go to higher order. Many people feel, however, that the sweet spot is the fourth-order rule, which offers a good balance of high accuracy and equations that are still relatively simple to program. The equations look like this:

$$k_1 = hf(x,t), \tag{8.17a}$$

$$k_2 = hf(x + \tfrac{1}{2}k_1, t + \tfrac{1}{2}h), \tag{8.17b}$$

$$k_3 = hf(x + \tfrac{1}{2}k_2, t + \tfrac{1}{2}h), \tag{8.17c}$$

$$k_4 = hf(x + k_3, t + h), \tag{8.17d}$$

$$x(t+h) = x(t) + \tfrac{1}{6}(k_1 + 2k_2 + 2k_3 + k_4). \tag{8.17e}$$

This is the *fourth-order Runge–Kutta method,* and it is by far the most common method for the numerical solution of ordinary differential equations. It is accurate to terms of order h^4 and carries an error of order h^5. Although its derivation is quite complicated (we'll not go over the algebra—it's very tedious), the final equations are relatively simple. There are just five of them, and yet the result is a method that is three orders of h more accurate than Euler's method for steps of the same size. In practice this can make the fourth-order method as much as a million times more accurate than Euler's method. Indeed the fourth-order method is significantly better even than the second-order method of Section 8.1.2. Alternatively, we can use the fourth-order Runge–Kutta method with much larger h and many fewer steps and still get accuracy just as good as Euler's method, giving a method that runs far faster yet gives comparable results.

For many professional physicists, the fourth-order Runge–Kutta method is the first method they turn to when they want to solve an ordinary differential equation on the computer. It is simple to program and gives excellent results. It is the workhorse of differential equation solvers and one of the best known computer algorithms of any kind anywhere.

EXAMPLE 8.3: THE FOURTH-ORDER RUNGE–KUTTA METHOD

Let us once more solve the differential equation from Eq. (8.7), this time using the fourth-order Runge–Kutta method. The program is again only a minor modification of our previous ones:

```
from math import sin                                    File: rk4.py
from numpy import arange
from pylab import plot,xlabel,ylabel,show

def f(x,t):
    return -x**3 + sin(t)

a = 0.0
b = 10.0
N = 10
h = (b-a)/N

tpoints = arange(a,b,h)
xpoints = []
x = 0.0
```

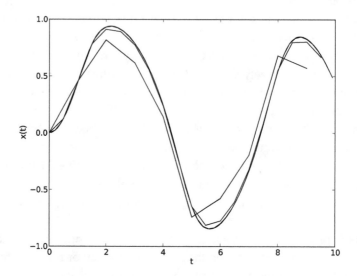

Figure 8.4: Solutions calculated with the fourth-order Runge–Kutta method. Solutions to Eq. (8.7) calculated using the fourth-order Runge–Kutta method with $N = 10$, 20, 50, and 100 steps.

```
for t in tpoints:
    xpoints.append(x)
    k1 = h*f(x,t)
    k2 = h*f(x+0.5*k1,t+0.5*h)
    k3 = h*f(x+0.5*k2,t+0.5*h)
    k4 = h*f(x+k3,t+h)
    x += (k1+2*k2+2*k3+k4)/6

plot(tpoints,xpoints)
xlabel("t")
ylabel("x(t)")
show()
```

Again we run the program repeatedly with $N = 10, 20, 50$, and 100. Figure 8.4 shows the results. Now we see that, remarkably, even the solution with 20 points is close to the final converged solution for the equation. With only 20 points we get quite a jagged curve—20 points is not enough to make the curve appear smooth in the plot—but the points nonetheless lie close to the final solution of the equation. With only 20 points the fourth-order method has calculated a solution almost as accurate as Euler's method with a thousand points.

One minor downside of the fourth-order Runge–Kutta method, and indeed of all Runge–Kutta methods, is that if you get the equations wrong, it may not be obvious in the solution they produce. If, for example, you miss one of the factors of $\frac{1}{2}$ or 2, or have a minus sign when you should have a plus, then the method will probably still produce a solution that looks approximately right. The solution will be *much* less accurate than the correct fourth-order method— if you don't use the equations exactly as in Eq. (8.17) you will probably only get a solution about as accurate as Euler's method, which, as we have seen, is much worse. This means that you must be careful when writing programs that use the Runge–Kutta method. Check your code in detail to make sure all the equations are exactly correct. If you make a mistake you may never realize it because your program will appear to give reasonable answers, but in fact there will be large errors. This contrasts with most other types of calculation in computational physics, where if you make even a small error in the program it is likely to produce ridiculous results that are so obviously wrong that the error is relatively easy to spot.

Exercise 8.1: A low-pass filter

Here is a simple electronic circuit with one resistor and one capacitor:

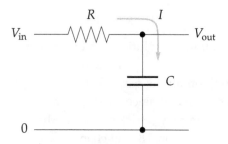

This circuit acts as a low-pass filter: you send a signal in on the left and it comes out filtered on the right.

Using Ohm's law and the capacitor law and assuming that the output load has very high impedance, so that a negligible amount of current flows through it, we can write down the equations governing this circuit as follows. Let I be the current that flows through R and into the capacitor, and let Q be the charge on the capacitor. Then:

$$IR = V_{\text{in}} - V_{\text{out}}, \qquad Q = CV_{\text{out}}, \qquad I = \frac{dQ}{dt}.$$

Substituting the second equation into the third, then substituting the result into the first equation, we find that $V_{in} - V_{out} = RC \, (dV_{out}/dt)$, or equivalently

$$\frac{dV_{out}}{dt} = \frac{1}{RC}(V_{in} - V_{out}).$$

a) Write a program (or modify a previous one) to solve this equation for $V_{out}(t)$ using the fourth-order Runge–Kutta method when the input signal is a square-wave with frequency 1 and amplitude 1:

$$V_{in}(t) = \begin{cases} 1 & \text{if } \lfloor 2t \rfloor \text{ is even,} \\ -1 & \text{if } \lfloor 2t \rfloor \text{ is odd,} \end{cases} \tag{8.18}$$

where $\lfloor x \rfloor$ means x rounded down to the next lowest integer. Use the program to make plots of the output of the filter circuit from $t = 0$ to $t = 10$ when $RC = 0.01$, 0.1, and 1, with initial condition $V_{out}(0) = 0$. You will have to make a decision about what value of h to use in your calculation. Small values give more accurate results, but the program will take longer to run. Try a variety of different values and choose one for your final calculations that seems sensible to you.

b) Based on the graphs produced by your program, describe what you see and explain what the circuit is doing.

A program similar to the one you wrote is running inside most stereos and music players, to create the effect of the "bass" control. In the old days, the bass control on a stereo would have been connected to a real electronic low-pass filter in the amplifier circuitry, but these days there is just a computer processor that simulates the behavior of the filter in a manner similar to your program.

8.1.4 SOLUTIONS OVER INFINITE RANGES

We have seen how to find the solution of a differential equation starting from a given initial condition and going a finite distance in t, but in some cases we want to find the solution all the way out to $t = \infty$. In that case we cannot use the method above directly, since we'd need an infinite number of steps to reach $t = \infty$, but we can play a trick similar to the one we played when we were doing integrals in Section 5.8, and change variables. We define

$$u = \frac{t}{1+t} \qquad \text{or equivalently} \qquad t = \frac{u}{1-u}, \tag{8.19}$$

so that as $t \to \infty$ we have $u \to 1$. Then, using the chain rule, we can rewrite our differential equation $dx/dt = f(x,t)$ as

$$\frac{dx}{du}\frac{du}{dt} = f(x,t), \tag{8.20}$$

or

$$\frac{dx}{du} = \frac{dt}{du} f\left(x, \frac{u}{1-u}\right).$$

(8.21)

But

$$\frac{dt}{du} = \frac{1}{(1-u)^2},$$

(8.22)

so

$$\frac{dx}{du} = (1-u)^{-2} f\left(x, \frac{u}{1-u}\right).$$

(8.23)

If we define a new function $g(x, u)$ by

$$g(x, u) = (1-u)^{-2} f\left(x, \frac{u}{1-u}\right)$$

(8.24)

then we have

$$\frac{dx}{du} = g(x, u),$$

(8.25)

which is a normal first-order differential equation again, as before. Solving this equation for values of u up to 1 is equivalent to solving the original equation for values of t up to infinity. The solution will give us $x(u)$ and we then map u back onto t using Eq. (8.19) to get $x(t)$.

EXAMPLE 8.4: SOLUTION OVER AN INFINITE RANGE

Suppose we want to solve the equation

$$\frac{dx}{dt} = \frac{1}{x^2 + t^2}$$

from $t = 0$ to $t = \infty$ with $x = 1$ at $t = 0$. What would be the equivalent differential equation in x and u that we would solve?

Applying Eq. (8.24), we have

$$g(x, u) = (1-u)^{-2} \frac{1}{x^2 + u^2/(1-u)^2} = \frac{1}{x^2(1-u)^2 + u^2}.$$

(8.26)

So we would solve the equation

$$\frac{dx}{du} = \frac{1}{x^2(1-u)^2 + u^2}$$

(8.27)

from $u = 0$ to $u = 1$, with an initial condition $x = 1$ at $u = 0$. We can calculate the solution with only a small modification of the program we used in Example 8.3:

File: odeinf.py

```
from numpy import arange
from pylab import plot,xlabel,ylabel,xlim,show

def g(x,u):
    return 1/(x**2*(1-u)**2+u**2)

a = 0.0
b = 1.0
N = 100
h = (b-a)/N

upoints = arange(a,b,h)
tpoints = []
xpoints = []

x = 1.0
for u in upoints:
    tpoints.append(u/(1-u))
    xpoints.append(x)
    k1 = h*g(x,u)
    k2 = h*g(x+0.5*k1,u+0.5*h)
    k3 = h*g(x+0.5*k2,u+0.5*h)
    k4 = h*g(x+k3,u+h)
    x += (k1+2*k2+2*k3+k4)/6

plot(tpoints,xpoints)
xlim(0,80)
xlabel("t")
ylabel("x(t)")
show()
```

Note how we made a list tpoints of the value of t at each step of the Runge–Kutta method, as we went along. Although we don't need these values for the solution itself, we use them at the end to make a plot of the final solution in terms of t rather than u. The resulting plot is shown in Fig. 8.5. (It only goes up to $t = 80$. Obviously it cannot go all the way out to infinity—one cannot draw an infinitely wide plot—but the solution itself does go out to infinity.)

As with the integrals of Section 5.8, there are other changes of variables that can be used in calculations like this, including transformations based on trigonometric functions, hyperbolic functions, and others. The transformation of Eq. (8.19) is often a good first guess—it works well in many cases—but other

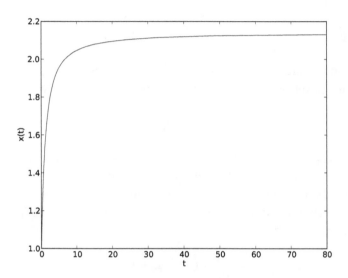

Figure 8.5: Solution of a differential equation to $t = \infty$**.** The solution of the differential equation in Eq. (8.4), calculated by solving all the way out to $t = \infty$ as described in the text, although only the part of the solution up to $t = 80$ is shown here.

choices can be appropriate too. A shrewd choice of variables can make the algebra easier, simplify the form of the function $g(x, u)$, or give the solution more accuracy in a region of particular interest.

8.2 DIFFERENTIAL EQUATIONS WITH MORE THAN ONE VARIABLE

So far we have considered ordinary differential equations with only one dependent variable x, but in many physics problems we have more than one variable. That is, we have *simultaneous differential equations*, where the derivative of each variable can depend on any or all of the variables, as well as the independent variable t. For example:

$$\frac{dx}{dt} = xy - x, \qquad \frac{dy}{dt} = y - xy + \sin^2 \omega t. \tag{8.28}$$

Note that there is still only one *independent* variable t. These are still ordinary differential equations, not partial differential equations.

A general form for two first-order simultaneous differential equations is[2]

$$\frac{dx}{dt} = f_x(x, y, t), \qquad \frac{dy}{dt} = f_y(x, y, t), \tag{8.29}$$

where f_x and f_y are general, possibly nonlinear, functions of x, y, and t. For an arbitrary number of variables the equations can be written using vector notation as

$$\frac{d\mathbf{r}}{dt} = \mathbf{f}(\mathbf{r}, t), \tag{8.30}$$

where $\mathbf{r} = (x, y, \ldots)$ and $\mathbf{f}$ is a vector of functions $\mathbf{f}(\mathbf{r}, t) = (f_x(\mathbf{r}, t), f_y(\mathbf{r}, t), \ldots)$.

Although simultaneous differential equations are often a lot harder to solve analytically than single equations, when solving computationally they are actually not much more difficult than the one-variable case. For instance, we can Taylor expand the vector $\mathbf{r}$ thus:

$$\mathbf{r}(t + h) = \mathbf{r}(t) + h\frac{d\mathbf{r}}{dt} + O(h^2) = \mathbf{r}(t) + h\mathbf{f}(\mathbf{r}, t) + O(h^2). \tag{8.31}$$

Dropping the terms of order h^2 and higher we get Euler's method for the multi-variable case:

$$\mathbf{r}(t + h) = \mathbf{r}(t) + h\mathbf{f}(\mathbf{r}, t). \tag{8.32}$$

The Taylor expansions used to derive the Runge–Kutta rules also generalize straightforwardly to the multi-variable case, and in particular the multi-variable version of the fourth-order Runge–Kutta method is an obvious vector generalization of the one-variable version:

$$\mathbf{k}_1 = h\mathbf{f}(\mathbf{r}, t), \tag{8.33a}$$

$$\mathbf{k}_2 = h\mathbf{f}(\mathbf{r} + \tfrac{1}{2}\mathbf{k}_1, t + \tfrac{1}{2}h), \tag{8.33b}$$

$$\mathbf{k}_3 = h\mathbf{f}(\mathbf{r} + \tfrac{1}{2}\mathbf{k}_2, t + \tfrac{1}{2}h), \tag{8.33c}$$

$$\mathbf{k}_4 = h\mathbf{f}(\mathbf{r} + \mathbf{k}_3, t + h), \tag{8.33d}$$

$$\mathbf{r}(t + h) = \mathbf{r}(t) + \tfrac{1}{6}(\mathbf{k}_1 + 2\mathbf{k}_2 + 2\mathbf{k}_3 + \mathbf{k}_4). \tag{8.33e}$$

These equations can be conveniently translated into Python using arrays to represent the vectors. Since Python allows us to do arithmetic with vectors di-

---

[2]Although it covers most cases of interest, this is not the most general possible form. In principle, dx/dt could also depend on dy/dt, possibly in nonlinear fashion. We'll assume that the equations have already been separated in the derivatives to remove such dependencies. It's worth noting, however, that such separation is not always possible—for instance, when the equations involve transcendental functions. In such cases, other methods, such as the relaxation methods discussed in Section 8.6.2, may be needed to find a solution.

rectly, and allows vectors to be both the arguments and the results of functions, the code is only slightly more complicated than for the one-variable case.

EXAMPLE 8.5: SIMULTANEOUS ORDINARY DIFFERENTIAL EQUATIONS

Let us calculate a solution to the equations given in Eq. (8.28) from $t = 0$ to $t = 10$, for the case $\omega = 1$ with initial condition $x = y = 1$ at $t = 0$. Here is a suitable program, again based on a slight modification of our earlier programs.

```python
from math import sin
from numpy import array,arange
from pylab import plot,xlabel,show

def f(r,t):
    x = r[0]
    y = r[1]
    fx = x*y - x
    fy = y - x*y + sin(t)**2
    return array([fx,fy],float)

a = 0.0
b = 10.0
N = 1000
h = (b-a)/N

tpoints = arange(a,b,h)
xpoints = []
ypoints = []

r = array([1.0,1.0],float)
for t in tpoints:
    xpoints.append(r[0])
    ypoints.append(r[1])
    k1 = h*f(r,t)
    k2 = h*f(r+0.5*k1,t+0.5*h)
    k3 = h*f(r+0.5*k2,t+0.5*h)
    k4 = h*f(r+k3,t+h)
    r += (k1+2*k2+2*k3+k4)/6
plot(tpoints,xpoints)
plot(tpoints,ypoints)
xlabel("t")
show()
```

File: odesim.py

Note in particular the definition of the function f(r,t), which takes a vector argument r, breaks it apart into its components x and y, forms the values of f_x and f_y from them, then puts those values together into an array and returns that array as the final output of the function. In fact, the construction of this function is really the only complicated part of the program; in other respects the program is almost identical to the program we used for the one-variable case in Example 8.3. The lines representing the Runge–Kutta method itself are unchanged except for the replacement of the scalar variable x by the new vector variable r.

This program will form the basis for the solution of many other problems in this chapter.

Exercise 8.2: The Lotka–Volterra equations

The Lotka–Volterra equations are a mathematical model of predator–prey interactions between biological species. Let two variables x and y be proportional to the size of the populations of two species, traditionally called "rabbits" (the prey) and "foxes" (the predators). You could think of x and y as being the population in thousands, say, so that $x = 2$ means there are 2000 rabbits. Strictly the only allowed values of x and y would then be multiples of 0.001, since you can only have whole numbers of rabbits or foxes. But 0.001 is a pretty close spacing of values, so it's a decent approximation to treat x and y as continuous real numbers so long as neither gets very close to zero.

In the Lotka–Volterra model the rabbits reproduce at a rate proportional to their population, but are eaten by the foxes at a rate proportional to both their own population and the population of foxes:

$$\frac{dx}{dt} = \alpha x - \beta xy,$$

where α and β are constants. At the same time the foxes reproduce at a rate proportional to the rate at which they eat rabbits—because they need food to grow and reproduce—but also die of old age at a rate proportional to their own population:

$$\frac{dy}{dt} = \gamma xy - \delta y,$$

where γ and δ are also constants.

a) Write a program to solve these equations using the fourth-order Runge–Kutta method for the case $\alpha = 1$, $\beta = \gamma = 0.5$, and $\delta = 2$, starting from the initial condition $x = y = 2$. Have the program make a graph showing both x and y as a function of time on the same axes from $t = 0$ to $t = 30$. (Hint: Notice that the differential equations in this case do not depend explicitly on time t—in vector notation, the right-hand side of each equation is a function $f(\mathbf{r})$ with no t dependence. You may nonetheless find it convenient to define a Python function $\mathtt{f(r,t)}$ including the time variable, so that your program takes the same form as programs given earlier in this chapter. You don't have to do it that way, but it can avoid some confusion. Several of the following exercises have a similar lack of explicit time-dependence.)

b) Describe in words what is going on in the system, in terms of rabbits and foxes.

Exercise 8.3: The Lorenz equations

One of the most celebrated sets of differential equations in physics is the Lorenz equations:

$$\frac{dx}{dt} = \sigma(y - x), \qquad \frac{dy}{dt} = rx - y - xz, \qquad \frac{dz}{dt} = xy - bz,$$

where σ, r, and b are constants. (The names σ, r, and b are odd, but traditional—they are always used in these equations for historical reasons.)

These equations were first studied by Edward Lorenz in 1963, who derived them from a simplified model of weather patterns. The reason for their fame is that they were one of the first incontrovertible examples of *deterministic chaos*, the occurrence of apparently random motion even though there is no randomness built into the equations. We encountered a different example of chaos in the logistic map of Exercise 3.6.

a) Write a program to solve the Lorenz equations for the case $\sigma = 10$, $r = 28$, and $b = \frac{8}{3}$ in the range from $t = 0$ to $t = 50$ with initial conditions $(x, y, z) = (0, 1, 0)$. Have your program make a plot of y as a function of time. Note the unpredictable nature of the motion. (Hint: If you base your program on previous ones, be careful. This problem has parameters r and b with the same names as variables in previous programs—make sure to give your variables new names, or use different names for the parameters, to avoid introducing errors into your code.)

b) Modify your program to produce a plot of z against x. You should see a picture of the famous "strange attractor" of the Lorenz equations, a lopsided butterfly-shaped plot that never repeats itself.

8.3 SECOND-ORDER DIFFERENTIAL EQUATIONS

So far we have looked at first-order differential equations, but first-order equations are in fact quite rare in physics. Many, perhaps most, of the equations encountered in physics are second-order or higher. Luckily, now that we know how to solve first-order equations, solving second-order ones is pretty easy, because of the following trick.

Consider first the simple case where there is only one dependent variable x. The general form for a second-order differential equation with one dependent variable is

$$\frac{d^2x}{dt^2} = f\left(x, \frac{dx}{dt}, t\right). \tag{8.34}$$

That is, the second derivative can be any arbitrary function, including possibly a nonlinear function, of x, t, and the derivative dx/dt. So we could have, for instance,

$$\frac{d^2x}{dt^2} = \frac{1}{x}\left(\frac{dx}{dt}\right)^2 + 2\frac{dx}{dt} - x^3 e^{-4t}. \tag{8.35}$$

Now here's the trick. We define a new quantity y by

$$\frac{dx}{dt} = y, \tag{8.36}$$

in terms of which Eq. (8.34) can be written

$$\frac{dy}{dt} = f(x, y, t). \tag{8.37}$$

Between them, Eqs. (8.36) and (8.37) are equivalent to the one second-order equation we started with, as we can prove by substituting (8.36) into (8.37) to recover (8.34) again. But (8.36) and (8.37) are both first order. So this process reduces our second-order equation to two simultaneous first-order equations. And we already know how to solve simultaneous first-order equations, so we can now use the techniques we have learned to solve our second-order equation as well.

We can do a similar trick for higher-order equations. For instance, the general form of a third-order equation is

$$\frac{d^3x}{dt^3} = f\left(x, \frac{dx}{dt}, \frac{d^2x}{dt^2}, t\right). \tag{8.38}$$

We define two additional variables y and z by

$$\frac{dx}{dt} = y, \qquad \frac{dy}{dt} = z, \tag{8.39}$$

so that Eq. (8.38) becomes

$$\frac{dz}{dt} = f(x, y, z, t). \tag{8.40}$$

Between them Eqs. (8.39) and (8.40) give us three first-order equations that are equivalent to our one third-order equation, so again we can solve using the methods we already know about for simultaneous first-order equations.

This approach can be generalized to equations of any order, although equations of order higher than three are rare in physics, so you probably won't need to solve them often.

The method can also be generalized in a straightforward manner to equations with more than one dependent variable—the variables become vectors but the basic equations are the same as above. Thus a set of simultaneous second-order equations can be written in vector form as

$$\frac{d^2\mathbf{r}}{dt^2} = \mathbf{f}\left(\mathbf{r}, \frac{d\mathbf{r}}{dt}, t\right), \tag{8.41}$$

which is equivalent to the first-order equations

$$\frac{d\mathbf{r}}{dt} = \mathbf{s}, \qquad \frac{d\mathbf{s}}{dt} = \mathbf{f}(\mathbf{r}, \mathbf{s}, t). \tag{8.42}$$

If we started off with two simultaneous second-order equations, for instance, then we would end up with *four* simultaneous first-order equations after applying the transformation above. More generally, an initial system of n equations of mth order becomes a system of $m \times n$ simultaneous first-order equations, which we can solve by the standard methods.

EXAMPLE 8.6: THE NONLINEAR PENDULUM

A standard problem in physics is the linear pendulum, where you approximate the behavior of a pendulum by a linear differential equation than can be solved exactly. But a real pendulum is nonlinear. Consider a pendulum with an arm of length ℓ holding a bob of mass m:

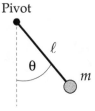

In terms of the angle θ of displacement of the arm from the vertical, the acceleration of the mass is $\ell \, d^2\theta/dt^2$ in the tangential direction. Meanwhile the force on the mass is vertically downward with magnitude mg, where $g = 9.81 \, \mathrm{m\,s^{-2}}$ is the acceleration due to gravity and, for the sake of simplicity, we are ignoring friction and assuming the arm to be massless. The component of this force in the tangential direction is $mg \sin \theta$, always toward the rest point at $\theta = 0$, and hence Newton's second law gives us an equation of motion for the pendulum of the form

$$ml \frac{d^2\theta}{dt^2} = -mg \sin \theta, \qquad (8.43)$$

or equivalently

$$\frac{d^2\theta}{dt^2} = -\frac{g}{\ell} \sin \theta. \qquad (8.44)$$

Because it is nonlinear it is not easy to solve this equation analytically, and no exact solution is known. But a solution on the computer is straightforward. We first use the trick described in the previous section to turn the second-order equation, Eq. (8.44), into two first-order equations. We define a new variable ω by

$$\frac{d\theta}{dt} = \omega. \qquad (8.45)$$

Then Eq. (8.44) becomes

$$\frac{d\omega}{dt} = -\frac{g}{\ell} \sin \theta. \qquad (8.46)$$

Between them, these two first-order equations are equivalent to the one second-order equation we started with. Now we combine the two variables θ and ω into a single vector $\mathbf{r} = (\theta, \omega)$ and apply the fourth-order Runge–Kutta method in vector form to solve the two equations simultaneously. We are only really interested in the solution for one of the variables, the variable θ. The method gives us the solution for both, but we can simply ignore the value of ω if we don't need it. The program will be similar to that of Example 8.5, except that the function f(r,t) must be redefined appropriately. If the arm of the pendulum were 10 cm long, for example, we would have

```
g = 9.81
l = 0.1

def f(r,t):
    theta = r[0]
    omega = r[1]
    ftheta = omega
    fomega = -(g/l)*sin(theta)
    return array([ftheta,fomega],float)
```

The rest of the program is left to you—see Exercise 8.4.

Exercise 8.4: Building on the results from Example 8.6 above, calculate the motion of a nonlinear pendulum as follows.

a) Write a program to solve the two first-order equations, Eqs. (8.45) and (8.46), using the fourth-order Runge–Kutta method for a pendulum with a 10 cm arm. Use your program to calculate the angle θ of displacement for several periods of the pendulum when it is released from a standstill at $\theta = 179°$ from the vertical. Make a graph of θ as a function of time.

b) Extend your program to create an animation of the motion of the pendulum. Your animation should, at a minimum, include a representation of the moving pendulum bob and the pendulum arm. (Hint: You will probably find the function `rate` discussed in Section 3.5 useful for making your animation run at a sensible speed. Also, you may want to make the step size for your Runge–Kutta calculation smaller than the frame-rate of your animation, i.e., do several Runge–Kutta steps per frame on screen. This is certainly allowed and may help to make your calculation more accurate.)

For a bigger challenge, take a look at Exercise 8.15 on page 398, which invites you to write a program to calculate the chaotic motion of the double pendulum.

Exercise 8.5: The driven pendulum

A pendulum like the one in Exercise 8.4 can be driven by, for example, exerting a small oscillating force horizontally on the mass. Then the equation of motion for the pendulum becomes

$$\frac{d^2\theta}{dt^2} = -\frac{g}{\ell}\sin\theta + C\cos\theta\sin\Omega t,$$

where C and Ω are constants.

a) Write a program to solve this equation for θ as a function of time with $\ell = 10$ cm, $C = 2\,\mathrm{s}^{-2}$ and $\Omega = 5\,\mathrm{s}^{-1}$ and make a plot of θ as a function of time from $t = 0$ to $t = 100$ s. Start the pendulum at rest with $\theta = 0$ and $d\theta/dt = 0$.

b) Now change the value of Ω, while keeping C the same, to find a value for which the pendulum resonates with the driving force and swings widely from side to side. Make a plot for this case also.

Exercise 8.6: Harmonic and anharmonic oscillators

The simple harmonic oscillator arises in many physical problems, in mechanics, electricity and magnetism, and condensed matter physics, among other areas. Consider the standard oscillator equation

$$\frac{d^2x}{dt^2} = -\omega^2 x.$$

a) Using the methods described in the preceding section, turn this second-order equation into two coupled first-order equations. Then write a program to solve them for the case $\omega = 1$ in the range from $t = 0$ to $t = 50$. A second-order equation requires two initial conditions, one on x and one on its derivative. For this problem use $x = 1$ and $dx/dt = 0$ as initial conditions. Have your program make a graph showing the value of x as a function of time.

b) Now increase the amplitude of the oscillations by making the initial value of x bigger—say $x = 2$—and confirm that the period of the oscillations stays roughly the same.

c) Modify your program to solve for the motion of the anharmonic oscillator described by the equation

$$\frac{d^2x}{dt^2} = -\omega^2 x^3.$$

Again take $\omega = 1$ and initial conditions $x = 1$ and $dx/dt = 0$ and make a plot of the motion of the oscillator. Again increase the amplitude. You should observe that the oscillator oscillates faster at higher amplitudes. (You can try lower amplitudes too if you like, which should be slower.) The variation of frequency with amplitude in an anharmonic oscillator was studied previously in Exercise 5.10.

d) Modify your program so that instead of plotting x against t, it plots dx/dt against x, i.e., the "velocity" of the oscillator against its "position." Such a plot is called a *phase space* plot.

e) The *van der Pol oscillator*, which appears in electronic circuits and in laser physics, is described by the equation

$$\frac{d^2x}{dt^2} - \mu(1 - x^2)\frac{dx}{dt} + \omega^2 x = 0.$$

Modify your program to solve this equation from $t = 0$ to $t = 20$ and hence make a phase space plot for the van der Pol oscillator with $\omega = 1$, $\mu = 1$, and initial conditions $x = 1$ and $dx/dt = 0$. Try it also for $\mu = 2$ and $\mu = 4$ (still with $\omega = 1$). Make sure you use a small enough value of the time interval h to get a smooth, accurate phase space plot.

Exercise 8.7: Trajectory with air resistance

Many elementary mechanics problems deal with the physics of objects moving or flying through the air, but they almost always ignore friction and air resistance to make the equations solvable. If we're using a computer, however, we don't need solvable equations.

Consider, for instance, a spherical cannonball shot from a cannon standing on level ground. The air resistance on a moving sphere is a force in the opposite direction to the motion with magnitude

$$F = \tfrac{1}{2}\pi R^2 \rho C v^2,$$

where R is the sphere's radius, ρ is the density of air, v is the velocity, and C is the so-called *coefficient of drag* (a property of the shape of the moving object, in this case a sphere).

a) Starting from Newton's second law, $F = ma$, show that the equations of motion for the position (x, y) of the cannonball are

$$\ddot{x} = -\frac{\pi R^2 \rho C}{2m} \dot{x}\sqrt{\dot{x}^2 + \dot{y}^2}, \qquad \ddot{y} = -g - \frac{\pi R^2 \rho C}{2m} \dot{y}\sqrt{\dot{x}^2 + \dot{y}^2},$$

where m is the mass of the cannonball, g is the acceleration due to gravity, and $\dot{x}$ and $\ddot{x}$ are the first and second derivatives of x with respect to time.

b) Change these two second-order equations into four first-order equations using the methods you have learned, then write a program that solves the equations for a cannonball of mass 1 kg and radius 8 cm, shot at 30° to the horizontal with initial velocity $100 \, \text{m s}^{-1}$. The density of air is $\rho = 1.22 \, \text{kg m}^{-3}$ and the coefficient of drag for a sphere is $C = 0.47$. Make a plot of the trajectory of the cannonball (i.e., a graph of y as a function of x).

c) When one ignores air resistance, the distance traveled by a projectile does not depend on the mass of the projectile. In real life, however, mass certainly does make a difference. Use your program to estimate the total distance traveled (over horizontal ground) by the cannonball above, and then experiment with the program to determine whether the cannonball travels further if it is heavier or lighter. You could, for instance, plot a series of trajectories for cannonballs of different masses, or you could make a graph of distance traveled as a function of mass. Describe briefly what you discover.

Exercise 8.8: Space garbage

A heavy steel rod and a spherical ball-bearing, discarded by a passing spaceship, are floating in zero gravity and the ball bearing is orbiting around the rod under the effect of its gravitational pull:

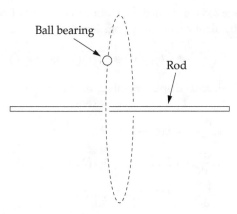

For simplicity we'll assume that the rod is of negligible cross-section and heavy enough that it doesn't move significantly, and that the ball bearing is orbiting around the rod's mid-point in a plane perpendicular to the rod.

a) Treating the rod as a line of mass M and length L and the ball bearing as a point mass m, show that the attractive force F felt by the ball bearing in the direction toward the center of the rod is given by

$$F = \frac{GMm}{L}\sqrt{x^2 + y^2} \int_{-L/2}^{L/2} \frac{dz}{(x^2 + y^2 + z^2)^{3/2}},$$

where G is Newton's gravitational constant and x and y are the coordinates of the ball bearing in the plane perpendicular to the rod. The integral can be done in closed form and gives

$$F = \frac{GMm}{\sqrt{(x^2 + y^2)(x^2 + y^2 + L^2/4)}}.$$

Hence show that the equations of motion for the position x, y of the ball bearing in the xy-plane are

$$\frac{d^2x}{dt^2} = -GM\frac{x}{r^2\sqrt{r^2 + L^2/4}}, \qquad \frac{d^2y}{dt^2} = -GM\frac{y}{r^2\sqrt{r^2 + L^2/4}},$$

where $r = \sqrt{x^2 + y^2}$.

b) Convert these two second-order equations into four first-order ones using the techniques of Section 8.3. Then, working in units where $G = 1$, write a program to solve them for $M = 10$, $L = 2$, and initial conditions $(x, y) = (1, 0)$ with velocity of $+1$ in the y direction. Calculate the orbit from $t = 0$ to $t = 10$ and make a plot of it, meaning a plot of y against x. You should find that the ball bearing does not orbit in a circle or ellipse as a planet does, but has a precessing orbit, which arises because the attractive force is not a simple $1/r^2$ force as it is for a planet orbiting the Sun.

Exercise 8.9: Vibration in a one-dimensional system

In Example 6.2 on page 235 we studied the motion of a system of N identical masses (in zero gravity) joined by identical linear springs like this:

As we showed, the horizontal displacements ξ_i of masses $i = 1 \ldots N$ satisfy equations of motion

$$m\frac{d^2\xi_1}{dt^2} = k(\xi_2 - \xi_1) + F_1,$$

$$m\frac{d^2\xi_i}{dt^2} = k(\xi_{i+1} - \xi_i) + k(\xi_{i-1} - \xi_i) + F_i,$$

$$m\frac{d^2\xi_N}{dt^2} = k(\xi_{N-1} - \xi_N) + F_N.$$

where m is the mass, k is the spring constant, and F_i is the external force on mass i. In Example 6.2 we showed how these equations could be solved by guessing a form for the solution and using a matrix method. Here we'll solve them more directly.

a) Write a program to solve for the motion of the masses using the fourth-order Runge–Kutta method for the case we studied previously where $m = 1$ and $k = 6$, and the driving forces are all zero except for $F_1 = \cos \omega t$ with $\omega = 2$. Plot your solutions for the displacements ξ_i of all the masses as a function of time from $t = 0$ to $t = 20$ on the same plot. Write your program to work with general N, but test it out for small values—$N = 5$ is a reasonable choice.

You will need first of all to convert the N second-order equations of motion into $2N$ first-order equations. Then combine all of the dependent variables in those equations into a single large vector $\mathbf{r}$ to which you can apply the Runge–Kutta method in the standard fashion.

b) Modify your program to create an animation of the movement of the masses, represented as spheres on the computer screen. You will probably find the `rate` function discussed in Section 3.5 useful for making your animation run at a sensible speed.

8.4 VARYING THE STEP SIZE

The methods we have seen so far in this chapter all use repeated steps of a fixed size h, the size being chosen by you, the programmer. In most situations, however, we can get better results if we allow the step size to vary during the running of the program, with the program choosing the best value at each step.

Suppose we are solving a first-order differential equation of the general form $dx/dt = f(x, t)$ and suppose as a function of time the solution looks something like Fig. 8.6. In some regions the function is slowly varying, in which case we can accurately capture its shape with only a few, widely spaced points. But in the central region of the figure the function varies rapidly and in this region we need points that are more closely spaced. If we are allowed to vary the size h of our steps, making them large in the regions where the solution varies little and small when we need more detail, then we can calculate the whole solution faster (because we need fewer points overall) but still very accurately (because we use small step sizes in the regions where they are needed). This type of scheme is called an *adaptive step size* method, and some version of it is used in most large-scale numerical solutions of differential equations.

The basic idea behind an adaptive step size scheme is to vary the step sizes h so that the error introduced per unit interval in t is roughly constant.

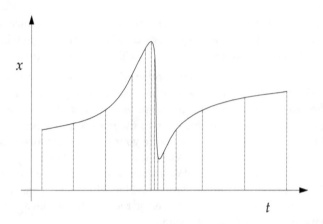

Figure 8.6: Adaptive step sizes. When solving a differential equation whose solution $x(t)$ varies slowly with t in some regions but more rapidly in others, it makes sense to use a varying step size. When the solution is slowly varying a large step size will give good results with less computational effort. When the solution is rapidly varying we must use smaller steps to get good accuracy.

For instance, we might specify that we want an error of 0.001 per unit time, or less, so that if we calculate a solution from say $t = 0$ to $t = 10$ we will get a total error of 0.01 or less. We achieve this by making the step size smaller in regions where the solution is tricky, but we must be careful because if we use smaller steps we will also need to take more steps and the errors pile up, so each individual step will have to be more accurate overall.

In practice the adaptive step size method has two parts. First we have to estimate the error on our steps, then we compare that error to our required accuracy and either increase or decrease the step size to achieve the accuracy we want. Here's how the approach works when applied to the fourth-order Runge–Kutta method.

We choose some initial value of h—typically very small, to be on the safe side—and, using our ordinary Runge–Kutta method, we first do *two* steps of the solution, each of size h, one after another—see Fig. 8.7. So if we start at time t, we will after two steps get to time $t + 2h$ and get an estimate of $x(t + 2h)$. Now here's the clever part: we go back to the start again, to time t, and we do one more Runge–Kutta step, but this time of twice the size, i.e., of size $2h$. This third larger step also takes us to time $t + 2h$ and gives us another estimate of $x(t + 2h)$, which will usually be close to but slightly different from the first estimate, since it was calculated in a different way. It turns out that by comparing

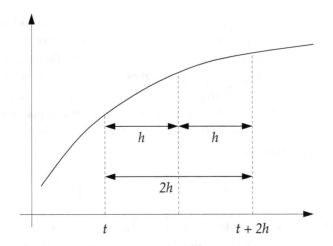

Figure 8.7: The adaptive step size method. Starting from some time t, the method involves first taking two steps of size h each, then going back to t again and taking a single step of size $2h$. Both give us estimates of the solution at time $t + 2h$ and by comparing these we can estimate the error.

the two estimates we can tell how accurate our calculation is.

The fourth-order Runge–Kutta method is *accurate* to fourth order but the *error* on the method is fifth order. That is, the size of the error on a single step is ch^5 to leading order for some constant c. So if we start at time t and do two steps of size h then the error will be roughly $2ch^5$. That is, the true value of $x(t + 2h)$ is related to our estimated value, call it x_1, by

$$x(t + 2h) = x_1 + 2ch^5. \tag{8.47}$$

On the other hand, when we do a single large step of size $2h$ the error is $c(2h)^5 = 32ch^5$, and so

$$x(t + 2h) = x_2 + 32ch^5, \tag{8.48}$$

where x_2 is our second estimate of $x(t + 2h)$. Equating these two expressions we get $x_1 = x_2 + 30ch^5$, which implies that the per-step error $\epsilon = ch^5$ on steps of size h is

$$\epsilon = ch^5 = \tfrac{1}{30}(x_1 - x_2). \tag{8.49}$$

Our goal is to make the size of this error exactly equal to some target accuracy that we choose. In general, unless we are very lucky, the two will not be exactly equal. Either Eq. (8.49) will be better than the target, which means we are performing steps that are smaller than they need to be and hence wasting time, or it will be worse than the target, which is unacceptable—the whole point here is to perform a calculation that meets the specified target accuracy.

So let us ask the following question: what size would our steps have to be to make the size of the error in Eq. (8.49) exactly equal to the target, to make our calculation exactly as accurate as we need it to be but not more? Let us denote this perfect step size h'. If we were to take steps of size h' then the error on a single step would be

$$\epsilon' = ch'^5 = ch^5\left(\frac{h'}{h}\right)^5 = \tfrac{1}{30}(x_1 - x_2)\left(\frac{h'}{h}\right)^5, \tag{8.50}$$

where we have used Eq. (8.49). At the same time suppose that the target accuracy per unit time for our calculation is δ, which means that the target accuracy for a single step of size h' would be $h'\delta$. We want to find the value of h' such that the actual accuracy (8.50) is equal to this target accuracy. We are only interested in the absolute magnitude of the error, not its sign, so we want the h' that satisfies

$$\tfrac{1}{30}|x_1 - x_2|\left(\frac{h'}{h}\right)^5 = h'\delta. \tag{8.51}$$

Rearranging for h' we then find that

$$h' = h\left(\frac{30h\delta}{|x_1 - x_2|}\right)^{1/4} = h\rho^{1/4}, \tag{8.52}$$

where

$$\rho = \frac{30h\delta}{|x_1 - x_2|} \tag{8.53}$$

which is precisely the ratio of the target accuracy $h\delta$ and the actual accuracy $\tfrac{1}{30}|x_1 - x_2|$ for steps of size h.

The complete method is now as follows. We perform two steps of size h and then, starting from the same starting point, one step of size $2h$. This gives us our two estimates x_1 and x_2 of $x(t + 2h)$. We use these to calculate the ratio ρ in Eq. (8.53). If $\rho > 1$ then we know that the actual accuracy of our Runge–Kutta steps is better than the target accuracy, so our calculation is fine, in the sense that it meets the target, but it is wasteful because it is using steps that are smaller than they need to be. So we keep the results and move on to time $t + 2h$ to continue our solution, but we make our steps bigger the next time around to avoid this waste. Plugging our value of ρ into Eq. (8.52) tells us exactly what the new larger value h' of the step size should be to achieve this.

Conversely, if $\rho < 1$ then the actual accuracy of our calculation is poorer than the target accuracy—we have missed our target and the current step of the calculation has failed. In this case we need to repeat the current step again, but with a smaller step size, and again Eq. (8.52) tells us what that step size should be.

Thus, after each step of the process, depending on the value of ρ, we either increase the value of h and move on to the next step or decrease the value of h and repeat the current step. Note that for the actual solution of our differential equation we always use the estimate x_1 for the value of x, not the estimate x_2, since x_1 was made using smaller steps and is thus, in general, more accurate. The estimate x_2 made with the larger step is used only for calculating the error and updating the step size, never for calculating the final solution.

The adaptive step size method involves more work for the computer than methods that use a fixed step size—we have to do at least three Runge–Kutta steps for every two that we actually use in calculating the solution, and sometimes more than three in cases where we have to repeat a step because we missed our target accuracy. However, the extra effort usually pays off because the method gets you an answer with the accuracy you require with very little waste. In the end the program almost always takes less time to run, and usually much less.

It is possible, by chance, for the two estimates x_1 and x_2 to coincidentally agree with one another—errors are inherently unpredictable and the two can occasionally be the same or roughly the same just by luck. If this happens, h' in Eq. (8.52) can erroneously become very large or diverge, causing the calculation to break down. To prevent this, one commonly places an upper limit on how much the value of h can increase from one step to another. For instance, a common rule of thumb is that it should not increase by more than a factor of two on any given pair of steps (pairs of successive steps being the fundamental unit in the method described here).

The adaptive step size method can be used to solve simultaneous differential equations as well as single equations. In such cases we need to decide how to generalize the formula (8.49) for the error, or equivalently the formula (8.53) for the ratio ρ, to the case of more than one dependent variable. The derivation leading to Eq. (8.49) can be duplicated for each variable to show that variables x, y, etc. have separate errors

$$\epsilon_x = \tfrac{1}{30}(x_1 - x_2), \qquad \epsilon_y = \tfrac{1}{30}(y_1 - y_2), \tag{8.54}$$

and so forth. There is, however, more than one way that these separate errors can be combined into a single overall error for use in Eq. (8.53), depending on the particular needs of the calculation. For instance, if we have variables x and y that represent coordinates of a point in a two-dimensional space, we might wish to perform a calculation that ensures that the Euclidean error in the position of the point meets a certain target, where by Euclidean error we mean $\sqrt{\epsilon_x^2 + \epsilon_y^2}$. In that case it is straightforward to see that we would use the same formulas for the adaptive method as before, except that $\frac{1}{30}|x_1 - x_2|$ in Eq. (8.53) should be replaced with $\sqrt{\epsilon_x^2 + \epsilon_y^2}$. On the other hand, suppose we are performing a calculation like that of Example 8.6 for the nonlinear pendulum, where we are solving a single second-order equation for θ but we introduce an additional variable ω to turn the problem into two first-order equations. In that case we don't really care about ω—it is introduced only for convenience—and its accuracy doesn't matter so long as θ is calculated accurately. In this situation we would use Eq. (8.53) directly, with x replaced by θ, and ignore ω in the calculation of the step sizes. (An example of such a calculation for the nonlinear pendulum is given below.) Thus it may take a little thought to determine, for any particular calculation, what the appropriate generalization of the adaptive method is to simultaneous equations, but the answer usually becomes clear once one determines the correct definition for the error on the calculation.

One further point is worth making about the adaptive step size method. It may seem unnecessarily strict to insist that we repeat the current step of the calculation if we miss our target accuracy. One might imagine that one could get reasonable answers if we always moved on to the next step, even when we miss our target: certainly there will be some steps where the error is a little bigger than the target value, but there will be others where it is a little smaller, and with luck it might all just wash out in the end—the total error at the end of the calculation would be roughly, if not exactly, where we want it to be. Unfortunately, however, this usually doesn't work. If one takes this approach, then one often ends up with a calculation that significantly misses the required accuracy target because there are a few steps that have unusually large errors. The problem is that the errors are cumulative—a large error on even one step makes all subsequent steps inaccurate too. If errors fluctuate from step to step then at some point you are going to get an undesirably large error which can doom the entire calculation. Thus it really is important to repeat steps that miss the target accuracy, rather than just letting them slip past, so that you can be certain no step has a very large error.

As an example of the adaptive step size method let us return once more to the nonlinear pendulum of Example 8.6. Figure 8.8 shows the results of a

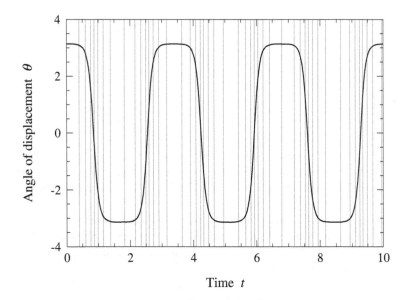

Figure 8.8: Motion of a nonlinear pendulum. This figure shows the angle θ of displacement of a nonlinear pendulum from the vertical as a function of time, calculated using the adaptive step size approach described in this section. The vertical dotted lines indicate the position of every twentieth Runge–Kutta step.

calculation of the motion of such a pendulum using adaptive step sizes. The solid curve shows the angle of displacement of the pendulum as a function of time—the wavelike form indicates that it's swinging back and forth. The vertical lines in the plot show the position of every twentieth Runge–Kutta step in the calculation (i.e., every tenth iteration of the adaptive method, since we always take two Runge–Kutta steps at once). As you can see from the figure, the method makes the step sizes longer in the flat portions of the curve at the top and bottom of each swing where little is happening, but in the steep portions where the pendulum is moving rapidly the step sizes are much smaller, which ensures accurate calculations of the motion.

Exercise 8.10: Cometary orbits

Many comets travel in highly elongated orbits around the Sun. For much of their lives they are far out in the solar system, moving very slowly, but on rare occasions their orbit brings them close to the Sun for a fly-by and for a brief period of time they move very fast indeed:

This is a classic example of a system for which an adaptive step size method is useful, because for the large periods of time when the comet is moving slowly we can use long time-steps, so that the program runs quickly, but short time-steps are crucial in the brief but fast-moving period close to the Sun.

The differential equation obeyed by a comet is straightforward to derive. The force between the Sun, with mass M at the origin, and a comet of mass m with position vector $\mathbf{r}$ is GMm/r^2 in direction $-\mathbf{r}/r$ (i.e., the direction towards the Sun), and hence Newton's second law tells us that

$$m\frac{d^2\mathbf{r}}{dt^2} = -\left(\frac{GMm}{r^2}\right)\frac{\mathbf{r}}{r}.$$

Canceling the m and taking the x component we have

$$\frac{d^2x}{dt^2} = -GM\frac{x}{r^3},$$

and similarly for the other two coordinates. We can, however, throw out one of the coordinates because the comet stays in a single plane as it orbits. If we orient our axes so that this plane is perpendicular to the z-axis, we can forget about the z coordinate and we are left with just two second-order equations to solve:

$$\frac{d^2x}{dt^2} = -GM\frac{x}{r^3}, \qquad \frac{d^2y}{dt^2} = -GM\frac{y}{r^3},$$

where $r = \sqrt{x^2 + y^2}$.

a) Turn these two second-order equations into four first-order equations, using the methods you have learned.

b) Write a program to solve your equations using the fourth-order Runge–Kutta method with a *fixed* step size. You will need to look up the mass of the Sun and Newton's gravitational constant G. As an initial condition, take a comet at coordinates $x = 4$ billion kilometers and $y = 0$ (which is somewhere out around the orbit of Neptune) with initial velocity $v_x = 0$ and $v_y = 500\,\mathrm{m\,s^{-1}}$. Make a graph showing the trajectory of the comet (i.e., a plot of y against x).

Choose a fixed step size h that allows you to accurately calculate at least two full orbits of the comet. Since orbits are periodic, a good indicator of an accurate calculation is that successive orbits of the comet lie on top of one another on your plot. If they do not then you need a smaller value of h. Give a short description of your findings. What value of h did you use? What did you observe in your simulation? How long did the calculation take?

c) Make a copy of your program and modify the copy to do the calculation using an adaptive step size. Set a target accuracy of $\delta = 1$ kilometer per year in the position of the comet and again plot the trajectory. What do you see? How do the speed, accuracy, and step size of the calculation compare with those in part (b)?

d) Modify your program to place dots on your graph showing the position of the comet at each Runge–Kutta step around a single orbit. You should see the steps getting closer together when the comet is close to the Sun and further apart when it is far out in the solar system.

Calculations like this can be extended to cases where we have more than one orbiting body—see Exercise 8.16 for an example. We can include planets, moons, asteroids, and others. Analytic calculations are impossible for such complex systems, but with careful numerical solution of differential equations we can calculate the motions of objects throughout the entire solar system.

Here's one further interesting wrinkle to the adaptive method. Recall Eq. (8.47), which relates the results of a "double step" of the method to the solution of the differential equation:

$$x(t + 2h) = x_1 + 2ch^5 + O(h^6). \tag{8.55}$$

(We have added the $O(h^6)$ here to remind us of the next term in the series.)

We also know from Eq. (8.49) that

$$ch^5 = \tfrac{1}{30}(x_1 - x_2), \tag{8.56}$$

where x_1 and x_2 are the two estimates of $x(t + 2h)$ calculated in the adaptive method. Substituting (8.56) into (8.55), we find that

$$x(t + 2h) = x_1 + \tfrac{1}{15}(x_1 - x_2) + O(h^6), \tag{8.57}$$

which is now accurate to order h^5 and has a error of order h^6—one order better than the standard fourth-order Runge–Kutta method. Equation (8.57) involves only quantities we have already computed in the course of the adaptive method, so it's essentially no extra work to calculate this more accurate estimate of the solution.

This trick is called *local extrapolation*. It is a kind of free bonus prize that comes along with the adaptive method, giving us a more accurate answer for no extra work. The only catch with it is that we don't know the size of the error on Eq. (8.57). It is, presumably, better than the error on the old fourth-order result (which is $2ch^5$, with ch^5 given by Eq. (8.56)), but we don't know by how much.

It is an easy extra step to incorporate local extrapolation into adaptive calculations. We could have used it in our solution of the motion of the pendulum in Fig. 8.8, for example. You could use it if you do Exercise 8.10 on calculating cometary orbits. It typically offers at least a modest improvement in the accuracy of your results.

The real interest in extrapolation, however, arises when we take the method further. It is possible to use methods similar to this not only to eliminate the leading-order error (the $O(h^5)$ term in Eq. (8.55)), but also any number of higher-order terms as well, resulting in impressively accurate solutions to differential equations even when using quite large values of h. The technique for doing this is called Richardson extrapolation and it's the basis of one of the most powerful methods for solving differential equations. Richardson extrapolation, however, is not usually used with the Runge–Kutta method, but rather with another method, called the "modified midpoint method," which we will examine in Section 8.5.4.

8.5 OTHER METHODS FOR DIFFERENTIAL EQUATIONS

So far in this chapter we have concentrated our attention on the Runge–Kutta method for solving differential equations. The Runge–Kutta method is a robust, accurate method that's easy to program and gives good results in most cases. It is, however, not the only method available. There are a number of other methods for solving differential equations that, while less widely used than the Runge–Kutta method, are nonetheless useful in certain situations. In this section we look at several additional methods, including the leapfrog and Verlet methods, and the Bulirsch–Stoer method, which combines a modified version of the leapfrog method with Richardson extrapolation to create one of the most accurate methods for solving differential equations (although it is also quite complex to program).

8.5.1 THE LEAPFROG METHOD

Consider a first-order differential equation in a single variable:

$$\frac{dx}{dt} = f(x, t). \tag{8.58}$$

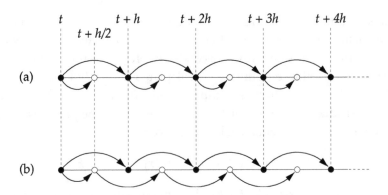

Figure 8.9: Second-order Runge–Kutta and the leapfrog method. (a) A diagrammatic representation of the calculations involved in the second-order Runge–Kutta method. On every step we use the starting position to calculate a value at the midpoint (open circle), then use that value to calculate the value at the end of the interval (filled circle). (b) The leapfrog method starts out the same, with a half step to the first midpoint and a full step to the end of the first interval. But thereafter each midpoint is calculated from the previous midpoint.

In Section 8.1.2 we introduced the second-order Runge–Kutta method (also sometimes called the midpoint method) in which, given the value of the dependent variable x at time t, one estimates its value at $t + h$ by using the slope at the midpoint $f(x(t + \frac{1}{2}h), t + \frac{1}{2}h)$. Because one doesn't normally know the value $x(t + \frac{1}{2}h)$, one first estimates it using Euler's method. The equations for the method can be written thus:

$$x(t + \tfrac{1}{2}h) = x(t) + \tfrac{1}{2}hf(x, t), \tag{8.59a}$$
$$x(t + h) = x(t) + hf(x(t + \tfrac{1}{2}h), t + \tfrac{1}{2}h). \tag{8.59b}$$

This is a slightly different way of writing the equations from the one we used previously (see Eq. (8.13)) but it is equivalent and it will be convenient for what follows.

The second-order Runge–Kutta method involves using these equations repeatedly to calculate the value of x at intervals of h as far as we wish to go. Each step is accurate to order h^2 and has an error of order h^3. When we combine many steps, one after another, the total error is one order of h worse (see Section 8.1.1), meaning it is of order h^2 in this case.

Figure 8.9a shows a simple graphical representation of what the Runge–Kutta method is doing. At each step we calculate the solution at the midpoint, and then use that solution as a stepping stone to calculate the value at $t + h$. The *leapfrog method* is a variant on this idea, as depicted in Fig. 8.9b. This method starts out the same way as second-order Runge–Kutta, with a half-step to the midpoint, follow by a full step to calculate $x(t + h)$. But then, for the next step, rather than calculating the midpoint value from $x(t + h)$ as we would in the Runge–Kutta method, we instead calculate it from the previous midpoint value $x(t + \frac{1}{2}h)$. In mathematical language we have

$$x\left(t + \tfrac{3}{2}h\right) = x\left(t + \tfrac{1}{2}h\right) + hf\left(x(t+h), t+h\right). \tag{8.60}$$

In this calculation $f(x(t + h), t + h)$ plays the role of the gradient at the midpoint between $t + \frac{1}{2}h$ and $t + \frac{3}{2}h$, so the calculation has second-order accuracy again and a third-order error. Moreover, once we have $x(t + \frac{3}{2}h)$ we can use it to do the next full step thus:

$$x(t + 2h) = x(t + h) + hf\left(x\left(t + \tfrac{3}{2}h\right), t + \tfrac{3}{2}h\right). \tag{8.61}$$

And we can go on repeating this process as long as we like. Given values of $x(t)$ and $x(t + \frac{1}{2}h)$, we repeatedly apply the equations

$$x(t + h) = x(t) + hf\left(x\left(t + \tfrac{1}{2}h\right), t + \tfrac{1}{2}h\right), \tag{8.62a}$$
$$x\left(t + \tfrac{3}{2}h\right) = x\left(t + \tfrac{1}{2}h\right) + hf\left(x(t+h), t+h\right). \tag{8.62b}$$

This is the leapfrog method, so called because each step "leaps over" the position of the previously calculated value. Like the second-order Runge–Kutta method, each step of the method is accurate to order h^2 and carries an error of order h^3. If we compound many steps of size h then the final result is accurate to order h and carries an h^2 error. The method can be extended to the solution of simultaneous differential equations just as the Runge–Kutta method can, by replacing the single variable x with a vector $\mathbf{r}$ and the function $f(x, t)$ with a vector function $\mathbf{f}(\mathbf{r}, t)$:

$$\mathbf{r}(t + h) = \mathbf{r}(t) + h\mathbf{f}\left(\mathbf{r}\left(t + \tfrac{1}{2}h\right), t + \tfrac{1}{2}h\right), \tag{8.63a}$$
$$\mathbf{r}\left(t + \tfrac{3}{2}h\right) = \mathbf{r}\left(t + \tfrac{1}{2}h\right) + h\mathbf{f}\left(\mathbf{r}(t+h), t+h\right). \tag{8.63b}$$

It can also be extended to the solution of second- or higher-order equations by converting the equations into simultaneous first-order equations, as shown in Section 8.3.

On the face of it, however, it's not immediately clear why we would want to use this method. It's true it is quite simple, but the fourth-order Runge–Kutta method is not much more complicated and significantly more accurate for al-most all calculations. But the leapfrog method has two significant virtues that make it worth considering. First, it is time-reversal symmetric, which makes it useful for physics problems where energy conservation is important. And second, its error is even in the step size h, which makes it ideal as a starting point for the Richardson extrapolation method mentioned at the end of Section 8.4. In the following sections we look at these issues in more detail.

8.5.2 TIME REVERSAL AND ENERGY CONSERVATION

The leapfrog method is time-reversal symmetric. When we use the method to solve a differential equation, the state of the calculation at any time t_1 is completely specified by giving the two values $x(t_1)$ and $x(t_1 + \frac{1}{2}h)$. Given only these values the rest of the solution going forward in time can be calculated by repeated application of Eq. (8.62). Suppose we continue the solution to a later time $t = t_2$, calculating values up to and including $x(t_2)$ and $x(t_2 + \frac{1}{2}h)$. Time-reversal symmetry means that if we take these values and use the leapfrog method backwards, with time interval $-h$ equal to minus the interval we used in the forward calculation, then we will retrace our steps and recover the exact values $x(t_1)$ and $x(t_1 + \frac{1}{2}h)$ at time t_1 (apart from any rounding error).

To see this let us set $h \to -h$ in Eq. (8.62):

$$x(t - h) = x(t) - hf\left(x\left(t - \tfrac{1}{2}h\right), t - \tfrac{1}{2}h\right), \tag{8.64a}$$
$$x\left(t - \tfrac{3}{2}h\right) = x\left(t - \tfrac{1}{2}h\right) - hf(x(t - h), t - h). \tag{8.64b}$$

Now put $t \to t + \frac{3}{2}h$ and we get

$$x\left(t + \tfrac{1}{2}h\right) = x\left(t + \tfrac{3}{2}h\right) - hf(x(t + h), t + h), \tag{8.65a}$$
$$x(t) = x(t + h) - hf\left(x\left(t + \tfrac{1}{2}h\right), t + \tfrac{1}{2}h\right). \tag{8.65b}$$

These equations give us the values of $x(t)$ and $x(t + \frac{1}{2}h)$ in terms of $x(t + h)$ and $x(t + \frac{3}{2}h)$. But if you compare these equations to Eq. (8.62), you'll see that they are simply performing the same mathematical operations as the forward calculation, only in reverse—everywhere we previously added a term $hf(x, t)$ we now subtract it again. Thus when we use the leapfrog method with step size $-h$ to solve a differential equation backwards, we get the exact same values $x(t)$ at every time-step that we get when we solve the equation forwards.

The same is not true of, for example, the second-order Runge–Kutta method. If you put $h \to -h$ in Eq. (8.59), the resulting equations do not give you the same mathematical operations as the forward Runge–Kutta method. The method will give you a solution in either the forward or backward direction, but the solutions will not agree exactly, in general, even after you allow for rounding error.

Why is time-reversal symmetry important? It turns out that it has a couple of useful implications. One concerns the conservation of energy.

Consider as an illustration the frictionless nonlinear pendulum, which we studied in Example 8.6. The motion of the pendulum is given by Eqs. (8.45) and (8.46), which read

$$\frac{d\theta}{dt} = \omega, \qquad \frac{d\omega}{dt} = -\frac{g}{\ell}\sin\theta. \qquad (8.66)$$

If we solve these equations using a Runge–Kutta method we can get a pretty good solution, as shown in Fig. 8.8 on page 361, but it is nonetheless only approximate, as nearly all computer calculations are. Among other things, this means that the total energy of the system, kinetic plus potential, is only approximately constant during the calculation. A frictionless pendulum should have constant energy, but the Runge–Kutta method isn't perfect and energies calculated using it tend to fluctuate and drift slightly over time. The top panel of Fig. 8.10 shows results from a solution of the equations above using the second-order Runge–Kutta method and the drift of the total energy with time is clearly visible. (We have deliberately used the less accurate second-order method in this case to make the drift larger and easier to see. With the fourth-order Runge–Kutta method, which is more accurate, the drift would be significantly smaller, though it would still be there.)

Now suppose we solve the same differential equations using the leapfrog method. Imagine doing so for one full swing of the pendulum. The pendulum starts at the furthest limit of its swing, swings all the way across, then all the way back again. In real life, the total energy of the system must remain constant throughout the motion, and in particular it must be the same when the pendulum returns to its initial point as it was when it started out. Our solution using the leapfrog method, on the other hand, is only approximate, so it's possible the energy might drift. Let us suppose for the sake of argument that it drifts upward, as it did for the Runge–Kutta method in the top panel of Fig. 8.10, so that its value at the end of the swing is slightly higher than at the beginning.

Now let us calculate the pendulum's motion once again, still using the leapfrog method but this time in reverse, starting at the end of the swing and solving backwards, with minus the step size that we used in our forward calculation. As we have shown, when we run the leapfrog method backwards in

We consider one full swing of the pendulum, starting on one side and swinging across then back.

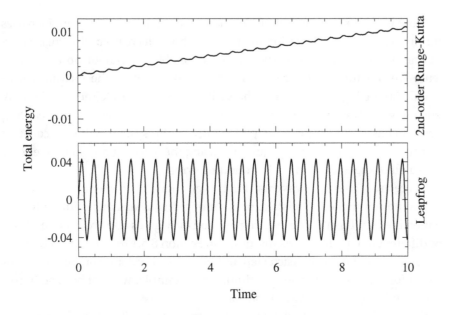

Figure 8.10: Total energy of the nonlinear pendulum. Top: The total energy, potential plus kinetic, of a nonlinear pendulum as a function of time, calculated using the second-order Runge–Kutta method. Bottom: The same energy calculated using the leapfrog method. Neither is constant, but the leapfrog method returns to the same value at the end of each swing of the pendulum and so conserves energy in the long run, while the energy calculated with Runge–Kutta drifts steadily away from the true value as time passes.

this fashion it will retrace its steps and end up exactly at the starting point of the motion again (apart from rounding error). Thus, if the energy increased during the forward calculation it must decrease when we do things in reverse.

But here's the thing. The physics of the pendulum is itself time-reversal symmetric. The motion of swinging across and back, the motion that the pendulum makes in a single period, is exactly the same backwards as it is forwards. Hence, when we perform the backward solution we are solving for the exact same motion and moreover doing it using the exact same method (since we are using the leapfrog method in both directions). This means that the values of the variables θ and ω will be exactly the same at each successive step of the solution in the reverse direction as they were going forward. Hence, if the energy increased during the forward solution it must also increase during the backward one.

Now we have a contradiction. We have shown that if the energy increases during the forward calculation then it must both decrease and increase during the backward one. Clearly this is impossible—it cannot do both—and hence we conclude that it cannot have increased during the forward calculation. An analogous argument shows it cannot decrease either, so the only remaining possibility is that it stays the same. In other words, the leapfrog method conserves energy. The total energy of the system will stay constant over time when we solve the equations using the leapfrog method, except for any small changes introduced by rounding error.

There are a couple of caveats. First, even though the energy is conserved we should not make the mistake of assuming this means our solution for the motion is exact. It isn't. The leapfrog method only gives approximate solutions for differential equations—as discussed in Section 8.5.1 the method is only accurate to second order on each step and has a third-order error. So the values we get for the angle θ for our pendulum, for example, will not be exactly correct, even though the energy is constant.

Second, the argument we have given applies to a full swing of the pendulum. It tells us that the energy at the end of a full swing will be the same as it was at the beginning. It does not tell us that the energy will be conserved throughout the swing, and indeed, as we will see, it is not. The energy may fluctuate during the course of the pendulum swing, but it will always come back to the correct value at the end of the swing. More generally, if the leapfrog method is used to solve equations of motion for any periodic system, such as a pendulum or a planet orbiting a star, then energy will be conserved over any full period of the system (or many full periods), but it will not, in general, be conserved over fractions of a period.

If we can live with these limitations, however, the leapfrog method can be useful for solving the equations of motion of energy conserving physical systems over long periods of time. If we wait long enough, a solution using a Runge–Kutta method will drift in energy—the pendulum might run down and stop swinging, or the planet might fall out of orbit and into its star. But a solution using the leapfrog method will run forever.

As an example look again at Fig. 8.10. The bottom panel shows the total energy of the nonlinear pendulum calculated using the leapfrog method and we can see that indeed it is constant on average over long periods of time—many swings of the pendulum—even though it oscillates over the course of individual swings. As the figure shows, the accuracy of the energy in the short term is actually poorer than the second-order Runge–Kutta method (notice that the vertical scales are different in the two panels), but in the long term the leapfrog method will be far better than the Runge–Kutta method, as the latter drifts further and further from the true value of the energy.

8.5.3 THE VERLET METHOD

Suppose, as in the previous section, that we are using the leapfrog method to solve the classical equations of motion for a physical system. Such equations, derived from Newton's second law $F = ma$, take the form of second-order differential equations

$$\frac{d^2x}{dt^2} = f(x, t),\tag{8.67}$$

or the vector equivalent when there is more than one dependent variable. Examples include the motion of projectiles, the pendulum of the previous section, and the cometary orbit of Exercise 8.10. As we have seen, we can convert such equations of motion into coupled first-order equations

$$\frac{dx}{dt} = v, \qquad \frac{dv}{dt} = f(x, t),\tag{8.68}$$

where we use the variable name v here as a reminder that, when we are talking about equations of motion, the quantity it represents is a velocity (or sometimes an angular velocity, as in the case of the pendulum).

If we want to apply the leapfrog method to these equations the normal strategy would be to define a vector $\mathbf{r} = (x, v)$, combine the two equations (8.68) into a single vector equation

$$\frac{d\mathbf{r}}{dt} = \mathbf{f}(\mathbf{r}, t),\tag{8.69}$$

and then solve this equation for $\mathbf{r}$ using the leapfrog method.

Rather than going this route, however, let us instead write out the leapfrog method in full, as applied to (8.68). If we are given the value of x at some time t and the value of v at time $t + \frac{1}{2}h$ then, applying the method, the value of x a time interval h later is

$$x(t + h) = x(t) + hv\left(t + \tfrac{1}{2}h\right).\tag{8.70}$$

And the value of v an interval h later is

$$v\left(t + \tfrac{3}{2}h\right) = v\left(t + \tfrac{1}{2}h\right) + hf\left(x(t + h), t + h\right).\tag{8.71}$$

We can derive a full solution to the problem by using just these two equations repeatedly, as many times as we wish. Notice that the equations involve the value of x only at time t plus integer multiples of h and the value of v only at half-integer multiples. We never need to calculate v at any of the integer points or x at the half integers. This is an improvement over the normal leapfrog method applied to the vector $\mathbf{r} = (x, v)$, which would involve solving for both x and v at all points, integer and half-integer. Equations (8.70) and (8.71) require only half as much work to evaluate as the full leapfrog method.

This simplification works only for equations of motion or other differential equations that have the special structure of Eq. (8.68), where the right-hand side of the first equation depends on v but not x and the right-hand side of the second equation depends on x but not v. Many physics problems, however, boil down to solving equations of motion, so the method is widely applicable.

A minor problem with the method arises if we want to calculate some quantity that depends on both x and v, such as the total energy of the system. Potential energy depends on position x while kinetic energy depends on velocity v, so calculating the total energy, potential plus kinetic, at any time t, requires us to know the values of both variables at that time. Unfortunately we know x only at the integer points and v only at the half-integer points, so we never know both at the same time.

But there's an easy solution to this problem. We can calculate the velocity at the integer points by doing an additional half step as follows. If we did know $v(t+h)$ then we could calculate $v(t + \frac{1}{2}h)$ from it by doing a half step backwards using Euler's method. That is, we would do Euler's method with a step size of $-\frac{1}{2}h$:

$$v(t + \tfrac{1}{2}h) = v(t+h) - \tfrac{1}{2}hf(x(t+h), t+h). \tag{8.72}$$

Alternatively, by rearranging this equation we can calculate $v(t+h)$ from $v(t + \frac{1}{2}h)$ like this:

$$v(t+h) = v(t + \tfrac{1}{2}h) + \tfrac{1}{2}hf(x(t+h), t+h). \tag{8.73}$$

This equation gives us v at integer steps in terms of quantities we already know from our leapfrog calculation, allowing us to calculate total energy (or any other quantity) at each step.

A complete calculation combines Eqs. (8.70) and (8.71) with Eq. (8.73), plus an initial half step at the very beginning to get everything started. Putting it all together, here's what we have.

We are given the initial values of x and v at some time t. Then

$$v(t + \tfrac{1}{2}h) = v(t) + \tfrac{1}{2}hf(x(t), t). \tag{8.74}$$

Then subsequent values of x and v are derived by repeatedly applying

$$x(t+h) = x(t) + hv\left(t+\tfrac{1}{2}h\right),$$ (8.75a)
$$k = hf(x(t+h), t+h),$$ (8.75b)
$$v(t+h) = v\left(t+\tfrac{1}{2}h\right) + \tfrac{1}{2}k,$$ (8.75c)
$$v\left(t+\tfrac{3}{2}h\right) = v\left(t+\tfrac{1}{2}h\right) + k.$$ (8.75d)

This variant of the leapfrog method is called the *Verlet method* after physicist Loup Verlet, who discovered it in the 1960s (although it was known to others long before that, as far back as the eighteenth century).

The method can be easily extended to equations of motion in more than one dimension. If we wish to solve an equation of motion of the form

$$\frac{d^2\mathbf{r}}{dt^2} = \mathbf{f}(\mathbf{r}, t),$$ (8.76)

where $\mathbf{r} = (x, y, \ldots)$ is a d-dimensional vector, then, given initial conditions on $\mathbf{r}$ and the velocity $\mathbf{v} = d\mathbf{r}/dt$, the appropriate generalization of the Verlet method involves first performing a half step to calculate $\mathbf{v}\left(t+\tfrac{1}{2}h\right)$:

$$\mathbf{v}\left(t+\tfrac{1}{2}h\right) = \mathbf{v}(t) + \tfrac{1}{2}h\mathbf{f}(\mathbf{r}(t), t),$$ (8.77)

then repeatedly applying the equations

$$\mathbf{r}(t+h) = \mathbf{r}(t) + h\mathbf{v}\left(t+\tfrac{1}{2}h\right),$$ (8.78a)
$$\mathbf{k} = h\mathbf{f}(\mathbf{r}(t+h), t+h),$$ (8.78b)
$$\mathbf{v}(t+h) = \mathbf{v}\left(t+\tfrac{1}{2}h\right) + \tfrac{1}{2}\mathbf{k},$$ (8.78c)
$$\mathbf{v}\left(t+\tfrac{3}{2}h\right) = \mathbf{v}\left(t+\tfrac{1}{2}h\right) + \mathbf{k}.$$ (8.78d)

Exercise 8.11: Write a program to solve the differential equation

$$\frac{d^2x}{dt^2} - \left(\frac{dx}{dt}\right)^2 + x + 5 = 0$$

using the leapfrog method. Solve from $t = 0$ to $t = 50$ in steps of $h = 0.001$ with initial condition $x = 1$ and $dx/dt = 0$. Make a plot of your solution showing x as a function of t.

Exercise 8.12: Orbit of the Earth

Use the Verlet method to calculate the orbit of the Earth around the Sun. The equations of motion for the position $\mathbf{r} = (x, y)$ of the planet in its orbital plane are the same as those for any orbiting body and are derived in Exercise 8.10 on page 361. In vector form, they are

$$\frac{d^2\mathbf{r}}{dt^2} = -GM\frac{\mathbf{r}}{r^3},$$

where $G = 6.6738 \times 10^{-11}\,\mathrm{m^3\,kg^{-1}\,s^{-2}}$ is Newton's gravitational constant and $M = 1.9891 \times 10^{30}\,\mathrm{kg}$ is the mass of the Sun.

The orbit of the Earth is not perfectly circular, the planet being sometimes closer to and sometimes further from the Sun. When it is at its closest point, or *perihelion*, it is moving precisely tangentially (i.e., perpendicular to the line between itself and the Sun) and it has distance $1.4710 \times 10^{11}\,\mathrm{m}$ from the Sun and linear velocity $3.0287 \times 10^4\,\mathrm{m\,s^{-1}}$.

a) Write a program to calculate the orbit of the Earth using the Verlet method, Eqs. (8.77) and (8.78), with a time-step of $h = 1$ hour. Make a plot of the orbit, showing several complete revolutions about the Sun. The orbit should be very slightly, but visibly, non-circular.

b) The gravitational potential energy of the Earth is $-GMm/r$, where $m = 5.9722 \times 10^{24}\,\mathrm{kg}$ is the mass of the planet, and its kinetic energy is $\frac{1}{2}mv^2$ as usual. Modify your program to calculate both of these quantities at each step, along with their sum (which is the total energy), and make a plot showing all three as a function of time on the same axes. You should find that the potential and kinetic energies vary visibly during the course of an orbit, but the total energy remains constant.

c) Now plot the total energy alone without the others and you should be able to see a slight variation over the course of an orbit. Because you're using the Verlet method, however, which conserves energy in the long term, the energy should always return to its starting value at the end of each complete orbit.

8.5.4 THE MODIFIED MIDPOINT METHOD

The leapfrog method offers another, more subtle, advantage over the Runge–Kutta method, namely that the total error on the method, after many steps, is an even function of the step size h. To put that another way, the expansion of the error in powers of h contains only even terms and no odd ones. This result will be crucial in the next section, where it forms the basis for a powerful solution method for differential equations called the Bulirsch–Stoer method.

The argument that the error on the leapfrog method is even in h relies once more on the fact that the method is time-reversal symmetric, as discussed in Section 8.5.2. Recall that a single step of the leapfrog method is accurate up to terms in h^2 and has an h^3 error to leading order. More generally, we can write the error on a single step as some function $\epsilon(h)$ of the step size, with the first term in that function being proportional to h^3. The question is what the other terms look like.

Imagine taking a small step using the leapfrog method, which gives the solution to our differential equation a short time later plus error $\epsilon(h)$. Now imagine taking the same step backwards, i.e., with step size $-h$. Given that the leapfrog method is time-reversal symmetric—the change in the solution going backwards is exactly the reverse of the change forwards—it follows that the backward error $\epsilon(-h)$ must be minus the forward one:

$$\epsilon(-h) = -\epsilon(h). \tag{8.79}$$

This equation tells us that $\epsilon(h)$ is an odd function. It is antisymmetric about the origin and its Taylor expansion in powers of h can contain only odd powers. We know the first term in the series is proportional to h^3, so in general $\epsilon(h)$ must take the form

$$\epsilon(h) = c_3 h^3 + c_5 h^5 + c_7 h^7 + \ldots \tag{8.80}$$

and so on, for some set of constants $c_3, c_5, \ldots$

But, as we've also seen, first for the Euler method in Section 8.1.1 and later for the Runge–Kutta and leapfrog methods as well, if you perform many steps of size h then the total, cumulative error over all of them is one order worse in h than it is for each single step. Roughly speaking, if the error on a single step is $\epsilon(h)$ and it takes Δ/h steps to cover an interval of time Δ, then the total error is of order $\epsilon(h) \times \Delta/h$, which is one order lower in h than $\epsilon(h)$ itself. Hence, given that in the present case $\epsilon(h)$ contains only odd powers of h starting with h^3, the total error on the leapfrog method, after many steps, must contain only even powers, starting with h^2.

This result is correct as far as it goes, but there is a catch. Recall that to get the leapfrog method started, we initially take one half-step using Euler's method, as in Eq. (8.59a) on page 365. This additional step introduces an error of its own. That error is of size h^2 to leading order (as always with Euler's method), which is the same as the overall error on the leapfrog method and hence makes the final error no worse in terms of the order of h. Unfortunately, however, the higher-order terms in Euler's method are not restricted to even powers of h—all powers from h^2 onward are present, meaning that this one extra step at the start of the calculation spoils our result about even powers above. The total error at the end of the whole calculation will now contain both even and odd powers.

There is, however, a solution for this problem. Suppose we wish to solve our differential equation from some initial time t forward to a later time $t + H$ (where H is not necessarily small) using n leapfrog steps of size $h = H/n$ each. Let us write the leapfrog method in a slightly different form from before. We define

$$x_0 = x(t),\tag{8.81a}$$

$$y_1 = x_0 + \tfrac{1}{2}hf(x_0, t).\tag{8.81b}$$

Then

$$x_1 = x_0 + hf(y_1, t + \tfrac{1}{2}h),\tag{8.82a}$$

$$y_2 = y_1 + hf(x_1, t + h),\tag{8.82b}$$

$$x_2 = x_1 + hf(y_2, t + \tfrac{3}{2}h),\tag{8.82c}$$

and so forth. The variables x_m here represent the solution at integer multiples of h and the variables y_m at half-integer multiples. In general, we have

$$y_{m+1} = y_m + hf(x_m, t + mh),\tag{8.83a}$$

$$x_{m+1} = x_m + hf\big(y_{m+1}, t + (m + \tfrac{1}{2})h\big).\tag{8.83b}$$

The last two points in the solution are $y_n = x(t + H - \tfrac{1}{2}h)$ and $x_n = x(t + H)$. Normally, we would take the value of x_n as our final solution for $x(t + H)$, but there is another possibility: we can also calculate a final value from y_n. Using the same trick that we used to derive Eq. (8.73), we can write

$$x(t + H) = y_n + \tfrac{1}{2}hf(x_n, t + H).\tag{8.84}$$

Thus we have two different ways to calculate a value for $x(t + H)$. Or we can combine the two, Eq. (8.84) and the estimate $x(t + H) = x_n$, taking their average thus:

$$x(t + H) = \tfrac{1}{2}\big[x_n + y_n + \tfrac{1}{2}hf(x_n, t + H)\big].\tag{8.85}$$

Miraculously, it turns out that if we calculate $x(t + H)$ from this equation then the odd-order error terms that arise from the Euler's method step at the start of the leapfrog calculation cancel out, giving a total error on Eq. (8.85) that once again contains only even powers of h. This result was originally proved by mathematician William Gragg in 1965 and the resulting method is sometimes called *Gragg's method* in his honor, although it is more commonly referred to as the *modified midpoint method*. The modified midpoint method combines the leapfrog method in the form of Eqs. (8.81) and (8.83) with Eq. (8.85) to make an estimate of $x(t + H)$ that carries a leading-order error of order h^2 and higher-order terms containing even powers of h only.

The modified midpoint method is rarely used alone, since it offers little advantage over either the ordinary leapfrog method (if you want an energy conserving solution) or the fourth-order Runge–Kutta method (which is significantly more accurate). It plays an important role, nonetheless, as the basis for the powerful Bulirsch–Stoer method, which we study in the next section.

8.5.5 THE BULIRSCH–STOER METHOD

The Bulirsch–Stoer method for solving differential equations combines two ideas we have seen already: the modified midpoint method and Richardson extrapolation. It's reminiscent in some ways of the Romberg method for evaluating integrals that we studied in Section 5.4, and the equations are similar. Here's how it works.

We are given a differential equation, and for now let's again assume the simplest case of a first-order, single-variable equation:

$$\frac{\mathrm{d}x}{\mathrm{d}t} = f(x, t). \tag{8.86}$$

We are also, as usual, given an initial condition at some time t, and, as in the previous section, let us solve the equation over an interval of time from t to some later time $t + H$.

We start by calculating a solution using the modified midpoint method and, in the first instance, we will use just a single step for the whole solution, from t to $t + H$ (or you can think of it as two half-steps—see Eq. (8.81)). In other words our step size for the modified midpoint method, which we'll call h_1, will just be equal to H. This gives an estimate of the value of $x(t + H)$, which we will denote $R_{1,1}$. (The R is for "Richardson extrapolation.") If H is a large interval of time then $R_{1,1}$ will be a rather crude estimate, but that need not worry us, as we'll see.

Once we have calculated $R_{1,1}$, we go back to the start at time t again and repeat our calculation, again using the modified midpoint method, but this time with two steps of size $h_2 = \frac{1}{2}H$. This gives us a second estimate of $x(t + H)$, which we'll call $R_{2,1}$.

We showed in the previous section that the total error on the modified midpoint method is an even function of the step size, so it follows that

$$x(t + H) = R_{2,1} + c_1 h_2^2 + \mathrm{O}(h_2^4), \tag{8.87}$$

where c_1 is an unknown constant. Similarly

$$x(t + H) = R_{1,1} + c_1 h_1^2 + O(h_1^4) = R_{1,1} + 4c_1 h_2^2 + O(h_2^4), \qquad (8.88)$$

where we have used the fact that $h_1 = 2h_2$. Since Eqs. (8.87) and (8.88) are both expressions for the same quantity we can equate them and, after rearranging, we find that

$$c_1 h_2^2 = \tfrac{1}{3}(R_{2,1} - R_{1,1}). \qquad (8.89)$$

Substituting this back into Eq. (8.87), we get

$$x(t + H) = R_{2,1} + \tfrac{1}{3}(R_{2,1} - R_{1,1}) + O(h_2^4). \qquad (8.90)$$

In other words, we have found a new estimate of $x(t + H)$ which is more accurate than either of the estimates that went into it—it has an error of order h^4, two orders in h better than the basic leapfrog method and as good as the fourth-order Runge–Kutta method (which, when you add up errors over more than one step, is accurate to order h^3 and carries an h^4 error). Let us call this new estimate $R_{2,2}$:

$$R_{2,2} = R_{2,1} + \tfrac{1}{3}(R_{2,1} - R_{1,1}). \qquad (8.91)$$

We can take this approach further. If we increase the number of steps to three, with step size $h_3 = \tfrac{1}{3}H$, and solve from t to $t + H$ again we get a new estimate $R_{3,1}$. Then, following the same line of argument as above, we can calculate a further estimate

$$R_{3,2} = R_{3,1} + \tfrac{4}{5}(R_{3,1} - R_{2,1}), \qquad (8.92)$$

which has an error of order h_3^4. This allows us to write

$$x(t + H) = R_{3,2} + c_2 h_3^4 + O(h_3^6), \qquad (8.93)$$

where c_2 is another constant. Combining Eqs. (8.90) and (8.91), we also have

$$x(t + H) = R_{2,2} + c_2 h_2^4 + O(h_2^6) = R_{2,2} + \tfrac{81}{16} c_2 h_3^4 + O(h_3^6), \qquad (8.94)$$

where we have made use of the fact that $h_2 = \tfrac{3}{2}h_3$. Equating this result with (8.93) and rearranging gives

$$c_2 h_3^4 = \tfrac{16}{65}(R_{3,2} - R_{2,2}), \qquad (8.95)$$

and substituting this into Eq. (8.93) gives

$$x(t + H) = R_{3,3} + O(h_3^6), \qquad (8.96)$$

where

$$R_{3,3} = R_{3,2} + \tfrac{16}{65}(R_{3,2} - R_{2,2}). \tag{8.97}$$

Now our error is of order h^6, and we've taken only three modified midpoint steps!

The power of this method lies in the way it cancels out the error terms to higher and higher orders on successive steps, along with the fact that the modified midpoint method has only even-order error terms, which means that every time we cancel out another term we gain two extra orders of accuracy in h.

We can take this process as far as we like. Each time around, we solve our differential equation again from t to $t + H$ using the modified midpoint method, but with one more step than last time. Suppose we denote the current number of steps by n and our modified midpoint estimate estimate of $x(t + H)$ by $R_{n,1}$. Then we can use the method above to cancel error terms and arrive at a series of further estimates $R_{n,2}$, $R_{n,3}$, and so on, where $R_{n,m}$ carries an error of order h^{2m}:

$$x(t + H) = R_{n,m} + c_m h_n^{2m} + O(h_n^{2m+2}), \tag{8.98}$$

where c_m is an unknown constant. The corresponding estimate $R_{n-1,m}$ made with one less step satisfies

$$x(t + H) = R_{n-1,m} + c_m h_{n-1}^{2m} + O(h_{n-1}^{2m+2}). \tag{8.99}$$

But $h_n = H/n$ and $h_{n-1} = H/(n-1)$, so

$$h_{n-1} = \frac{n}{n-1} h_n. \tag{8.100}$$

Substituting this into (8.99), equating with (8.98), and rearranging, we then find that

$$c_m h_n^{2m} = \frac{R_{n,m} - R_{n-1,m}}{[n/(n-1)]^{2m} - 1}. \tag{8.101}$$

And putting this in Eq. (8.98) gives us a new estimate of $x(t + H)$ two orders of h more accurate:

$$x(t + H) = R_{n,m+1} + O(h_n^{2m+2}), \tag{8.102}$$

where

$$R_{n,m+1} = R_{n,m} + \frac{R_{n,m} - R_{n-1,m}}{[n/(n-1)]^{2m} - 1}. \tag{8.103}$$

Equation (8.103) is the fundamental equation of Richardson extrapolation, and the heart of the Bulirsch–Stoer solution method. It allows us to calculate remarkably accurate estimates of $x(t + H)$ while only using a very few steps of the modified midpoint method.

A diagram may help to make the structure of the method clearer:

$$
\begin{array}{ll}
n = 1: & R_{1,1} \\
& \searrow \\
n = 2: & R_{2,1} \rightarrow R_{2,2} \\
& \searrow \quad \searrow \\
n = 3: & R_{3,1} \rightarrow R_{3,2} \rightarrow R_{3,3} \\
& \searrow \quad \searrow \quad \searrow \\
n = 4: & R_{4,1} \rightarrow R_{4,2} \rightarrow R_{4,3} \rightarrow R_{4,4}
\end{array}
$$

$$\underbrace{\qquad}_{\substack{\text{Modified} \\ \text{midpoint}}} \quad \overbrace{\qquad\qquad\qquad}^{\text{Richardson extrapolation}}$$

For each value of n we calculate a basic modified midpoint estimate $R_{n,1}$ with n steps, and then a series of further extrapolation estimates, working along a row of the diagram. Each extrapolation estimate depends on two previous estimates, as indicated by the arrows, and the last estimate in each row is the highest-order estimate for that value of n.

The method also gives us estimates of the error each time around. The quantity $c_m h_n^{2m}$ in Eq. (8.101) is precisely the (leading-order) error on the current estimate of $x(t + H)$. The Bulirsch–Stoer method involves increasing the number of steps n until the error on our best estimate of $x(t + H)$ is as small as we want it to be. As with the adaptive Runge–Kutta method of Section 8.4, we typically specify the accuracy we want in terms of the error per unit time δ, in which case the required accuracy for a solution over the interval H is $H\delta$. When the error falls below this value, the calculation is finished. Thus the Bulirsch–Stoer method is actually an adaptive method—it performs only as many steps as are needed to give us the accuracy we require.[3]

If you previously read Section 5.4, on the technique for calculating integrals known as Romberg integration, the diagram above may look familiar—it is similar to the one on page 161 in that section. This is no coincidence. Romberg integration and the Bulirsch–Stoer method are both applications of the same idea of Richardson extrapolation, to two different problems, and Eq. (8.103)

---

[3] Actually it goes one step further than we really need, since the error calculated in Eq. (8.101) is the error on the *previous* estimate $R_{n,m}$ and not the error on the new estimate $R_{n,m+1}$. So, just as with the local extrapolation method discussed at the end of Section 8.4, the results will usually be a little more accurate than our target accuracy.

for the Bulirsch–Stoer method embodies essentially the same idea as Eq. (5.51) for Romberg integration, though there are some differences. In particular, in Romberg integration we doubled the number of steps each time around, instead of just increasing it by one. This is a convenient choice when doing integrals because it gives us "nested" sample points that improve the speed of the calculations by allowing us to reuse previous results, as discussed in Section 5.3. There is no equivalent speed improvement to be had when solving differential equations, which is a shame in a sense, but does leave us free to choose the number of steps however we like. Various choices have been investigated and the results seem to indicate that the simple choice used here, of increasing n by one each time around, is a good one—better in most cases than doubling the value of n.

There are some limitations to the Bulirsch–Stoer method. One is that it only calculates a really accurate answer for the final value $x(t + H)$. At all the intermediate points we only get the raw midpoint-method estimates, which are not particularly accurate. (They carry an error of order h^2.)

Furthermore, we are, in effect, calculating the terms in a series expansion of $x(t + H)$ in powers of h and the method is only worthwhile if the series converges reasonably quickly. If you need hundreds or thousands of terms to get a good answer, then the Bulirsch–Stoer method is not a good choice. This means in practice that the interval H over which we are solving has to be kept reasonably small. Practical experience suggests that the method works best if the number of modified midpoint steps is never greater than about eight or ten, which limits the size of the time interval H to relatively modest values.

Both of these problems can be overcome with the same simple technique: if we want a solution from say $t = a$ to $t = b$, we divide that time into some number N of smaller intervals of size $H = (b - a)/N$ and apply the Bulirsch–Stoer method separately to each one in turn. We should choose N large enough that we get a complete picture of the solution without having to rely on the modified midpoint estimates in the interior of the intervals. And H should be small enough that in any one interval the number of modified midpoint steps needed to reach the target accuracy is never too large.

The complete Bulirsch–Stoer method is then as follows. Let δ be the desired accuracy of your solution per unit time. Divide the entire solution into N equal intervals of length H each and apply the following steps to solve your differential equation in each one in turn:

1. Set $n = 1$ and use the modified midpoint method of Section 8.5.4 to calculate an estimate $R_{1,1}$ of the solution from t to $t + H$ using just one step.

2. Increase n by one and calculate a new modified midpoint estimate $R_{n,1}$ with that many steps.

3. Use Eq. (8.103) to calculate further estimates $R_{n,2} \ldots R_{n,n}$—a complete row in the diagram on page 380.

4. After calculating the whole row, compare the error given by Eq. (8.101) with the target accuracy $H\delta$. If the error is larger than the target accuracy, go back to step 2 again. Otherwise, move on to the next time interval H.

When the entire solution has been covered in this way, the calculation ends.

The Bulirsch–Stoer method can easily be extended to the solution of simultaneous differential equations by replacing the single dependent variable x with a vector $\mathbf{r}$ of two or more variables, as in Section 8.2. Then the estimates $R_{n,m}$ also become vectors $\mathbf{R}_{n,m}$ but the equations for the method remain otherwise the same. We can also apply the Bulirsch–Stoer method to second- or higher-order differential equations by first transforming those equations into first-order ones, as described in Section 8.3.

Although it is somewhat more complicated to program than the Runge–Kutta method, the Bulirsch–Stoer method can work significantly better even than the adaptive version of Runge–Kutta, giving more accurate solutions with less work. Because, as we have said, it relies on a series expansion, the method is mainly useful for equations whose solutions are relatively smooth, so that expansions work well. Differential equations with pathological behaviors—large fluctuations, divergences, and so forth—are not suitable candidates. If you have a differential equation that displays such behaviors then the adaptive Runge–Kutta method of Section 8.4 is a better choice. But in cases where it is applicable, the Bulirsch–Stoer method is considered by many to be the best method available for solving ordinary differential equations, the king of differential equation solvers.

EXAMPLE 8.7: BULIRSCH–STOER METHOD FOR THE NONLINEAR PENDULUM

Let us return to the nonlinear pendulum, which we examined previously in Example 8.6 and Section 8.4. The equations of motion were given in Eqs. (8.45) and (8.46), which we repeat here:

$$\frac{d\theta}{dt} = \omega, \qquad \frac{d\omega}{dt} = -\frac{g}{\ell} \sin\theta. \qquad (8.104)$$

Let us solve these equations for the case of a pendulum with an arm $\ell = 10\,\text{cm}$ long, initially at rest with $\theta = 179°$, i.e., pointing almost, but not quite, vertically upward. (These are the same conditions as in Exercise 8.4.) Here is a

complete program to solve for the motion of the pendulum using the Bulirsch–
Stoer method:

```
from math import sin,pi                              File: bulirsch.py
from numpy import empty,array,arange
from pylab import plot,show

g = 9.81
l = 0.1
theta0 = 179*pi/180
a = 0.0
b = 10.0
N = 100           # Number of "big steps"
H = (b-a)/N       # Size of "big steps"
delta = 1e-8      # Required position accuracy per unit time

def f(r):
    theta = r[0]
    omega = r[1]
    ftheta = omega
    fomega = -(g/l)*sin(theta)
    return array([ftheta,fomega],float)

tpoints = arange(a,b,H)
thetapoints = []
r = array([theta0,0.0],float)

# Do the "big steps" of size H
for t in tpoints:

    thetapoints.append(r[0])

    # Do one modified midpoint step of size H
    # to get things started
    n = 1
    r1 = r + 0.5*H*f(r)
    r2 = r + H*f(r1)

    # The array R1 stores the first row of the
    # extrapolation table, which contains only the single
    # modified midpoint estimate of the solution at the
    # end of the interval
    R1 = empty([1,2],float)
    R1[0] = 0.5*(r1 + r2 + 0.5*H*f(r2))
```

```
    # Now increase n until the required accuracy is reached
    error = 2*H*delta
    while error>H*delta:

        n += 1
        h = H/n

        # Modified midpoint method
        r1 = r + 0.5*h*f(r)
        r2 = r + h*f(r1)
        for i in range(n-1):
            r1 += h*f(r2)
            r2 += h*f(r1)

        # Calculate extrapolation estimates.  Arrays R1 and R2
        # hold the two most recent lines of the table
        R2 = R1
        R1 = empty([n,2],float)
        R1[0] = 0.5*(r1 + r2 + 0.5*h*f(r2))
        for m in range(1,n):
            epsilon = (R1[m-1]-R2[m-1])/((n/(n-1))**(2*m)-1)
            R1[m] = R1[m-1] + epsilon
        error = abs(epsilon[0])

    # Set r equal to the most accurate estimate we have,
    # before moving on to the next big step
    r = R1[n-1]

# Plot the results
plot(tpoints,thetapoints)
plot(tpoints,thetapoints,"b.")
show()
```

There are a couple of points worth noting about this program. Notice, for instance, how we gave the variable error an initial value of $2H\delta$, thus ensuring that we go around the while loop at least once. Notice also how we have used the two arrays R1 and R2 to store the most recent two rows of extrapolation estimates $R_{n,m}$. Since the calculation of each row requires only the values in the current and previous rows, and since, ultimately, we are only interested in the final value of the final row, there is no need to retain more than two rows of estimates at any time. The rest can be safely discarded.

If we run the program it produces a solution essentially identical to our previous solution of the same problem using the adaptive Runge–Kutta method, Fig. 8.8. The real difference between the two methods lies in the time it takes them to reach a solution. The Bulirsch–Stoer method in this case takes about 3800 modified midpoint steps in total (including the steps performed for each individual value of n). The Runge–Kutta method takes about 4200 steps to calculate a solution to the same accuracy, which at first glance doesn't seem very different from 3800. But a Runge–Kutta step takes more computer time than a modified midpoint step, requiring four evaluations of the function f where the modified midpoint method requires only two. This means that the total number of operations for the Runge–Kutta solution is around 16 800, compared with only about 7600 for the Bulirsch–Stoer solution.

The Bulirsch–Stoer method does require us to do some additional work to perform the Richardson extrapolation, but the computational effort involved is typically small by comparison with the rest of the calculation. So to a good approximation we expect to arrive at a solution about twice as fast with the Bulirsch–Stoer method as with adaptive Runge–Kutta. The running time of neither calculation was very great in this case—they both finished in seconds— but for a larger calculation, something more taxing than this modest example, an improvement of a factor of two could make a great deal of difference. The difference between a program that runs in a week and one that runs in two weeks is significant. Moreover, the advantages of the Bulirsch–Stoer method become more pronounced if we demand a more accurate solution by reducing the value of the accuracy parameter δ. This makes the method particularly attractive for cases where solutions of very high precision are required.

Before ending this section, we should mention that the method described here is not exactly the original Bulirsch–Stoer method as invented by Bulirsch and Stoer. There are two ways in which it differs from the original. First, the original method used a number n of modified midpoint steps that increased exponentially, doubling on successive steps rather than increasing by one.[4] As mentioned above, however, the current belief (based primarily on accumulated experience rather than any rigorous result) is that the method is more efficient when n just goes up by one each time. Second, the original Bulirsch–Stoer method used a different extrapolation scheme, based on rational approximants rather than the Richardson extrapolation described in this section, which uses

---

[4]Specifically, it used a sequence of values $n = 2, 3, 4, 6, 8, 12, 16 \ldots$ so that each value was twice the next-to-last one.

polynomial approximants. It was originally thought that rational approximants gave more accurate results, but again experience has shown this not to be the case, and current thought favors the polynomial scheme.

Exercise 8.13: Planetary orbits

This exercise asks you to calculate the orbits of two of the planets using the Bulirsch–Stoer method. The method gives results significantly more accurate than the Verlet method used to calculate the Earth's orbit in Exercise 8.12.

The equations of motion for the position x, y of a planet in its orbital plane are the same as those for any orbiting body and are derived in Exercise 8.10 on page 361:

$$\frac{d^2x}{dt^2} = -GM\frac{x}{r^3}, \qquad \frac{d^2y}{dt^2} = -GM\frac{y}{r^3},$$

where $G = 6.6738 \times 10^{-11}\,\mathrm{m^3\,kg^{-1}\,s^{-2}}$ is Newton's gravitational constant, $M = 1.9891 \times 10^{30}\,\mathrm{kg}$ is the mass of the Sun, and $r = \sqrt{x^2 + y^2}$.

Let us first solve these equations for the orbit of the Earth, duplicating the results of Exercise 8.12, though with greater accuracy. The Earth's orbit is not perfectly circular, but rather slightly elliptical. When it is at its closest approach to the Sun, its perihelion, it is moving precisely tangentially (i.e., perpendicular to the line between itself and the Sun) and it has distance $1.4710 \times 10^{11}\,\mathrm{m}$ from the Sun and linear velocity $3.0287 \times 10^4\,\mathrm{m\,s^{-1}}$.

a) Write a program, or modify the one from Example 8.7, to calculate the orbit of the Earth using the Bulirsch–Stoer method to a positional accuracy of $1\,\mathrm{km}$ per year. Divide the orbit into intervals of length $H = 1$ week and then calculate the solution for each interval using the combined modified midpoint/Richardson extrapolation method described in this section. Make a plot of the orbit, showing at least one complete revolution about the Sun.

b) Modify your program to calculate the orbit of the dwarf planet Pluto. The distance between the Sun and Pluto at perihelion is $4.4368 \times 10^{12}\,\mathrm{m}$ and the linear velocity is $6.1218 \times 10^3\,\mathrm{m\,s^{-1}}$. Choose a suitable value for H to make your calculation run in reasonable time, while once again giving a solution accurate to $1\,\mathrm{km}$ per year.

You should find that the orbit of Pluto is significantly elliptical—much more so than the orbit of the Earth. Pluto is a Kuiper belt object, similar to a comet, and (unlike true planets) it's typical for such objects to have quite elliptical orbits.

8.5.6 INTERVAL SIZE FOR THE BULIRSCH–STOER METHOD

As we have said, the Bulirsch–Stoer method works best if we keep the number n of modified midpoint steps small, rising to at most eight or ten on any round of the calculation. To ensure this we must choose a suitable value of H, small enough that the extrapolation process converges to the required accuracy quickly. In Example 8.7 we set H manually and this works fine for simple problems. One can just use trial and error to find a suitable value.

For large-scale calculations, however, it's better to have the value of H chosen automatically by the computer. It saves time and ensures that the value used is always a good one. There are various adaptive schemes one can use to calculate a good H, but here's one that is robust and relatively simple.

Suppose, as previously, that we are solving our differential equation from time $t = a$ to $t = b$ and let us choose some initial number of intervals N into which we divide this time, so that the length of each interval is $H = (b-a)/N$. Normally the initial value of N will be a small number, like four, or two, or even just one.

Now we carry out the operations of the Bulirsch–Stoer method on each of the intervals in turn as normal, subdividing each one into n modified midpoint steps and then extrapolating the results as we increase the value of n. For each value of n we calculate the error, Eq. (8.101), on our results and if this error meets our required accuracy level then the calculation for the current interval is finished.

However, if n reaches a predetermined maximum value—typically around eight—and we have not yet met our accuracy goal, then we abandon the calculation and instead subdivide the current time interval of size H into two smaller intervals of size $\frac{1}{2}H$ each. Then we apply the Bulirsch–Stoer method to each of these smaller intervals in turn.

We continue this process as long as necessary, repeatedly subdividing intervals until we reach the required accuracy. Any interval that fails to meet our accuracy goal before the number of steps n reaches the allowed maximum, is subdivided into two parts and the method applied to the parts separately.

Note that different intervals may be subdivided different numbers of times. This means that the ultimate size of the intervals used for different parts of the solution may not be the same. The division of the complete solution into intervals might end up looking something like this:

In portions of the solution where a good result can be obtained with a larger value of H the method will take advantage of that fact. In other portions, where necessary, it will use a smaller value of H (as in the central portions above). Exercises 8.17 and 8.18 give you an opportunity to try out this scheme.

8.6 BOUNDARY VALUE PROBLEMS

All the examples we have considered so far in this chapter have been *initial value problems*, meaning that we are solving differential equations given the initial values of the variables. This is the most common form of differential equation problem encountered in physics, but it is not the only one. There are also *boundary value problems*.

Consider, for instance, the differential equation governing the height above the ground of a ball thrown in the air:

$$\frac{d^2x}{dt^2} = -g,$$ (8.105)

where g is the acceleration due to gravity and we're ignoring friction. To fix the solution of this equation we could specify initial conditions, two initial conditions being required for a second-order equation. We could, for instance, specify the initial height of the ball and its initial upward velocity. However there is another possibility: we could specify our two conditions by giving one initial condition and one *ending* condition. We could, for instance, specify that the ball has height $x = 0$ at $t = 0$ and that $x = 0$ again at some later time $t = t_1$. In other words, we are specifying the time at which the ball is thrown and the time at which it lands. Then our goal would be to find the solution that satisfies these conditions. This problem might arise for instance if we wished to calculate the trajectory of a projectile necessary to make it land at a specific point, which is a classic problem in artillery fire.

Problems of this kind are called boundary value problems. They are somewhat harder to solve computationally than the initial value problems we have looked at previously, but a solution can be achieved by combining two techniques that we have already seen, as follows.

8.6.1 THE SHOOTING METHOD

A fundamental technique for solving boundary value problems is the *shooting method*. The shooting method is a trial-and-error method that searches for the correct values of the initial conditions that match a given set of boundary

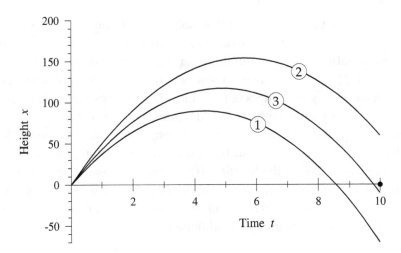

Figure 8.11: The shooting method. The shooting method allows us to match boundary conditions at the beginning and end of a solution. In this example we are solving for the trajectory of a thrown ball and require that its height x be zero—i.e., that it lands—at time $t = 10$. The shooting method involves making a guess as to the initial conditions that will achieve this. In this example, we undershoot our target (represented by the dot on the right) on our first guess (trajectory 1). On the second guess we overshoot. On the third we are closer, but still not perfect.

conditions, in effect turning the calculation back into an initial value problem. Consider the problem of the thrown ball above. In this problem we are given the initial position but not the velocity of the ball. All we know is that the ball lands a certain time later; the velocity is whatever it has to be to make this happen.

So we start by guessing a value for the initial upward velocity. Using this value we solve the differential equation and follow the ball until the time t_1 at which it is supposed to land and we ask whether it had height zero at that time—see Fig. 8.11. Probably it did not. Probably we overshot or undershot. In that case we change our guess for the initial velocity and try again.

The question is how exactly we should modify our guesses to converge to the correct value for the initial velocity. To understand how to do this, let us look at the problem in a slightly different way. In principle there is some function $x = f(v)$ which gives the height x of the ball at time t_1 as a function of the initial vertical velocity v. We don't know what this function is, but we can calculate it for any given value of v by solving the differential equation with that initial velocity. Our goal in solving the boundary value problem is to find the value of v that makes the function zero. That is, we want to solve the equation $f(v) = 0$. But this is simply a matter of finding the root of the function $f(v)$ and we already know how to do that. We saw a number of methods for finding roots in Section 6.3, including binary search and the secant method. Either of these methods would work for the current problem.

So the shooting method involves using one of the standard methods for solving differential equations, such as the fourth-order Runge–Kutta method, to calculate the value of the function $f(v)$, which relates the unknown initial conditions to the final boundary condition(s). Then we use a root finding method such as binary search to find the value of this function that matches the given value of the boundary condition(s).

EXAMPLE 8.8: VERTICAL POSITION OF A THROWN BALL

Let us solve the problem above with the thrown ball for the case where the ball lands back at $x = 0$ after $t = 10$ seconds. The first step, as is normal for second-order equations, is to convert Eq. (8.105) into two first-order equations:

$$\frac{dx}{dt} = y, \qquad \frac{dy}{dt} = -g. \tag{8.106}$$

We will solve these using fourth-order Runge–Kutta, then perform a binary search to find the value of the initial velocity that matches the boundary conditions. Here is a program to accomplish the calculation:

File: `throw.py`

```
from numpy import array,arange

g = 9.81        # Acceleration due to gravity
a = 0.0         # Initial time
b = 10.0        # Final time
N = 1000        # Number of Runge-Kutta steps
h = (b-a)/N     # Size of Runge-Kutta steps
target = 1e-10  # Target accuracy for binary search

def f(r):
    x = r[0]
    y = r[1]
    fx = y
    fy = -g
    return array([fx,fy],float)
```

```
# Function to solve the equation and calculate the final height
def height(v):
    r = array([0.0,v],float)
    for t in arange(a,b,h):
        k1 = h*f(r)
        k2 = h*f(r+0.5*k1)
        k3 = h*f(r+0.5*k2)
        k4 = h*f(r+k3)
        r += (k1+2*k2+2*k3+k4)/6
    return r[0]

# Main program performs a binary search
v1 = 0.01
v2 = 1000.0
h1 = height(v1)
h2 = height(v2)

while abs(h2-h1)>target:
    vp = (v1+v2)/2
    hp = height(vp)
    if h1*hp>0:
        v1 = vp
        h1 = hp
    else:
        v2 = vp
        h2 = hp

v = (v1+v2)/2
print("The required initial velocity is",v,"m/s")
```

One point to notice about this program is that the condition for the binary search to stop is a condition on the accuracy of the height of the ball at the final time $t = 10$, not a condition on the initial velocity. In most cases we care about matching the boundary conditions accurately, not calculating the initial conditions accurately.

If we run the program it prints the following:

```
The required initial velocity is 49.05 m/s
```

In principle, we could now take this value and use it to solve the differential equation once again, to compute the actual trajectory that the ball follows, verifying in the process that indeed it lands back on the ground at the allotted time $t = 10$.

8.6.2   THE RELAXATION METHOD

There is another method for solving boundary value problems that finds some use in physics, the *relaxation method*.[5] This method involves defining a shape for the entire solution, one that matches the boundary conditions but may not be a correct solution of the differential equation in the region between the boundaries. Then one successively modifies this shape to bring it closer and closer to a solution of the differential equation, while making sure that it continues to satisfy the boundary conditions.

In a way, the relaxation method is the opposite of the shooting method. The shooting method starts with a correct solution to the differential equation that may not match the boundary conditions and modifies it until it does. The relaxation method starts with a solution that matches the boundary conditions, but may not be a correct solution to the equation.

In fact the relaxation method is most often used not for solving boundary value problems for ordinary differential equations, but for partial differential equations, and so we will delay our discussion of it until the next chapter, which focuses on partial differential equations. Exercise 9.7 at the end of that chapter gives you an opportunity to apply the relaxation method to an ordinary differential equation problem. If you do that exercise you will see that the method is essentially the same for ordinary differential equations as it is for partial ones.

8.6.3   EIGENVALUE PROBLEMS

A special type of boundary value problem arises when the equation (or equations) being solved are linear and homogeneous, meaning that every term in the equation is linear in the dependent variable. A good example is the Schrödinger equation. For a single particle of mass m in one dimension, the time-independent Schrödinger equation is

$$-\frac{\hbar^2}{2m}\frac{d^2\psi}{dx^2} + V(x)\psi(x) = E\psi(x), \tag{8.107}$$

where $\psi(x)$ is the wavefunction, $V(x)$ is the potential energy at position x, and E is the total energy of the particle, potential plus kinetic. Note how every term

---

[5]The relaxation method has the same name as the method for solving nonlinear equations introduced in Section 6.3.1, which is no coincidence. The two are in fact the same method. The only difference is that in Section 6.3.1 we applied the relaxation method to solutions for single variables, whereas we are solving for entire functions in the case of differential equations.

in the equation is linear in ψ.

Consider the problem of a particle in a square potential well with infinitely high walls. That is,

$$V(x) = \begin{cases} 0 & \text{for } 0 < x < L, \\ \infty & \text{elsewhere,} \end{cases} \qquad (8.108)$$

where L is the width of the well. This problem is solvable analytically, but it is instructive to see how we would solve it numerically as well.

The probability of finding the particle in the region with $V(x) = \infty$ is zero, so the wavefunction has to go to zero at $x = 0$ and $x = L$. Thus this appears to be a standard boundary-value problem which we could solve, for instance, using the shooting method. Since the differential equation is second-order, we would start by turning it into two first-order ones, thus:

$$\frac{d\psi}{dx} = \phi, \qquad \frac{d\phi}{dx} = \frac{2m}{\hbar^2}\big[V(x) - E\big]\psi. \qquad (8.109)$$

To calculate a solution we need two initial conditions, one for each of the variables ψ and ϕ. We know the value of ψ is zero at $x = 0$ but we don't know the value of $\phi = d\psi/dx$, so we guess an initial value then calculate the solution from $x = 0$ to $x = L$. Figure 8.12 shows an example calculated using fourth-order Runge–Kutta (the solid curve in the figure).

In principle the solution should equal zero again at $x = L$, but in this case it doesn't. Based on our experience with the shooting method in Section 8.6.1, we might guess that we can fix this problem by changing the initial condition on the derivative $\phi = d\psi/dx$. Using a root-finding method such as binary search we should be able to find the value of the derivative that makes the wavefunction exactly zero at $x = L$. Unfortunately in the present case this will not work.

To see why, consider what happens when we change the initial condition for our solution. If, for example, we double the initial value of $d\psi/dx$, that doubles the value of the wavefunction close to $x = 0$, as shown by the dashed line in Fig. 8.12. But the Schrödinger equation is a linear equation, meaning that if $\psi(x)$ is a solution then so also is $2\psi(x)$. Thus if we double the initial values of $\psi(x)$ the corresponding solution to the Schrödinger equation is just the same solution that we had before, but times two. This means that if the solution previously failed to pass through zero at $x = L$, it will still fail to pass through zero. And indeed no amount of adjustment of the initial condition will ever make the solution pass through zero. The initial condition only affects the overall magnitude of the solution but does not change its essential shape.

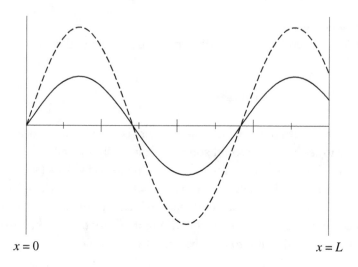

$x = 0$ $x = L$

Figure 8.12: Solution of the Schrödinger equation in a square well. In attempting to solve the Schrödinger equation in a square well using the shooting method the initial value of the wavefunction at position $x = 0$ is known to be zero but we must guess a value for the initial gradient. The result will almost certainly be a wavefunction that fails to correctly return to zero at the far wall of the well, at $x = L$ (solid curve). Changing our guess for the initial slope does not help because the equation is linear, meaning the solution merely gets rescaled by a constant (dashed curve). If the solution did not previously pass through zero at $x = L$ then it still won't, and indeed will not for any choice of initial conditions.

The fundamental issue underlying our problem in this case is that there simply is no solution to the equation that is zero at both $x = 0$ and $x = L$. There is no solution that satisfies the boundary conditions. In fact, there are in this case solutions to the Schrödinger equation only for certain special values of the energy parameter E, the so-called allowed values or eigenvalues; for other values no solutions exist. It is precisely because of this phenomenon that the energies of quantum particles are quantized, meaning that there are discrete allowed energy levels and all other energies are forbidden.

So how do we find the allowed energies and solve the Schrödinger equation? One way to do it is to use a method akin to the shooting method, but instead of varying the initial conditions we vary the energy E. For a particular set of initial conditions, we vary E to find the value for which $\psi = 0$ at $x = L$. That value is an energy eigenstate of the system, an allowed quantum energy, and the corresponding solution to the Schrödinger equation is the wavefunction in that eigenstate. We can think of the solution to the Schrödinger equation as giving us a function $f(E)$ equal to the value of the wavefunction at $x = L$ and we want to find the value of E that makes this function zero. That is, we want to find a root of the function. As with the shooting method, we can find that root using any of the methods we learned about in Section 6.3, such as binary search or the secant method.

And what about the unknown boundary condition on $\phi = d\psi/dx$? How is that to be fixed in this scheme? The answer is that it doesn't matter. Since, as we have said, the only effect of changing this boundary condition is to multiply the whole wavefunction by a constant, we can give it any value we like. The end result will be a solution for the wavefunction that is correct except for a possible overall multiplying factor. Traditionally this factor is fixed by requiring that the wavefunction be normalized so that $\int |\psi(x)|^2 \, dx = 1$. If we need a wavefunction normalized in this fashion, then we can normalize it after the rest of the solution has been computed by calculating the value of the integral $\int |\psi(x)|^2 \, dx$ using any integration method we like, and then dividing the wavefunction throughout by the square root of that value.

EXAMPLE 8.9: GROUND STATE ENERGY IN A SQUARE WELL

Let us calculate the ground state energy of an electron in a square potential well with infinitely high walls separated by a distance L equal to the Bohr radius $a_0 = 5.292 \times 10^{-11}$ m. Here's a program to do the calculation using the secant method:

File: squarewell.py

```
from numpy import array,arange

# Constants
m = 9.1094e-31      # Mass of electron
hbar = 1.0546e-34   # Planck's constant over 2*pi
e = 1.6022e-19      # Electron charge
L = 5.2918e-11      # Bohr radius
N = 1000
h = L/N

# Potential function
def V(x):
    return 0.0

def f(r,x,E):
    psi = r[0]
    phi = r[1]
```

```
    fpsi = phi
    fphi = (2*m/hbar**2)*(V(x)-E)*psi
    return array([fpsi,fphi],float)

# Calculate the wavefunction for a particular energy
def solve(E):
    psi = 0.0
    phi = 1.0
    r = array([psi,phi],float)

    for x in arange(0,L,h):
        k1 = h*f(r,x,E)
        k2 = h*f(r+0.5*k1,x+0.5*h,E)
        k3 = h*f(r+0.5*k2,x+0.5*h,E)
        k4 = h*f(r+k3,x+h,E)
        r += (k1+2*k2+2*k3+k4)/6

    return r[0]

# Main program to find the energy using the secant method
E1 = 0.0
E2 = e
psi2 = solve(E1)

target = e/1000
while abs(E1-E2)>target:
    psi1,psi2 = psi2,solve(E2)
    E1,E2 = E2,E2-psi2*(E2-E1)/(psi2-psi1)

print("E =",E2/e,"eV")
```

If we run the program, it prints the energy of the ground state thus:

```
E = 134.286371694 eV
```

which is indeed the correct answer. Note that the function V(x) does nothing in this program—since the potential everywhere in the well is zero, it plays no role in the calculation. However, by including the function in the program we make it easier to solve other, less trivial potential well problems. For instance, suppose the potential inside the well is not zero but varies as

$$V(x) = V_0 \frac{x}{L}\left(\frac{x}{L} - 1\right),$$ (8.110)

where $V_0 = 100\,\text{eV}$. It's a simple matter to solve for the ground-state energy of this problem also. We have only to change the function V(x), thus:

```
V0 = 100*e
def V(x):
    return V0*(x/L)*(x/L-1)
```

Then we run the program again and this time find that the ground state energy is

```
E = 112.540107208 eV
```

Though the original square well problem is relatively easy to solve analytically, this second version of the problem with a varying potential would be much harder, and yet a solution is achieved easily using the computer.

Exercise 8.14: Quantum oscillators

Consider the one-dimensional, time-independent Schrödinger equation in a harmonic (i.e., quadratic) potential $V(x) = V_0 x^2 / a^2$, where V_0 and a are constants.

a) Write down the Schrödinger equation for this problem and convert it from a second-order equation to two first-order ones, as in Example 8.9. Write a program, or modify the one from Example 8.9, to find the energies of the ground state and the first two excited states for these equations when m is the electron mass, $V_0 = 50\,\text{eV}$, and $a = 10^{-11}\,\text{m}$. Note that in theory the wavefunction goes all the way out to $x = \pm\infty$, but you can get good answers by using a large but finite interval. Try using $x = -10a$ to $+10a$, with the wavefunction $\psi = 0$ at both boundaries. (In effect, you are putting the harmonic oscillator in a box with impenetrable walls.) The wavefunction is real everywhere, so you don't need to use complex variables, and you can use evenly spaced points for the solution—there is no need to use an adaptive method for this problem.

   The quantum harmonic oscillator is known to have energy states that are equally spaced. Check that this is true, to the precision of your calculation, for your answers. (Hint: The ground state has energy in the range 100 to 200 eV.)

b) Now modify your program to calculate the same three energies for the anharmonic oscillator with $V(x) = V_0 x^4 / a^4$, with the same parameter values.

c) Modify your program further to calculate the properly normalized wavefunctions of the anharmonic oscillator for the three states and make a plot of them, all on the same axes, as a function of x over a modest range near the origin—say $x = -5a$ to $x = 5a$.

   To normalize the wavefunctions you will have to calculate the value of the integral $\int_{-\infty}^{\infty} |\psi(x)|^2 \, dx$ and then rescale ψ appropriately to ensure that the area under the square of each of the wavefunctions is 1. Either the trapezoidal rule or Simpson's rule will give you a reasonable value for the integral. Note, however, that you may find a few very large values at the end of the array holding the wavefunction. Where do these large values come from? Are they real, or spurious?

One simple way to deal with the large values is to make use of the fact that the system is symmetric about its midpoint and calculate the integral of the wavefunction over only the left-hand half of the system, then double the result. This neatly misses out the large values.

The methods described in this section allow us to calculate solutions of the Schrödinger equation in one dimension, but the real world, of course, is three-dimensional. In three dimensions the Schrödinger equation becomes a partial differential equation, whose solution requires a different set of techniques. Partial differential equations are the topic of the next chapter.

FURTHER EXERCISES

8.15 The double pendulum: If you did Exercise 8.4 you will have created a program to calculate the movement of a nonlinear pendulum. Although it is nonlinear, the nonlinear pendulum's movement is nonetheless perfectly regular and periodic—there are no surprises. A *double pendulum*, on the other hand, is completely the opposite—chaotic and unpredictable. A double pendulum consists of a normal pendulum with another pendulum hanging from its end. For simplicity let us ignore friction, and assume that both pendulums have bobs of the same mass m and massless arms of the same length ℓ. Thus the setup looks like this:

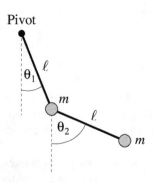

The position of the arms at any moment in time is uniquely specified by the two angles θ_1 and θ_2. The equations of motion for the angles are most easily derived using the Lagrangian formalism, as follows.

The heights of the two bobs, measured from the level of the pivot are

$$h_1 = -\ell \cos\theta_1, \qquad h_2 = -\ell(\cos\theta_1 + \cos\theta_2),$$

so the potential energy of the system is

$$V = mgh_1 + mgh_2 = -mg\ell(2\cos\theta_1 + \cos\theta_2),$$

where g is the acceleration due to gravity. (The potential energy is negative because we have chosen to measure it downwards from the level of the pivot.)

The velocities of the two bobs are given by

$$v_1 = \ell\dot{\theta}_1, \qquad v_2^2 = \ell^2\big[\dot{\theta}_1^2 + \dot{\theta}_2^2 + 2\dot{\theta}_1\dot{\theta}_2 \cos(\theta_1 - \theta_2)\big],$$

where $\dot{\theta}$ means the derivative of θ with respect to time t. (If you don't see where the second velocity equation comes from, it's a good exercise to derive it for yourself from the geometry of the pendulum.) Then the total kinetic energy is

$$T = \tfrac{1}{2}mv_1^2 + \tfrac{1}{2}mv_2^2 = m\ell^2\big[\dot{\theta}_1^2 + \tfrac{1}{2}\dot{\theta}_2^2 + \dot{\theta}_1\dot{\theta}_2\cos(\theta_1 - \theta_2)\big],$$

and the Lagrangian of the system is

$$\mathcal{L} = T - V = m\ell^2\big[\dot{\theta}_1^2 + \tfrac{1}{2}\dot{\theta}_2^2 + \dot{\theta}_1\dot{\theta}_2\cos(\theta_1 - \theta_2)\big] + mg\ell(2\cos\theta_1 + \cos\theta_2).$$

Then the equations of motion are given by the Euler–Lagrange equations

$$\frac{\mathrm{d}}{\mathrm{d}t}\left(\frac{\partial\mathcal{L}}{\partial\dot{\theta}_1}\right) = \frac{\partial\mathcal{L}}{\partial\theta_1}, \qquad \frac{\mathrm{d}}{\mathrm{d}t}\left(\frac{\partial\mathcal{L}}{\partial\dot{\theta}_2}\right) = \frac{\partial\mathcal{L}}{\partial\theta_2},$$

which in this case give

$$2\ddot{\theta}_1 + \ddot{\theta}_2\cos(\theta_1 - \theta_2) + \dot{\theta}_2^2\sin(\theta_1 - \theta_2) + 2\frac{g}{\ell}\sin\theta_1 = 0,$$

$$\ddot{\theta}_2 + \ddot{\theta}_1\cos(\theta_1 - \theta_2) - \dot{\theta}_1^2\sin(\theta_1 - \theta_2) + \frac{g}{\ell}\sin\theta_2 = 0,$$

where the mass m has canceled out.

These are second-order equations, but we can convert them into first-order ones by the usual method, defining two new variables, ω_1 and ω_2, thus:

$$\dot{\theta}_1 = \omega_1, \qquad \dot{\theta}_2 = \omega_2.$$

In terms of these variables our equations of motion become

$$2\dot{\omega}_1 + \dot{\omega}_2\cos(\theta_1 - \theta_2) + \omega_2^2\sin(\theta_1 - \theta_2) + 2\frac{g}{\ell}\sin\theta_1 = 0,$$

$$\dot{\omega}_2 + \dot{\omega}_1\cos(\theta_1 - \theta_2) - \omega_1^2\sin(\theta_1 - \theta_2) + \frac{g}{\ell}\sin\theta_2 = 0.$$

Finally we have to rearrange these into the standard form of Eq. (8.29) with a single derivative on the left-hand side of each one, which gives

$$\dot{\omega}_1 = -\frac{\omega_1^2 \sin(2\theta_1 - 2\theta_2) + 2\omega_2^2 \sin(\theta_1 - \theta_2) + (g/\ell)\left[\sin(\theta_1 - 2\theta_2) + 3\sin\theta_1\right]}{3 - \cos(2\theta_1 - 2\theta_2)},$$

$$\dot{\omega}_2 = \frac{4\omega_1^2 \sin(\theta_1 - \theta_2) + \omega_2^2 \sin(2\theta_1 - 2\theta_2) + 2(g/\ell)\left[\sin(2\theta_1 - \theta_2) - \sin\theta_2\right]}{3 - \cos(2\theta_1 - 2\theta_2)}.$$

(This last step is quite tricky and involves some trigonometric identities. If you have not seen the derivation before, you may find it useful to go through it for yourself.)

These two equations, along with the equations $\dot{\theta}_1 = \omega_1$ and $\dot{\theta}_2 = \omega_2$, give us four first-order equations which between them define the motion of the double pendulum.

a) Derive an expression for the total energy $E = T + V$ of the system in terms of the variables θ_1, θ_2, ω_1, and ω_2, plus the constants g, ℓ, and m.

b) Write a program using the fourth-order Runge–Kutta method to solve the equations of motion for the case where $\ell = 40$ cm, with the initial conditions $\theta_1 = \theta_2 = 90°$ and $\omega_1 = \omega_2 = 0$. Use your program to calculate the total energy of the system assuming that the mass of the bobs is 1 kg each, and make a graph of energy as a function of time from $t = 0$ to $t = 100$ seconds.

Because of energy conservation, the total energy should be constant over time (actually it should be zero for these particular initial conditions), but you will find that it is not perfectly constant because of the approximate nature of the solution of the differential equations. Choose a suitable value of the step size h to ensure that the variation in energy is less than 10^{-5} joules over the course of the calculation.

c) Make a copy of your program and modify the copy to create a second program that does not produce a graph, but instead makes an animation of the motion of the double pendulum over time. At a minimum, the animation should show the two arms and the two bobs.

Hint: As in Exercise 8.4 you will probably find the function rate useful in order to make your program run at a steady speed. You will probably also find that the value of h needed to get the required accuracy in your solution gives a frame-rate much faster than any that can reasonably be displayed in your animation, so you won't be able to display every time-step of the calculation in the animation. Instead you will have to arrange the program so that it updates the animation only once every several Runge–Kutta steps.

8.16   The three-body problem: If you mastered Exercise 8.10 on cometary orbits, here's a more challenging problem in celestial mechanics—and a classic in the field—the *three-body problem*.

Three stars, in otherwise empty space, are initially at rest, with the following masses and positions, in arbitrary units:

	Mass	x	y
Star 1	150	3	1
Star 2	200	−1	−2
Star 3	250	−1	1

(All the z coordinates are zero, so the three stars lie in the xy plane.)

a) Show that the equation of motion governing the position $\mathbf{r}_1$ of the first star is

$$\frac{d^2\mathbf{r}_1}{dt^2} = Gm_2 \frac{\mathbf{r}_2 - \mathbf{r}_1}{|\mathbf{r}_2 - \mathbf{r}_1|^3} + Gm_3 \frac{\mathbf{r}_3 - \mathbf{r}_1}{|\mathbf{r}_3 - \mathbf{r}_1|^3}$$

and derive two similar equations for the positions $\mathbf{r}_2$ and $\mathbf{r}_3$ of the other two stars. Then convert the three second-order equations into six equivalent first-order equations, using the techniques you have learned.

b) Working in units where $G = 1$, write a program to solve your equations and hence calculate the motion of the stars from $t = 0$ to $t = 2$. Make a plot showing the trails of all three stars (i.e., a graph of y against x for each star).

c) Modify your program to make an animation of the motion on the screen from $t = 0$ to $t = 10$. You may wish to make the three stars different sizes or colors (or both) so that you can tell which is which.

To do this calculation properly you will need to use an adaptive step size method, for the same reasons as in Exercise 8.10—the stars move very rapidly when they are close together and very slowly when they are far apart. An adaptive method is the only way to get the accuracy you need in the fast-moving parts of the motion without wasting hours uselessly calculating the slow parts with a tiny step size. Construct your program so that it introduces an error of no more than 10^{-3} in the position of any star per unit time.

Creating an animation with an adaptive step size can be challenging, since the steps do not all correspond to the same amount of real time. The simplest thing to do is just to ignore the varying step sizes and make an animation as if they were all equal, updating the positions of the stars on the screen at every step or every several steps. This will give you a reasonable visualization of the motion, but it will look a little odd because the stars will slow down, rather than speed up, as they come close together, because the adaptive calculation will automatically take more steps in this region.

A better solution is to vary the frame-rate of your animation so that the frames run proportionally faster when h is smaller, meaning that the frame-rate needs to be equal to C/h for some constant C. You can achieve this by using the `rate` function from the `visual` package to set a different frame-rate on each step, equal to C/h. If you do this, it's a good idea to not let the value of h grow too large, or the animation will make some large jumps that look uneven on the screen. Insert extra program lines to ensure that h never exceeds a value h_{max} that you choose. Values for the constants of around $C = 0.1$ and $h_{max} = 10^{-3}$ seem to give reasonable results.

8.17 Cometary orbits and the Bulirsch–Stoer method: Repeat the calculation of the cometary orbit in Exercise 8.10 (page 361) using the adaptive Bulirsch–Stoer method of

Section 8.5.6 to calculate a solution accurate to $\delta = 1$ kilometer per year in the position of the comet. Calculate the solution from $t = 0$ to $t = 2 \times 10^9$ s, initially using just a single time interval of size $H = 2 \times 10^9$ s and allowing a maximum of $n = 8$ modified midpoint steps before dividing the interval in half and trying again. Then these intervals may be subdivided again, as described in Section 8.5.6, as many times as necessary until the method converges in eight steps or less in each interval.

Make a plot of the orbit (i.e., a plot of y against x) and have your program add dots to the trajectory to show where the ends of the time intervals lie. You should see the time intervals getting shorter in the part of the trajectory close to the Sun, where the comet is moving rapidly.

Hint: The simplest way to do this calculation is to make use of recursion, the ability of a Python function to call itself. (If you're not familiar with the idea of recursion you might like to look at Exercise 2.13 on page 83 before doing this exercise.) Write a user-defined function called, say, `step(r,t,H)` that takes as arguments the position vector $\mathbf{r} = (x, y)$ at a starting time t and an interval length H, and returns the new value of $\mathbf{r}$ at time $t + H$. This function should perform the modified midpoint/Richardson extrapolation calculation described in Section 8.5.5 until either the calculation converges to the required accuracy or you reach the maximum number $n = 8$ of modified midpoint steps. If it fails to converge in eight steps, have your function call itself, twice, to calculate separately the solution for the first then the second half of the interval from t to $t + H$, something like this:

```
r1 = step(r,t,H/2)
r2 = step(r1,t+H/2,H/2)
```

(Then *these* functions can call themselves, and so forth, subdividing the interval as many times as necessary to reach the required accuracy.)

8.18 Oscillating chemical reactions: The *Belousov–Zhabotinsky reaction* is a chemical oscillator, a cocktail of chemicals which, when heated, undergoes a series of reactions that cause the chemical concentrations in the mixture to oscillate between two extremes. You can add an indicator dye to the reaction which changes color depending on the concentrations and watch the mixture switch back and forth between two different colors for as long as you go on heating the mixture.

Physicist Ilya Prigogine formulated a mathematical model of this type of chemical oscillator, which he called the "Brusselator" after his home town of Brussels. The equations for the Brusselator are

$$\frac{dx}{dt} = 1 - (b+1)x + ax^2y, \qquad \frac{dy}{dt} = bx - ax^2y.$$

Here x and y represent concentrations of chemicals and a and b are positive constants.

Write a program to solve these equations for the case $a = 1$, $b = 3$ with initial conditions $x = y = 0$, to an accuracy of at least $\delta = 10^{-10}$ per unit time in both x and y, using the adaptive Bulirsch–Stoer method described in Section 8.5.6. Calculate a solution from $t = 0$ to $t = 20$, initially using a single time interval of size $H = 20$. Allow a maximum of $n = 8$ modified midpoint steps in an interval before you divide in half and try again.

Make a plot of your solutions for x and y as a function of time, both on the same graph, and have your program add dots to the curves to show where the boundaries of the time intervals lie. You should find that the points are significantly closer together in parts of the solution where the variables are changing rapidly.

Hint: The simplest way to perform the calculation is to make use of recursion, as described in Exercise 8.17.

CHAPTER **9**

PARTIAL DIFFERENTIAL EQUATIONS

$\mathbb{S}$OME OF the most challenging problems in computational physics—and also some of the most interesting—involve the solution of partial differential equations, such as the wave equation, the diffusion equation, the Laplace and Poisson equations, Maxwell's equations, and the Schrödinger equation. Although the basic numerical methods for solving these problems are straightforward, they are also computationally demanding, so a lot of effort has been invested in finding ways to arrive at solutions quickly.

As with the ordinary differential equations of Chapter 8, problems involving partial differential equations can be divided into initial value problems and boundary value problems. By contrast with the situation in Chapter 8, however, boundary value problems for partial differential equations are usually simpler to solve than initial value problems, so we'll study them first. With boundary value problems we have a partial differential equation describing the behavior of a variable in a space and we are given some constraint on the variable around the boundary of the space. The goal is to solve for the values inside the rest of the space. A classic example of a boundary value problem is the problem of solving for the electric potential in electrostatics. Suppose, for instance, we have a box as shown in Fig. 9.1. The box is empty and has one wall at voltage V and the others at voltage zero. Our goal is to determine the value of the electrostatic potential at points within the box. Problems like this arise in electronics and in many physics experiments.

The electrostatic potential ϕ is related to the vector electric field $\mathbf{E}$ by $\mathbf{E} = -\nabla\phi$. And Maxwell's equations tell us that $\nabla \cdot \mathbf{E} = 0$ in the absence of any electric charges, so $\nabla \cdot \nabla\phi = 0$ or

$$\nabla^2\phi = 0, \tag{9.1}$$

which is *Laplace's equation*. If we write out the Laplacian operator ∇^2 in full,

404

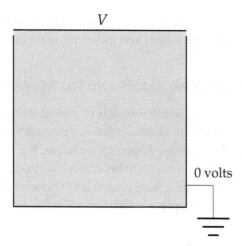

V

0 volts

Figure 9.1: A simple electrostatics problem. An empty box has conducting walls, all of which are grounded at 0 volts except for the wall at the top, which is at some other voltage V. The small gaps between the top wall and the others are intended to show that they are insulated from one another. We'll assume in our calculations that these gaps have negligible width.

the equation takes the form

$$\frac{\partial^2 \phi}{\partial x^2} + \frac{\partial^2 \phi}{\partial y^2} + \frac{\partial^2 \phi}{\partial z^2} = 0. \tag{9.2}$$

The challenge is to solve this equation inside the box, subject to the boundary conditions that $\phi = V$ on the top wall and $\phi = 0$ on the other walls.

The other broad class of partial differential equation problems are the initial value problems. In an initial value problem, the field or other variable of interest is not only subject to boundary conditions, but is also time varying. Typically we are given the initial state of the field at some time $t = 0$, in a manner similar to the initial conditions we used for ordinary differential equations in Chapter 8, and we want to calculate how the field evolves over time thereafter, subject to that initial condition plus the boundary conditions. An example of an initial value problem would be the wave equation for the vibration of a string, like a string in a musical instrument, for which there are boundary conditions (the string is usually pinned and stationary at its ends) but also an initial condition (the string is plucked or struck at $t = 0$) and the challenge is to solve for the motion of the string.

In the following sections of this chapter we will look at techniques for solving both boundary value problems and initial value problems.

9.1 BOUNDARY VALUE PROBLEMS AND THE RELAXATION METHOD

A fundamental technique for the solution of partial differential equations is the *method of finite differences*. Consider the electrostatics example of Fig. 9.1, whose solution involves solving Laplace's equation, Eq. (9.2), for the electric potential ϕ, subject to the appropriate boundary conditions. To make the pictures manageable (and the programs easier) let's consider the problem in two-dimensional space, where Laplace's equation takes the simpler form

$$\frac{\partial^2 \phi}{\partial x^2} + \frac{\partial^2 \phi}{\partial y^2} = 0. \tag{9.3}$$

Real problems are in three dimensions, of course, but the techniques for two and three dimensions are fundamentally the same; once we can do the two-dimensional version, the three-dimensional one is a straightforward generalization.

The method of finite differences involves dividing space into a grid of discrete points. Many kinds of grids are used, depending on the particular problem in hand, but in the simplest and commonest case we use a regular, square grid as shown in Fig. 9.2. Note that we put points on the boundaries of the space (where we already know the value of ϕ) as well as in the interior (where we want to calculate the solution).

Suppose the spacing of the grid points is a and consider a point at position (x, y). As we saw in Section 5.10.5, second derivatives can be approximated using a finite difference formula, just as first derivatives can, and, as we saw in Section 5.10.6, so can partial derivatives. Applying the results we learned in those sections (particularly Eq. (5.109) on page 197), we can write the second derivative of ϕ in the x-direction as

$$\frac{\partial^2 \phi}{\partial x^2} = \frac{\phi(x+a, y) + \phi(x-a, y) - 2\phi(x, y)}{a^2}. \tag{9.4}$$

This formula gives us an expression for the second derivative in terms of the values $\phi(x, y)$, $\phi(x-a, y)$, and $\phi(x+a, y)$ at three adjacent points on our grid. Similarly the derivative in the y-direction is

$$\frac{\partial^2 \phi}{\partial y^2} = \frac{\phi(x, y+a) + \phi(x, y-a) - 2\phi(x, y)}{a^2}, \tag{9.5}$$

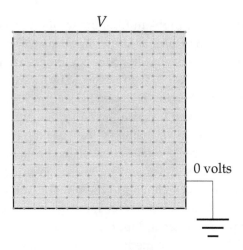

Figure 9.2: Finite differences. In the method of finite differences we divide space into a grid of points. Notice that we put points on the boundaries of the space as well as in the interior.

and the Laplacian operator in two dimensions is:

$$\frac{\partial^2 \phi}{\partial x^2} + \frac{\partial^2 \phi}{\partial y^2} = \frac{\phi(x+a,y) + \phi(x-a,y) + \phi(x,y+a) + \phi(x,y-a) - 4\phi(x,y)}{a^2}.$$

$$(9.6)$$

In other words, we add together the values of ϕ at all the grid points immediately adjacent to (x,y) and subtract four times the value at (x,y), then divide by a^2. Sometimes this rule is represented visually by a diagram like this:

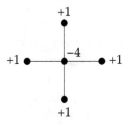

Combining Eqs. (9.3) and (9.6) and canceling the factor of a^2, our Laplace equation now takes the form

$$\phi(x+a,y) + \phi(x-a,y) + \phi(x,y+a) + \phi(x,y-a) - 4\phi(x,y) = 0. \quad (9.7)$$

We have one equation like this for every grid point (x,y) and the solution to the entire set of equations gives us the value of $\phi(x,y)$ at every grid point. In other words, we have turned our problem from the solution of a partial differential equation into the solution of a set of linear simultaneous equations.

We studied the solution of simultaneous equations in Chapter 6, where we saw that sets of equations like (9.7) can be solved by a variety of methods, such as Gaussian elimination (Section 6.1.1) or LU decomposition (Section 6.1.4). In the present case, however, because the equations are particularly simple, there is a quicker way to solve them: we can use the relaxation method of Section 6.3.1. We introduced the relaxation method as a method for the solution of nonlinear equations, but there's no reason in principle why it cannot be applied to linear ones as well, and in the present case it's actually a good choice—better than Gaussian elimination, for instance, because it involves less work, both for the programmer and for the computer.

To make use of the relaxation method, we rearrange Eq. (9.7) to read

$$\phi(x,y) = \tfrac{1}{4}\big[\phi(x+a,y) + \phi(x-a,y) + \phi(x,y+a) + \phi(x,y-a)\big], \qquad (9.8)$$

which tells us that $\phi(x,y)$ is simply the average of the values on the immediately adjacent points of the grid. Now we fix $\phi(x,y)$ at the boundaries of the system to have the value that we know it must have, which for the current electrostatic problem means $\phi = V$ on the top wall of the box and $\phi = 0$ on the other three walls. Then we guess some initial values for $\phi(x,y)$ at the grid points in the interior of the box. The guesses don't have to be good ones. For instance, we could just guess that the values are all zero, which is obviously wrong, but it doesn't much matter. Now we apply Eq. (9.8) to calculate new values ϕ' on all the grid points in the interior space. That is, we calculate

$$\phi'(x,y) = \tfrac{1}{4}\big[\phi(x+a,y) + \phi(x-a,y) + \phi(x,y+a) + \phi(x,y-a)\big], \qquad (9.9)$$

Then we take these values and feed them back in on the right-hand side again to calculate another new set of values, and so on until ϕ settles down to a fixed value. If it does that means that we have found a solution to Eq. (9.8).

This approach for solving the Laplace equation is called the *Jacobi method*. Like all relaxation methods, it does require that the iteration process actually converge to a solution, rather than diverging; as we saw in Section 6.3.1, this is not always guaranteed for relaxation methods. Situations where the calculation diverges are said to be *numerically unstable*, and there are some relaxation methods for partial differential equations that can be numerically unstable. But it turns out that the Jacobi method is not one of them. It can be proved that the Jacobi method is always stable and so always gives a solution.

As described in Section 6.3.1, there are a couple of different ways to decide when to stop the iteration process for the relaxation method, of which the simplest is just to wait until the values of ϕ stop changing, at least within whatever limits of accuracy you have set for yourself. Thus, having calculated $\phi'(x,y)$ from Eq. (9.9), you could calculate the change $\phi'(x,y) - \phi(x,y)$ and when the magnitude of this change is smaller than your target accuracy on every grid point, the calculation ends.

EXAMPLE 9.1: SOLUTION OF LAPLACE'S EQUATION

Let us compute a solution to the two-dimensional electrostatics problem of Fig. 9.1 using the Jacobi method, for the case where the box is 1 m along each side, $V = 1$ volt, and the grid spacing $a = 1$ cm, so that there are 100 grid points on a side, or 101 if we count the points at both the beginning and the end. Here's a program to calculate the solution and make a density plot of the result:

File: laplace.py

```
from numpy import empty,zeros,max
from pylab import imshow,gray,show

# Constants
M = 100          # Grid squares on a side
V = 1.0          # Voltage at top wall
target = 1e-6    # Target accuracy

# Create arrays to hold potential values
phi = zeros([M+1,M+1],float)
phi[0,:] = V
phiprime = empty([M+1,M+1],float)

# Main loop
delta = 1.0
while delta>target:

    # Calculate new values of the potential
    for i in range(M+1):
        for j in range(M+1):
            if i==0 or i==M or j==0 or j==M:
                phiprime[i,j] = phi[i,j]
            else:
                phiprime[i,j] = (phi[i+1,j] + phi[i-1,j] \
                               + phi[i,j+1] + phi[i,j-1])/4
```

```
# Calculate maximum difference from old values
delta = max(abs(phi-phiprime))

# Swap the two arrays around
phi,phiprime = phiprime,phi

# Make a plot
imshow(phi)
gray()
show()
```

A couple of things are worth noting in this program. First, observe that neither the system size L nor the grid spacing a ever enters into the calculation. We perform the whole calculation in terms of the indices i and j of the grid points, integers that run from 0 to 100 and which we can conveniently use as the indices of the arrays phi and phiprime that hold the values of the electric potential. Notice also how when we calculate the new values of ϕ, which are stored in the array phiprime, we run through each point i,j and first check whether it is on the boundary of the box. If it is, then the new value ϕ' of the electric potential is equal to the old one ϕ—the electric potential never changes on the boundaries. If, on the other hand, the point is in the interior of the box then we calculate a new value using Eq. (9.9). Finally, notice the line "phi,phiprime = phiprime,phi", which swaps around the two arrays phi and phiprime, so that the new values of electric potential get used as the old values the next time around the loop.

Figure 9.3 shows the density plot that results from running the program. As we can see, there is a region of high electric potential around the top wall of the box, as we would expect, and low potential around the other three walls.

It's important to appreciate that, since we have approximated our derivatives with finite differences, the Jacobi method is only going to give us an approximate solution to our problem. Even if we make the target accuracy for the iteration very small, the solution will still contain relatively large errors because the finite difference approximation to the second derivative is not very accurate. One can use higher-order derivative approximations, of the kind we studied in Section 5.10.4, to improve the accuracy of the calculation—the generalization of the method is straightforward—or one can just increase the number of grid points, making them closer together, which improves the accuracy of the current approximation. Increasing the number of grid points,

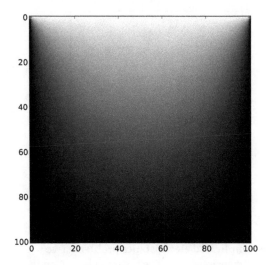

Figure 9.3: Solution of the Laplace equation. A density plot of the electric potential for the two-dimensional electrostatics problem shown in Fig. 9.1, calculated using the Jacobi method.

however, will also make the program run more slowly.

Another point to notice is that the calculation gives us the value of ϕ only at the grid points and not elsewhere. If we need values in between the grid points we could calculate them using an interpolation scheme such as the linear interpolation of Section 5.11, although this introduces a further approximation. This approximation can also be improved by making the grid size smaller, though again the resulting program will run more slowly.

A more technical point to be aware of is that the boundaries around our space may not always be simple and square the way they are in the exercise above. We may have diagonal surfaces or round surfaces or holes cut out of the space. Such things can make it difficult to divide up the space with a square grid because the grid points don't fall exactly on the boundaries. We can always approximate an inconvenient shape by moving the boundaries to the grid points closest to them, applying the boundary conditions at these points rather than exactly on the true boundaries. This, however, introduces more approximations into the calculation. There exist more complicated finite difference methods that get around these issues by using grids whose spacing

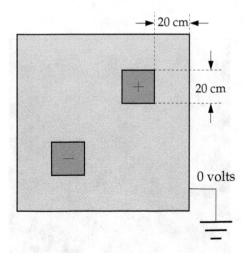

Figure 9.4: A more complicated electrostatics problem. Two square charges are placed inside a square two-dimensional box. The potential is zero on the walls and the charges have charge densities $+1\,\mathrm{Cm}^{-2}$ and $-1\,\mathrm{Cm}^{-2}$.

varies from place to place or non-rectilinear grids. With such methods we may be able to arrange for points to fall exactly on the boundaries of the space, or at least closer to them than with more basic methods, but the calculations also become more complicated—the simple equation (9.9) for calculating the electric potential will no longer apply and we must derive alternate and more complicated ones, which may not even be the same for every point on the grid. We will not need such methods for the calculations we do here, however.

EXAMPLE 9.2: SOLUTION OF THE POISSON EQUATION

As a more complex example of the use of the relaxation method, let us consider the solution of the Poisson equation of electrostatics

$$\nabla^2\phi = -\frac{\rho}{\epsilon_0},\tag{9.10}$$

which governs the electric potential in the presence of a charge density ρ. Here ϵ_0 is the permittivity of empty space (and we are assuming that we are in empty space). Let us consider this equation in two dimensions again, for simplicity, and again in a square box 1 meter along each side. This time, however, all the walls of the box will be at voltage zero, but there will be two square charges in the box, one positive, one negative, as depicted in Fig. 9.4. The two

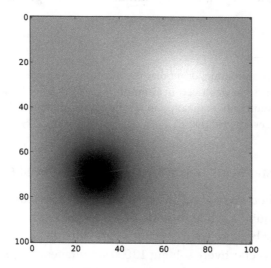

Figure 9.5: Solution for the electrostatic potential of Fig. 9.4. Solution for the electrostatic potential calculated by iteration of Eq. (9.12).

charges are each 20 cm on a side and 20 cm from the walls of the box and have charge density $\pm 1\,\mathrm{Cm}^{-2}$.

To solve this problem using the relaxation method we make use of Eq. (9.6) to rewrite Eq. (9.10) as

$$\frac{\phi(x+a,y)+\phi(x-a,y)+\phi(x,y+a)+\phi(x,y-a)-4\phi(x,y)}{a^2}=-\frac{\rho(x,y)}{\epsilon_0}.$$

$$(9.11)$$

Now we rearrange this expression for $\phi(x,y)$ and get

$$\phi(x,y)=\tfrac{1}{4}\left[\phi(x+a,y)+\phi(x-a,y)+\phi(x,y+a)+\phi(x,y-a)\right]+\frac{a^2}{4\epsilon_0}\rho(x,y).$$

$$(9.12)$$

This is similar to Eq. (9.8) for the Laplace equation, but with the addition of the term in $\rho(x,y)$. Note that the spacing a of the grid points no longer drops out of the calculation. We can, however, still solve the problem as we did before by choosing initial values for $\phi(x,y)$ on a grid and then iterating to convergence. The program is only slightly modified from the one for the Laplace equation on page 409—Exercise 9.1 below gives you the opportunity to work out the details for yourself. Figure 9.5 shows the solution you should get for the electrostatic potential.

Exercise 9.1: Write a program, or modify the one from Example 9.1, to solve Poisson's equation for the system described in Example 9.2. Work in units where $\epsilon_0 = 1$ and continue the iteration until your solution for the electric potential changes by less than 10^{-6} V per step at every grid point.

9.2   FASTER METHODS FOR BOUNDARY VALUE PROBLEMS

The Jacobi method gives good answers, but it has a serious shortcoming: it's slow. It takes quite a while to settle down to the final solution. To be fair this is partly just because we're solving a very difficult problem. The program of Example 9.1 is effectively solving 10 000 simultaneous equations for 10 000 unknowns, one for every point on the 100×100 grid, which is no mean feat. Still, the calculation that produced Fig. 9.3 took about ten minutes, and bigger calculations on larger lattices could take a very long time indeed. If we could find a faster method of solution, it would certainly be a good thing. In fact, it turns out that with two relatively simple modifications we can make the Jacobi method much faster.

9.2.1   OVERRELAXATION

On each iteration of the Jacobi method the values of ϕ on all the grid points converge a little closer to their final values. One way to make this convergence faster is on each step to "overshoot" the new value a little. For instance, suppose on a particular grid point the value goes from 0.1 to 0.3 on one iteration, and eventually converges to 0.5 if we wait long enough. Then we could observe those values 0.1 and 0.3 and tell ourselves, "Instead of changing to 0.3 on this step, let's overshoot and go to, say, 0.4." In this case, doing so will get us closer to the final value of 0.5. This method is called *overrelaxation*. In detail it works like this.[1]

Consider again the solution of Laplace's equation, as in Example 9.1, and suppose at some point in the operation of our program we have a set of values $\phi(x, y)$ on our grid points and at the next iteration we have a new set $\phi'(x, y)$

---

[1] If you did Exercise 6.11 on page 260, then you have seen the basic idea behind overrelaxation before, albeit in the context of ordinary algebraic equations rather than differential equations.

calculated from Eq. (9.9). Then the new set can be written in terms of the old as

$$\phi'(x,y) = \phi(x,y) + \Delta\phi(x,y), \tag{9.13}$$

where $\Delta\phi(x,y)$ is the change in ϕ on this step, given by $\Delta\phi(x,y) = \phi'(x,y) - \phi(x,y)$. Now we define a set of overrelaxed values $\phi_\omega(x,y)$ by

$$\phi_\omega(x,y) = \phi(x,y) + (1+\omega)\Delta\phi(x,y), \tag{9.14}$$

where $\omega > 0$. In other words we change each $\phi(x,y)$ by a little more than we normally would, the exact amount being controlled by the parameter ω. Substituting $\Delta\phi(x,y) = \phi'(x,y) - \phi(x,y)$ into Eq. (9.14) gives

$$\begin{aligned} \phi_\omega(x,y) &= \phi(x,y) + (1+\omega)\left[\phi'(x,y) - \phi(x,y)\right] \\ &= (1+\omega)\phi'(x,y) - \omega\phi(x,y) \\ &= \frac{1+\omega}{4}\left[\phi(x+a,y) + \phi(x-a,y) + \phi(x,y+a) + \phi(x,y-a)\right] - \omega\phi(x,y), \end{aligned} \tag{9.15}$$

where we have used Eq. (9.9) for $\phi'(x,y)$.

The overrelaxation method involves using this equation instead of Eq. (9.9) to calculate new values of ϕ on each step, but in other respects is the same as the Jacobi method. In order to use the equation we must choose a value for the parameter ω. We discuss in a moment how this is done.

9.2.2 THE GAUSS–SEIDEL METHOD

A second trick for speeding up the Jacobi method is as follows. In our program to solve Laplace's equation in Example 9.1 we calculated the new values $\phi'(x,y)$ of the potential at each grid point one by one, working along the rows of the grid in turn, so that when we come to each new point we have already calculated new values at two of the neighbors of that point, though not at the other two. Assuming the new values are better than the old ones, i.e., closer to the true solution we are searching for, it makes sense to use these better values to calculate $\phi'(x,y)$ rather than using the old values at those grid points. If we do this, we get a variant of the Jacobi method called the *Gauss–Seidel method*.[2] In the Gauss–Seidel method we never use the old values of $\phi(x,y)$ if we have new and better ones for the same grid points. Among other things,

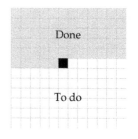

Done

To do

---

[2]Yes, it's *Carl Friedrich Gauss* again.

this means that we can throw away the old value for any grid square as soon as we calculate the new one. A simple way of doing this in practice is to store the new values in the same array as the old ones, overwriting the old ones in the process. The Gauss–Seidel method can be written (in a notation commonly used in computer science) as

$$\phi(x,y) \leftarrow \tfrac{1}{4}\big[\phi(x+a,y) + \phi(x-a,y) + \phi(x,y+a) + \phi(x,y-a)\big], \qquad (9.16)$$

meaning that we calculate the value on the right-hand side and use it to replace the value on the left. This technique has the nice feature that we only need one array to store the values of $\phi(x,y)$, and not two as in the traditional Jacobi method.

We can also combine the Gauss–Seidel method with the overrelaxation method of Section 9.2.1 to get a method that is faster than either, which we would write thus:

$$\phi(x,y) \leftarrow \frac{1+\omega}{4}\big[\phi(x+a,y) + \phi(x-a,y) + \phi(x,y+a) + \phi(x,y-a)\big]$$
$$- \omega\phi(x,y). \qquad (9.17)$$

In fact most of the speed-up in this combined overrelaxation/Gauss–Seidel method comes from the overrelaxation and not from Gauss and Seidel, so you might think we could just forget about Gauss–Seidel. There are, however, two reasons not to do this. First, the Gauss–Seidel method is not just faster, but also uses only one array instead of two, so it uses less memory. Given that it is faster *and* uses less memory, there is no reason to use the Jacobi method in most cases. Second, and more importantly, it turns out that in fact the simple Jacobi overrelaxation method as we described it in Section 9.2.1 *does not work*. It is numerically unstable (at least on square grids), whereas the Gauss–Seidel version is stable. The reasons for the instability are subtle but suffice it to say that the Gauss–Seidel method is better in essentially all respects than the Jacobi method.

Before we can use the method described here, we need to choose a value for the overrelaxation parameter ω. There is no general optimal value that gives the best performance of the method for all problems. The choice of value depends both on the specific equations we are solving and on the shape of the grid we use. However, there are some guidelines that can be helpful. First, larger values of ω in general give faster calculations, but only up to a point. And if we use too large a value the calculation can become numerically unstable. It has been proved that the method is in general stable for $\omega < 1$ but unstable otherwise, so we should confine ourselves to values less than one. Other than this, the best way to find a good value is usually by experimentation. For the solution of Laplace's equation using a square grid of points the best value, the value that gives the fastest solution, is somewhere in the vicinity of $\omega = 0.9$, or a little larger. A solution of the problem from Example 9.1 using $\omega = 0.9$ took the author's computer just 38 seconds—much better than the 10 minutes taken by the Jacobi method. Exercise 9.2 gives you an opportunity to develop a program of your own to do the calculation.

Exercise 9.2: Use the combined overrelaxation/Gauss–Seidel method to solve Laplace's equation for the two-dimensional problem in Example 9.1—a square box 1 m on each side, at voltage $V = 1$ volt along the top wall and zero volts along the other three. Use a grid of spacing $a = 1\,\text{cm}$, so that there are 100 grid points along each wall, or 101 if you count the points at both ends. Continue the iteration of the method until the value of the electric potential changes by no more than $\delta = 10^{-6}\,\text{V}$ at any grid point on any step, then make a density plot of the final solution, similar to that shown in Fig. 9.3. Experiment with different values of ω to find which value gives the fastest solution. As mentioned above, you should find that a value around 0.9 does well. In general larger values cause the calculation to run faster, but if you choose too large a value the speed drops off and for values above 1 the calculation becomes unstable.

Exercise 9.3: Consider the following simple model of an electronic capacitor, consisting of two flat metal plates enclosed in a square metal box:

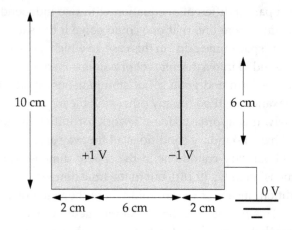

For simplicity let us model the system in two dimensions. Using any of the methods we have studied, write a program to calculate the electrostatic potential in the box on a grid of 100×100 points, where the walls of the box are at voltage zero and the two plates (which are of negligible thickness) are at voltages ± 1 V as shown. Have your program calculate the value of the potential at each grid point to a precision of 10^{-6} volts and then make a density plot of the result.

Hint: Notice that the capacitor plates are at fixed *voltage*, not fixed charge, so this problem differs from the problem with the two charges in Exercise 9.1. In effect, the capacitor plates are part of the boundary condition in this case: they behave the same way as the walls of the box, with potentials that are fixed at a certain value and cannot change.

9.3 INITIAL VALUE PROBLEMS

In the first part of this chapter we looked at the solution of boundary value problems. The other main class of partial differential equation problems in physics is initial value problems, where we are told the starting conditions for a variable and our goal is to predict future variation as a function of time. A simple example of such a problem is the diffusion equation in one (spatial) dimension:

$$\frac{\partial \phi}{\partial t} = D \frac{\partial^2 \phi}{\partial x^2}, \qquad (9.18)$$

where D is a diffusion coefficient. The diffusion equation is used to calculate the motion of diffusing gases and liquids, as well as the flow of heat in thermal conductors. (When used to study heat, it is sometimes called the *heat equation*.)

The variable $\phi(x, t)$ in Eq. (9.18) depends on both position x and time t. So, like the two-dimensional Laplace equation that we solved in the previous section, this is a partial differential equation with two independent variables. You might imagine, therefore, that one could solve it the same way: create a grid of points—a "space-time grid" in this case, in which the two dimensions of space-time are divided into a discrete set of points—then write the derivatives in finite difference form and get a set of simultaneous equations that can be solved by a relaxation method (or any other suitable method).

Unfortunately, this approach doesn't work for initial value problems because we don't have boundary conditions all the way around the space. Typically we have boundary conditions in the spatial dimension or dimensions (e.g., the x dimension in Eq. (9.18)), but in the time dimension we have instead an *initial* condition, meaning we are told where the value of ϕ starts but not where it ends. This means that the relaxation method breaks down because we don't know what value to use for ϕ at the time-like end of the grid. Instead of a relaxation method, therefore, we use a different approach for solving initial value problems, a *forward integration method*, as follows.

9.3.1 THE FTCS METHOD

Consider the solution of the diffusion equation, Eq. (9.18). As with our earlier solution of Laplace's equation, we start by dividing the *spatial* dimension (or dimensions) into a grid of points. In the case of Eq. (9.18) there is only one spatial dimension, so the "grid" is actually just a line of points along the x-axis. We will use evenly spaced points, although unevenly spaced ones are sometimes used in special cases. Let the spacing of the points be a. Then the derivative on the right-hand side of (9.18) can be written, as previously, using Eq. (5.109):

$$\frac{\partial^2 \phi}{\partial x^2} = \frac{\phi(x+a,t) + \phi(x-a,t) - 2\phi(x,t)}{a^2}, \tag{9.19}$$

and so our diffusion equation, Eq. (9.18), becomes

$$\frac{d\phi}{dt} = \frac{D}{a^2}\left[\phi(x+a,t) + \phi(x-a,t) - 2\phi(x,t)\right]. \tag{9.20}$$

If we think of the value of ϕ at the different grid points as separate variables, we now have a set of simultaneous *ordinary* differential equations in those variables, equations of exactly the kind we studied in Chapter 8, which can be solved by the methods we already know about. The catch is that there can be a lot of equations to solve—hundreds, thousands, or even millions, depending on how fine a grid we use and how many spatial dimensions we have. A few thousand equations is quite feasible with modern computers; a few million is pushing the limits, unless you have a supercomputer.

The most common method for solving the differential equations in Eq. (9.20) is Euler's method, which we studied in Section 8.1.1. This may seem like a strange choice: we said in Chapter 8 that one never uses Euler's method because higher-order Runge–Kutta methods are not much harder to program and give better results. That's true here too, but the point to notice is that the approximation to the second derivative on the right-hand side of Eq. (9.20) is not very accurate—it introduces a second-order error as shown in Section 5.10.5. There is not much point expending a lot of effort to get a very accurate solution for the differential equation in t if the inputs to that solution—the right-hand side of the equation—aren't accurate in the first place. Euler's method also has a second-order error and gives errors typically of comparable size to those coming from the second derivative, so there is little harm in using it in this case, and its relative simplicity gives it an advantage over the Runge–Kutta method.

Recall that Euler's method for solving a differential equation of the form

$$\frac{d\phi}{dt} = f(\phi, t), \tag{9.21}$$

is to Taylor expand $\phi(t)$ about time t and write

$$\phi(t+h) \simeq \phi(t) + h\frac{d\phi}{dt} = \phi(t) + hf(\phi, t). \tag{9.22}$$

Applying the same approach to Eq. (9.20) gives us

$$\phi(x, t+h) = \phi(x, t) + h\frac{D}{a^2}\left[\phi(x+a, t) + \phi(x-a, t) - 2\phi(x, t)\right]. \tag{9.23}$$

If we know the value of ϕ at every grid point x at some time t, then this equation tells us the value at every grid point at time $t + h$, a short interval later. Hence, by repeatedly using this equation at every grid point, we can derive the solution to our partial differential equation at a succession of time points a distance h apart, in a manner similar the way we solved ordinary differential equations in Chapter 8. This is called the *forward-time centered-space method* for solving partial differential equations, or FTCS for short.

EXAMPLE 9.3: THE HEAT EQUATION

The flat base of a container made of 1 cm thick stainless steel is initially at a uniform temperature of 20° Celsius everywhere. The container is placed in a bath of cold water at 0°C and filled with hot water at 50°C:

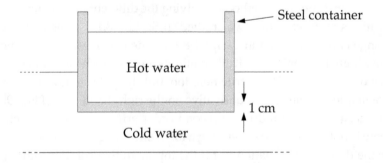

Our goal is to calculate the temperature profile of the steel as a function of distance x from the hot side to the cold side, and as a function of time. For simplicity let us treat the base of the container as being arbitrarily wide, so that its temperature profile is the same everywhere, and assume that neither the hot nor the cold water changes temperature appreciably.

Thermal conduction is governed by the diffusion equation (also called the heat equation in this context), so solving this problem is a matter of solving the one-dimensional diffusion equation for the temperature T:

$$\frac{\partial T}{\partial t} = D\frac{\partial^2 T}{\partial x^2}. \tag{9.24}$$

Let us divide the x-axis into 100 equal grid intervals, meaning that there will be 101 grid points in total, counting the first and last ones. The first and last points have fixed temperatures of $50°$ and $0°$C, respectively, while the intermediate points are initially all at $20°$C. We also need the heat diffusion coefficient for stainless steel, also called the *thermal diffusivity*, which is $D = 4.25 \times 10^{-6}\,\mathrm{m^2\,s^{-1}}$.

Here is a program to perform the calculation using the FTCS method and make a plot of the temperature profile at times $t = 0.01\,\mathrm{s}$, $0.1\,\mathrm{s}$, $0.4\,\mathrm{s}$, $1\,\mathrm{s}$, and $10\,\mathrm{s}$, all on the same graph:

```
from numpy import empty                                    File: heat.py
from pylab import plot,xlabel,ylabel,show

# Constants
L = 0.01        # Thickness of steel in meters
D = 4.25e-6     # Thermal diffusivity
N = 100         # Number of divisions in grid
a = L/N         # Grid spacing
h = 1e-4        # Time-step
epsilon = h/1000

Tlo = 0.0       # Low temperature in Celsius
Tmid = 20.0     # Intermediate temperature in Celsius
Thi = 50.0      # Hi temperature in Celsius

t1 = 0.01
t2 = 0.1
t3 = 0.4
t4 = 1.0
t5 = 10.0
tend = t5 + epsilon
```

```
# Create arrays
T = empty(N+1,float)
T[0] = Thi
T[N] = Tlo
T[1:N] = Tmid
Tp = empty(N+1,float)
Tp[0] = Thi
Tp[N] = Tlo

# Main loop
t = 0.0
c = h*D/(a*a)
while t<tend:

    # Calculate the new values of T
    for i in range(1,N):
        Tp[i] = T[i] + c*(T[i+1]+T[i-1]-2*T[i])
    T,Tp = Tp,T
    t += h

    # Make plots at the given times
    if abs(t-t1)<epsilon:
        plot(T)
    if abs(t-t2)<epsilon:
        plot(T)
    if abs(t-t3)<epsilon:
        plot(T)
    if abs(t-t4)<epsilon:
        plot(T)
    if abs(t-t5)<epsilon:
        plot(T)

xlabel("x")
ylabel("T")
show()
```

This is a straightforward use of Eq. (9.23). The only slightly subtle point arises in making the plots. We need to make a plot when the time variable t is equal to any of the times 0.01, 0.1, 0.4, and so forth, which are called t1, t2, t3 in the program. But because the time is a floating-point variable it is subject to numerical rounding error, as discussed in Section 4.2, so we cannot simply write "if t==t1". As we said in Section 4.2, one should never compare two floats for equality because numerical error may make them different in practice even when in theory they are really the same. So instead we compute the absolute value of the difference "abs(t-t1)" and check if it is smaller than the small number epsilon; if it is, then t is very close to t1, which is all we need.

Here's one further nice trick. In the program above we calculate the new values of the temperature at each time-step with the lines

```
for i in range(1,N):
    Tp[i] = T[i] + c*(T[i+1]+T[i-1]-2*T[i])
```

but we can if we wish do this more simply—and more quickly—by recalling that Python can do arithmetic with entire arrays in a single step. Using the "slicing" methods that we introduced in Section 2.4.5 we can calculate the new temperatures with a single line:

```
Tp[1:N] = T[1:N] + c*(T[2:N+1]+T[0:N-1]-2*T[1:N])
```

Because Python can do operations with entire arrays almost as fast as it can with single variables, this makes a huge difference to the speed of the program. On the author's computer the first, pedestrian program takes about 60 seconds to finish. The modified program using the trick above takes only 3 seconds— faster by about a factor of twenty.

Figure 9.6 shows the figure produced by the program. As you can see, the temperature profile starts off with a large region of the system at 20°C and much smaller regions of hotter and colder temperatures near the ends. But as time goes by the profile becomes smoother until by $t = 10$ it is a single straight line from the hot temperature of 50°C to the cold temperature of 0°C.

In this example, we assumed that the boundary conditions on the system, the temperatures at the beginning and end of the grid, were constant in time. But there is no reason why this has to be the case. There are many interesting physics problems where the boundary conditions vary in time to drive the system through a set of different states, and the FTCS method can be used to solve these problems as well. The method is exactly the same as for the case of fixed boundary conditions; the only change we have to make is varying the array elements that hold the boundary conditions as time goes by.

For instance, we might want to study the thermal diffusion problem above, but allow the hot water inside the container to cool down over time. It's a simple matter to vary the array element representing the temperature at the hot boundary so as to incorporate this change. In other respects the program would be exactly the same.

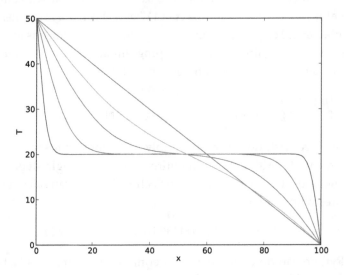

Figure 9.6: Solution of the heat equation. Solution of the heat equation for the problem in Example 9.3. The five curves represent the temperature profile as a function of position at times $t = 0.01, 0.1, 0.4, 1$, and 10.

Exercise 9.4 gives you an opportunity to try your hand at such a problem in a calculation of the temperature inside the Earth's crust, as the temperature at the surface varies with the seasons.

Exercise 9.4: Thermal diffusion in the Earth's crust

A classic example of a diffusion problem with a time-varying boundary condition is the diffusion of heat into the crust of the Earth, as surface temperature varies with the seasons. Suppose the mean daily temperature at a particular point on the surface varies as:

$$T_0(t) = A + B \sin \frac{2\pi t}{\tau},$$

where $\tau = 365$ days, $A = 10°C$ and $B = 12°C$. At a depth of 20 m below the surface almost all annual temperature variation is ironed out and the temperature is, to a good approximation, a constant $11°C$ (which is higher than the mean surface temperature of $10°C$—temperature increases with depth, due to heating from the hot core of the planet). The thermal diffusivity of the Earth's crust varies somewhat from place to place, but for our purposes we will treat it as constant with value $D = 0.1\,\mathrm{m^2\,day^{-1}}$.

Write a program, or modify one of the ones given in this chapter, to calculate the temperature profile of the crust as a function of depth up to 20 m and time up to 10 years. Start with temperature everywhere equal to 10°C, except at the surface and the deepest point, choose values for the number of grid points and the time-step h, then run your program for the first nine simulated years, to allow it to settle down into whatever pattern it reaches. Then for the tenth and final year plot four temperature profiles taken at 3-month intervals on a single graph to illustrate how the temperature changes as a function of depth and time.

9.3.2 NUMERICAL STABILITY

The FTCS method works well for the diffusion equation, but there are other cases in which it works less well. An example of the latter is the wave equation, one of the most important partial differential equations in physics, describing sound waves and vibrations, electromagnetic waves, water waves, and a host of other phenomena. The FTCS method fails badly when applied to the wave equation, and it does so for interesting reasons.

In one-dimension the wave equation is usually written as

$$\frac{\partial^2 \phi}{\partial x^2} - \frac{1}{v^2} \frac{\partial^2 \phi}{\partial t^2} = 0, \tag{9.25}$$

but for our purposes it will be useful to rearrange it in the form

$$\frac{\partial^2 \phi}{\partial t^2} = v^2 \frac{\partial^2 \phi}{\partial x^2}. \tag{9.26}$$

We could use this equation to model, for instance, the movement of a vibrating string of length L, held fixed at both ends. To solve the equation using the FTCS method we would start by dividing the string into discrete points with spacing a. Then we would replace the second derivative on the right-hand side with a discrete difference, using Eq. (5.109), and hence derive a set of second-order ordinary differential equations, one for each grid point:

$$\frac{d^2 \phi}{dt^2} = \frac{v^2}{a^2} \left[\phi(x+a, t) + \phi(x-a, t) - 2\phi(x, t) \right]. \tag{9.27}$$

Now, using the techniques we learned in Section 8.3, we can change this second-order equation into two first-order equations by defining a new variable ψ thus:

$$\frac{d\phi}{dt} = \psi(x, t), \qquad \frac{d\psi}{dt} = \frac{v^2}{a^2} \left[\phi(x+a, t) + \phi(x-a, t) - 2\phi(x, t) \right]. \tag{9.28}$$

425

Then, applying Euler's method, these equations become two FTCS equations thus:

$$\phi(x, t+h) = \phi(x, t) + h\psi(x, t), \tag{9.29a}$$

$$\psi(x, t+h) = \psi(x, t) + h\frac{v^2}{a^2}\left[\phi(x+a, t) + \phi(x-a, t) - 2\phi(x, t)\right]. \tag{9.29b}$$

Now we simply iterate these equations from any given starting condition to get our solution.

Figure 9.7 shows three snapshots from a typical solution, depicting the displacement ϕ of the string at three different times t during the run of the program. As we can see, the solution starts off looking fine—we have a normal looking wave on our string (Fig. 9.7a)—but around about the 50 millisecond mark errors start to creep in (Fig. 9.7b), and by 100 milliseconds the errors have grown to dominate the calculation and the results are meaningless (Fig. 9.7c). If we were to let the calculation run longer, we would find that the errors keep on growing until they become so large that they overflow the largest numbers the computer can store. This is not mere rounding error of the kind discussed in Section 4.2. This is something more serious. The calculation has become *numerically unstable.*

What is going on here? Why did the FTCS method work so well for the diffusion equation, but break down for the wave equation? To answer this question let us return for a moment to the diffusion equation and consider again the basic FTCS form:

$$\phi(x, t+h) = \phi(x, t) + h\frac{D}{a^2}\left[\phi(x+a, t) + \phi(x-a, t) - 2\phi(x, t)\right]. \tag{9.30}$$

(We derived this previously in Eq. (9.23).)

Consider now the following argument, which is known as a *von Neumann stability analysis,* after its inventor John von Neumann, one of the early pioneers of computational physics. We know that the spatial variation of ϕ at any time t can always be expressed in the form of a Fourier series $\phi(x, t) = \sum_k c_k(t)\, e^{ikx}$ for some suitable set of wavevectors k and (time-varying and potentially complex) coefficients $c_k(t)$. Given such an expression let us then ask what form the solution takes at the next time-step of the FTCS calculation, as given by Eq. (9.30) above. Since the equation is linear, we can answer this question by studying what happens to each term in the Fourier series separately, and then adding them up at the end. Plugging a single term $\phi(x, t) = c_k(t)\, e^{ikx}$ into the

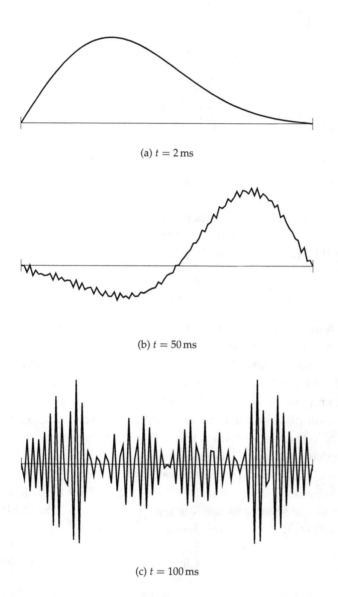

(a) $t = 2\,\text{ms}$

(b) $t = 50\,\text{ms}$

(c) $t = 100\,\text{ms}$

Figure 9.7: FTCS solution of the wave equation. A solution of the wave equation, Eq. (9.26), representing, for example, the displacement of a vibrating string, at three different times t, as calculated using the FTCS method of Eq. (9.29). While panel (a) looks promising, visible problems have appeared by panel (b), and by panel (c) they have grown so large that the solution is unworkable.

equation, we find that

$$\phi(x, t+h) = c_k(t)e^{ikx} + h\frac{D}{a^2}c_k(t)\left[e^{ik(x+a)} + e^{ik(x-a)} - 2e^{ikx}\right]$$

$$= \left[1 + h\frac{D}{a^2}\left(e^{ika} + e^{-ika} - 2\right)\right]c_k(t)e^{ikx}$$

$$= \left[1 - h\frac{4D}{a^2}\sin^2 \tfrac{1}{2}ka\right]c_k(t)e^{ikx}, \tag{9.31}$$

where we have made use of the fact that $e^{i\theta} + e^{-i\theta} = 2\cos\theta$, as well as the half-angle formula $1 - \cos\theta = 2\sin^2 \tfrac{1}{2}\theta$.

This equation tells us that each term in our Fourier series transforms independently under the FTCS equations—the term for wavevector k is multiplied by a k-dependent factor, but does not mix with any of the other wavevectors.[3] Knowing this allows us to read off the coefficient of e^{ikx} at time $t + h$ from Eq. (9.31):

$$c_k(t+h) = \left[1 - h\frac{4D}{a^2}\sin^2 \tfrac{1}{2}ka\right]c_k(t). \tag{9.32}$$

So on each time-step of the calculation, $c_k(t)$ is just multiplied by the factor in brackets, which depends on k but not on either x or t.

Now we can see how the calculation could become unstable: if the magnitude of the factor in brackets exceeds unity for any wavevector k, then the Fourier component with that value of k will grow exponentially as it gets repeatedly multiplied by the same factor on each time-step. Luckily the factor in brackets in this case is always less than one. However, its *magnitude* could become greater than one if it became negative and fell below -1. This happens if the term $h(4D/a^2)\sin^2 \tfrac{1}{2}ka$ ever becomes greater than 2. To put that another way, the solution is stable if $h(4D/a^2)\sin^2 \tfrac{1}{2}ka \le 2$ for all k. But the largest possible value of $\sin^2 \tfrac{1}{2}ka$ for any k is 1, so the solution will be stable for all k if we have $h(4D/a^2) \le 2$, or equivalently

$$h \le \frac{a^2}{2D}. \tag{9.33}$$

If h is larger than this limit then the solution can diverge, which is an unphysical behavior for the diffusion equation—thermal diffusion never results in a cool object becoming infinitely hot, for example.

---

[3] Technically, this is because the wave equation is diagonal in k-space. Analyses like the present one can be performed for equations that are not diagonal in k-space (so long as they are still linear), but one needs to use a different functional basis instead of the Fourier basis, one that makes the equation of interest diagonal.

On the other hand, if $h < a^2/2D$ then all terms in the Fourier series will decay exponentially and eventually go to zero, except for the $k = 0$ term, for which the multiplicative factor in Eq. (9.32) is always exactly 1. And this is precisely the physical behavior we expect of the diffusion equation. We expect all Fourier components except the one for $k = 0$ to decay, leaving a solution that is uniform in space (unless there are time-varying boundary conditions that are preventing the system from reaching a steady state).

The von Neumann analysis method doesn't work for all equations. For instance, it doesn't work for nonlinear differential equations, because it requires that the full solution of the equation, written as a Fourier composition, can be analyzed by treating each of its Fourier components separately, which is not possible for nonlinear equations. Nonetheless it is applicable to many of the equations we encounter in physics and is one of the simplest ways of demonstrating numerical stability or instability.

Having used the von Neumann analysis to analyze the stability of the diffusion equation, let us now return to the wave equation, for which the FTCS equations are given in Eq. (9.29). Applying the von Neumann analysis to these equations is more complicated than for the diffusion equation because there are now two FTCS equations, one for each of the two variables ϕ and ψ, but the analysis can still be done. We consider the two variables to be the elements of a two element vector (ϕ, ψ), and write a single term in the Fourier series for the solution as

$$\begin{pmatrix} \phi(x,t) \\ \psi(x,t) \end{pmatrix} = \begin{pmatrix} c_\phi(t) \\ c_\psi(t) \end{pmatrix} e^{ikx}. \tag{9.34}$$

Substituting into Eq. (9.29) and following the same line of argument as before, we find that the coefficients of e^{ikx} at the next time-step are

$$c_\phi(t + h) = c_\phi(t) + hc_\psi(t), \tag{9.35a}$$

$$c_\psi(t + h) = c_\psi(t) - hc_\phi(t)\frac{4v^2}{a^2}\sin^2 \tfrac{1}{2}ka, \tag{9.35b}$$

or in vector form

$$c(t + h) = Ac(t), \tag{9.36}$$

where $c(t)$ is the vector (c_ϕ, c_ψ) and A is the matrix

$$A = \begin{pmatrix} 1 & h \\ -hr^2 & 1 \end{pmatrix} \quad \text{with} \quad r = \frac{2v}{a}\sin \tfrac{1}{2}ka. \tag{9.37}$$

In other words the vector $c(t)$ gets multiplied on each time-step by a 2×2 matrix that depends on k but not on x or t.

429

Now let us write $\mathbf{c}(t)$ as a linear combination of the two (right) eigenvectors of $\mathbf{A}$, which we will call $\mathbf{v}_1$ and $\mathbf{v}_2$, so that $\mathbf{c}(t) = \alpha_1\mathbf{v}_1 + \alpha_2\mathbf{v}_2$ for some constants α_1 and α_2. Substituting this form into Eq. (9.36) gives

$$\mathbf{c}(t+h) = \mathbf{A}(\alpha_1\mathbf{v}_1 + \alpha_2\mathbf{v}_2) = \alpha_1\lambda_1\mathbf{v}_1 + \alpha_2\lambda_2\mathbf{v}_2, \tag{9.38}$$

where λ_1 and λ_2 are the eigenvalues corresponding to the two eigenvectors. We can then repeat the process and multiply by the matrix again to get the value on the next time-step:

$$\mathbf{c}(t+2h) = \alpha_1\lambda_1^2\mathbf{v}_1 + \alpha_2\lambda_2^2\mathbf{v}_2. \tag{9.39}$$

And after m time-steps

$$\mathbf{c}(t+mh) = \alpha_1\lambda_1^m\mathbf{v}_1 + \alpha_2\lambda_2^m\mathbf{v}_2. \tag{9.40}$$

If either (or both) of the eigenvalues have magnitude greater than unity, this solution will diverge to infinity as we take higher and higher powers of the eigenvalues. Or, to put that another way, the solution is numerically stable only if both eigenvalues have magnitude less than or equal to one.

The eigenvalues of the matrix are given by the characteristic determinant equation $\det(\mathbf{A} - \lambda\mathbf{I}) = 0$, where $\mathbf{I}$ is the identity matrix. Making use of Eq. (9.37) for $\mathbf{A}$ and writing out the determinant in full, the equation becomes $(1-\lambda)^2 + h^2r^2 = 0$, which has solutions $\lambda = 1 \pm ihr$. Thus the eigenvalues are complex in this case and both have the same magnitude,

$$|\lambda| = \sqrt{1 + h^2r^2} = \sqrt{1 + \frac{4h^2v^2}{a^2}\sin^2\tfrac{1}{2}ka}. \tag{9.41}$$

But this magnitude is never less than unity. No matter how small we make the value of h, the value of $|\lambda|$ will always be greater than one (except for the $k = 0$ Fourier component), and hence the FTCS method is *never stable* for the wave equation.

Unfortunately, this means that the FTCS method simply will not work for solving the wave equation, except over very short time intervals—short enough that the errors don't grow large and swamp the solution. For any but the shortest of intervals, however, we are going to need a different method to solve the wave equation.

Exercise 9.5: FTCS solution of the wave equation

Consider a piano string of length L, initially at rest. At time $t = 0$ the string is struck by the piano hammer a distance d from the end of the string:

The string vibrates as a result of being struck, except at the ends, $x = 0$ and $x = L$, where it is held fixed.

a) Write a program that uses the FTCS method to solve the complete set of simultaneous first-order equations, Eq. (9.28), for the case $v = 100\,\mathrm{m\,s^{-1}}$, with the initial condition that $\phi(x) = 0$ everywhere but the velocity $\psi(x)$ is nonzero, with profile

$$\psi(x) = C\frac{x(L-x)}{L^2}\exp\left[-\frac{(x-d)^2}{2\sigma^2}\right],$$

where $L = 1\,\mathrm{m}$, $d = 10\,\mathrm{cm}$, $C = 1\,\mathrm{m\,s^{-1}}$, and $\sigma = 0.3\,\mathrm{m}$. You will also need to choose a value for the time-step h. A reasonable choice is $h = 10^{-6}\,\mathrm{s}$.

b) Make an animation of the motion of the piano string using the facilities provided by the visual package, which we studied in Section 3.4. There are various ways you could do this. A simple one would be to just place a small sphere at the location of each grid point on the string. A more sophisticated approach would be to use the curve object in the visual package—see the on-line documentation at www.vpython.org for details. A convenient feature of the curve object is that you can specify its set of x positions and y positions separately as arrays. In this exercise the x positions only need to specified once, since they never change, while the y positions will need to be specified anew each time you take a time-step. Also, since the vertical displacement of the string is much less than its horizontal length, you will probably need to multiply the vertical displacement by a fairly large factor to make it visible on the screen.

Allow your animation to run for some time, until numerical instabilities start to appear.

9.3.3 THE IMPLICIT AND CRANK–NICOLSON METHODS

One possible approach for remedying the stability problems of the previous section is to use the so-called *implicit method*. Consider Eq. (9.29) again, and let us make the substitution $h \to -h$ in the equations, giving

$$\phi(x, t - h) = \phi(x, t) - h\psi(x, t), \tag{9.42a}$$

$$\psi(x, t - h) = \psi(x, t) - h\frac{v^2}{a^2}\big[\phi(x + a, t) + \phi(x - a, t) - 2\phi(x, t)\big]. \tag{9.42b}$$

These equations now tell us how to go back, not forward, in time by an interval h. But if we make a second substitution $t \to t + h$ then, after rearranging, we get

$$\phi(x, t + h) - h\psi(x, t + h) = \phi(x, t) \tag{9.43a}$$

$$\psi(x, t + h) - h\frac{v^2}{a^2}\big[\phi(x + a, t + h) + \phi(x - a, t + h) - 2\phi(x, t + h)\big] = \psi(x, t). \tag{9.43b}$$

Now the equations give us ϕ and ψ at $t + h$ again, in terms of the values at t, but they don't do so directly—we don't have an explicit expression for $\phi(x, t + h)$ as we did in Eq. (9.29). However, we can regard the equations as a set of simultaneous equations in the values of ϕ and ψ at each grid point, which can be solved by standard methods of the kind we looked at in Chapter 6, such as Gaussian elimination. Programs using these equations are going to be more complicated than our FTCS program, and will presumably run slower, but they have the advantage of being numerically stable, as we can show by again performing a von Neumann stability analysis. Using the same form as before for the solution, Eq. (9.34), and retracing our earlier steps, the equivalent of Eq. (9.36) is now

$$\mathbf{B}\mathbf{c}(t + h) = \mathbf{c}(t), \tag{9.44}$$

where $\mathbf{B}$ is the matrix

$$\mathbf{B} = \begin{pmatrix} 1 & -h \\ hr^2 & 1 \end{pmatrix}, \tag{9.45}$$

with $r = (2v/a)\sin\frac{1}{2}ka$ as before. Multiplying (9.44) on both sides by $\mathbf{B}^{-1}$, this then implies that $\mathbf{c}(t + h) = \mathbf{B}^{-1}\mathbf{c}(t)$, with

$$\mathbf{B}^{-1} = \frac{1}{1 + h^2 r^2}\begin{pmatrix} 1 & h \\ -hr^2 & 1 \end{pmatrix}, \tag{9.46}$$

where we have made use of the standard formula for the inverse of a 2×2 matrix.

Following the argument we made before, the calculation will be stable only if both eigenvalues of this matrix have magnitude less than or equal to one. The eigenvalues are again solutions of the characteristic equation, which in this case reads $[1 - (1 + h^2 r^2)\lambda]^2 + h^2 r^2 = 0$, giving eigenvalues

$$\lambda = \frac{1 \pm ihr}{1 + h^2 r^2} \tag{9.47}$$

and again the two eigenvalues have the same magnitude:

$$|\lambda| = \frac{1}{\sqrt{1 + h^2 r^2}}. \tag{9.48}$$

But this value is always less than or equal to one, no matter what the values of h and r. Thus the calculation is always numerically stable. We say that the implicit method for the wave equation is *unconditionally stable*.

Unfortunately, we are not finished yet. Being stable is not enough. We have just shown that all Fourier components of our solution will decay exponentially as time passes, which means any initial wave will die away (except for the $k = 0$ component, which is constant and so doesn't give any wave-like behavior). Thus our solution is stable, but it's still unphysical: as we know from elementary physics, waves described by the wave equation normally propagate forever without growing either larger or smaller. In a sense the implicit method goes too far. It overcorrects for the exponential growth of the FTCS method, creating a method with exponential decay instead. What we need is some method that lies in between the FTCS and implicit methods, a method where solutions neither decay or grow out of hand. Such a method is the *Crank–Nicolson method*.

The Crank–Nicolson method is precisely a hybrid of our two previous methods. The equations are derived by taking the average of Eqs. (9.29) and (9.43) to get

$$\phi(x, t + h) - \tfrac{1}{2}h\psi(x, t + h) = \phi(x, t) + \tfrac{1}{2}h\psi(x, t), \tag{9.49a}$$

$$\psi(x, t + h) - h\frac{v^2}{2a^2}\left[\phi(x + a, t + h) + \phi(x - a, t + h) - 2\phi(x, t + h)\right]$$

$$= \psi(x, t) + h\frac{v^2}{2a^2}\left[\phi(x + a, t) + \phi(x - a, t) - 2\phi(x, t)\right]. \tag{9.49b}$$

These equations are again indirect—they do not give us an explicit expression for ϕ at $t + h$ but must instead be solved as a set of simultaneous equations to give us the values we want.

If we apply our von Neumann analysis to the Crank–Nicolson method, using again a solution of the form (9.34), we get $\mathbf{B}\mathbf{c}(t + h) = \mathbf{A}\mathbf{c}(t)$, where the matrices $\mathbf{A}$ and $\mathbf{B}$ are the same as before, except that the constant r now takes the value $r = (v/a)\sin \frac{1}{2}ka$, (i.e., it is a factor of two smaller, because of the extra factors of $\frac{1}{2}$ in Eq. (9.49)). Rearranging, we get

$$\mathbf{c}(t + h) = \mathbf{B}^{-1}\mathbf{A}\,\mathbf{c}(t), \tag{9.50}$$

where

$$\mathbf{B}^{-1}\mathbf{A} = \frac{1}{1 + h^2 r^2}\begin{pmatrix} 1 & h \\ -hr^2 & 1 \end{pmatrix}\begin{pmatrix} 1 & h \\ -hr^2 & 1 \end{pmatrix} = \frac{1}{1 + h^2 r^2}\begin{pmatrix} 1 - h^2 r^2 & 2h \\ -2hr^2 & 1 - h^2 r^2 \end{pmatrix}. \tag{9.51}$$

The characteristic equation for the eigenvalues of this matrix is

$$\left[1 - h^2 r^2 - (1 + h^2 r^2)\lambda\right]^2 + 4h^2 r^2 = 0, \tag{9.52}$$

which has solutions

$$\lambda = \frac{1 - h^2 r^2 \pm 2ihr}{1 + h^2 r^2}, \tag{9.53}$$

and, once again, the two eigenvectors have the same magnitude:

$$|\lambda| = \frac{\sqrt{(1 - h^2 r^2 + 2ihr)(1 - h^2 r^2 - 2ihr)}}{1 + h^2 r^2} = 1. \tag{9.54}$$

In other words, the Crank–Nicolson method falls exactly on the boundary between the unstable (FTCS) and stable (implicit) methods, with wave amplitudes that neither grow uncontrollably nor die out as time passes, but instead remain exactly constant (apart from rounding error, presumably). And this is exactly the physical behavior we expect of the wave equation: waves present in the initial conditions should, in theory, persist forever. Thus one should be able, with the help of the Crank–Nicolson method, to solve the wave equation for its behavior over long periods of time.

Although the Crank–Nicolson method is more complicated than the FTCS method, it is still relatively fast. An important point to notice is that in Eq. (9.49) the variables $\phi(x, t)$ and $\psi(x, t)$ on each grid point depend only on the values on the points immediately to the left and right of them. This means that when viewed as a matrix problem the simultaneous equations we need to solve involve tridiagonal matrices, and we saw in Section 6.1.6 that such problems can be solved quickly using Gaussian elimination. Exercise 9.8 at the end of this chapter gives you the opportunity to apply the Crank–Nicolson method to the calculation of a solution to the Schrödinger equation of quantum mechanics, to make an animation showing how the wavefunction of a particle evolves over time.

9.3.4 SPECTRAL METHODS

Finite difference methods are not the only approach for solving partial differential equations. There are a number of other methods available, some of which have fewer stability problems, run faster, or give more accurate solutions, though typically at the cost of substantially more difficult analysis and programming. One popular method is the *finite element method* (not to be confused with the finite difference methods studied in this chapter). In this method one solves the differential equation of interest approximately within small elements of space and time and then stitches those solutions together at the boundaries of the elements to get a complete solution. The finite element method is widely used for large-scale solution of partial differential equations, but is also highly nontrivial—it is the subject of entire books just on its own. We will not look at the finite element method here, but we will look at another method that is invaluable for many of the partial differential equations encountered in physics, is not particularly complicated, and—when it's applicable—gives results much better than either finite difference or finite element methods. This is the *spectral method*, also sometimes called the *Fourier transform method*.

Consider again the wave equation, Eq. (9.26), for a wave on a string of length L, fixed at both ends so that $\phi = 0$ at $x = 0$ and at $x = L$. Then consider the trial solution

$$\phi_k(x, t) = \sin\left(\frac{\pi k x}{L}\right) e^{i\omega t}. \tag{9.55}$$

Assuming ϕ is supposed to be real, we should really take the real part as our solution, but, as is often the case in physics problems, it will be more convenient to do the math using the complex form and take the real part at the end. So for the moment we will retain the full complex form.

So long as the constant k is an integer, Eq. (9.55) satisfies the requirement of being zero at $x = 0$ and $x = L$, and, substituting into Eq. (9.26), we find that it is indeed a solution of the wave equation provided

$$\omega = \frac{\pi v k}{L}. \tag{9.56}$$

Now let us divide the string into N equal intervals, bounded by $N+1$ grid points, counting the ones at either end, both of which have $\phi = 0$. The positions of the points are

$$x_n = \frac{n}{N}L, \tag{9.57}$$

and the value of our solution at these points is

$$\phi_k(x_n, t) = \sin\left(\frac{\pi kn}{N}\right) \exp\left(i\frac{\pi vkt}{L}\right). \tag{9.58}$$

Since the wave equation is linear, any linear combination of solutions like these for different values of k is also a solution. Thus

$$\phi(x_n, t) = \frac{1}{N} \sum_{k=1}^{N-1} b_k \sin\left(\frac{\pi kn}{N}\right) \exp\left(i\frac{\pi vkt}{L}\right) \tag{9.59}$$

is a solution for any choice of coefficients b_k, which may be complex. Note that the sum starts at $k = 1$, because the $k = 0$ term vanishes. (The factor of $1/N$ in front of the sum is optional, but it's convenient for the developments that follow.)

Now notice that at time $t = 0$ this solution takes the form

$$\phi(x_n, 0) = \frac{1}{N} \sum_{k=1}^{N-1} b_k \sin\left(\frac{\pi kn}{N}\right). \tag{9.60}$$

If we now write the complex coefficients in the form $b_k = \alpha_k + i\eta_k$ and take the real part of (9.60) we get

$$\phi(x_n, 0) = \frac{1}{N} \sum_{k=1}^{N-1} \alpha_k \sin\left(\frac{\pi kn}{N}\right), \tag{9.61}$$

which is a standard Fourier sine series with coefficients α_k, of the kind that we looked at in Chapter 7, and particularly in Section 7.3. Such a series can represent any set of samples $\phi(x_n)$ and hence we can match any given initial value of ϕ with the solution (9.59) on the grid points.

Similarly, the time derivative of the real part of the solution at $t = 0$ is

$$\frac{\partial \phi}{\partial t} = -\left(\frac{\pi v}{L}\right)\frac{1}{N} \sum_{k=1}^{N-1} k\eta_k \sin\left(\frac{\pi kn}{N}\right), \tag{9.62}$$

which, apart from the leading numerical factor, is another sine series, but with coefficients $k\eta_k$. Hence, by a suitable choice of η_k this series allows us to match any given initial derivative of ϕ.

To put it another way, if we are given both the initial value and the initial derivative of ϕ at all points, then that fixes both the real and imaginary parts of the coefficients b_k, and hence our entire solution, Eq. (9.59), is determined, not just at $t = 0$ but at all times. Note that it is normal that we need both the value and the derivative of ϕ because the wave equation is second-order in time, requiring two initial conditions to fix the solution.

There are a couple of nice features of this approach. First, since Eqs. (9.61) and (9.62) are both standard Fourier sine series, we can calculate the coefficients α_k and η_k using the fast Fourier transform of Section 7.4, or more correctly the fast discrete sine transform. Doing this is usually a lot faster than directly evaluating the sums in the equations. Moreover, once we have the coefficients, we can also evaluate the solution, Eq. (9.59), using fast transforms. Again writing $b_k = \alpha_k + i\eta_k$ and taking the real part of (9.59), we have

$$\phi(x_n, t) = \frac{1}{N} \sum_{k=1}^{N-1} \left[\alpha_k \cos\left(\frac{\pi v k t}{L} \right) - \eta_k \sin\left(\frac{\pi v k t}{L} \right) \right] \sin\left(\frac{\pi k n}{N} \right). \qquad (9.63)$$

But this is just another sine series, with coefficients equal to the quantity in square brackets. Thus we can evaluate the sum using the fast inverse discrete sine transform.

Second, notice that, unlike finite difference methods, this method doesn't require us to "step through" one time-step after another in order to calculate the solution at some specified time t. Equation (9.63) gives the solution at any time directly, without passing through previous times. Thus, if we want to know the solution at just a single time in the future, we can go straight there. Or if we want to create an animation of the system at intervals of Δt, we can just calculate the solution at those intervals and nowhere else.

In principle, the spectral method is slower than FTCS for calculating the solution in a single time slice. As we saw in Section 7.4, the fast Fourier transform takes time proportional to $N \log N$, whereas one step of the FTCS method takes time proportional to N. For large enough values of N, therefore, one FTCS step will always be quicker than one spectral method step. But the fact that we don't have to calculate the solution at all time-steps often outweighs the difference in speed of individual steps. We might have to do a million steps of the FTCS method to reach the time we want, whereas we only have to do a single step of the spectral method. (And of course FTCS is numerically unstable, while the spectral method has no such problems. The Crank–Nicolson method of Section 9.3.3 is stable, but somewhat slower than FTCS, and it still requires us to go through many time-steps to get to the time we want.)

The spectral method does have its limitations. In particular, it only works for problems such as this one where the boundary conditions are rather simple, such as $\phi = 0$ at the edges of a simply shaped region or box. There is no straightforward way to adapt the method to more strangely shaped boundary conditions. Also the method is applicable only to linear differential equations because for nonlinear equations we cannot add together a set of individual solutions to get a complete solution as we did above. The finite difference methods suffer from neither of these limitations.

Exercise 9.9 gives you an opportunity to develop a program to apply the spectral method to the solution of the same Schrödinger equation problem studied with the Crank–Nicolson method in Exercise 9.8. If you do both exercises you can compare the answers to see how well they agree and find which method is faster for this problem.

FURTHER EXERCISES

9.6 What would the equivalent of Eq. (9.7) be in three dimensions?

9.7 The relaxation method for ordinary differential equations: There is no reason why the relaxation method must be restricted to the solution of differential equations with two or more independent variables. It can also be applied to those with one independent variable, i.e., to ordinary differential equations. In this context, as with partial differential equations, it is a technique for solving boundary value problems, which are less common with ordinary differential equations but do occur—we discussed them in Section 8.6.

Consider the problem we looked at in Example 8.8 on page 390, in which a ball of mass $m = 1\,\text{kg}$ is thrown from height $x = 0$ into the air and lands back at $x = 0$ ten seconds later. The problem is to calculate the trajectory of the ball, but we cannot do it using initial value methods like the ordinary Runge–Kutta method because we are not told the initial velocity of the ball. One approach to finding a solution is the shooting method of Section 8.6.1. Another is the relaxation method.

Ignoring friction effects, the trajectory is the solution of the ordinary differential equation

$$\frac{d^2x}{dt^2} = -g,$$

where g is the acceleration due to gravity.

    a) Replacing the second derivative in this equation with its finite-difference approximation, Eq. (5.109), derive a relaxation-method equation for solving this problem on a time-like "grid" of points with separation h.

b) Taking the boundary conditions to be that $x = 0$ at $t = 0$ and $t = 10$, write a program to solve for the height of the ball as a function of time using the relaxation method with 100 points and make a plot of the result from $t = 0$ to $t = 10$. Run the relaxation method until the answers change by 10^{-6} or less at every point on each step.

Note that, unlike the shooting method, the relaxation method does not give us the initial value of the velocity needed to achieve the required solution. It gives us only the solution itself, although one could get an approximation to the initial velocity by calculating a numerical derivative of the solution at time $t = 0$. On balance, however, the relaxation method for ordinary differential equations is most useful when one wants to know the details of the solution itself, but not the initial conditions needed to achieve it.

9.8   The Schrödinger equation and the Crank–Nicolson method: Perhaps the most important partial differential equation, at least for physicists, is the Schrödinger equation. This exercise uses the Crank–Nicolson method to solve the full time-dependent Schrödinger equation and hence develop a picture of how a wavefunction evolves over time. The following exercise, Exercise 9.9, solves the same problem again, but using the spectral method.

We will look at the Schrödinger equation in one dimension. The techniques for calculating solutions in two or three dimensions are basically the same as for one dimension, but the calculations take much longer on the computer, so in the interests of speed we'll stick with one dimension. In one dimension the Schrödinger equation for a particle of mass M with no potential energy reads

$$-\frac{\hbar^2}{2M}\frac{\partial^2\psi}{\partial x^2} = i\hbar\frac{\partial\psi}{\partial t}.$$

For simplicity, let's put our particle in a box with impenetrable walls, so that we only have to solve the equation in a finite-sized space. The box forces the wavefunction ψ to be zero at the walls, which we'll put at $x = 0$ and $x = L$.

Replacing the second derivative in the Schrödinger equation with a finite difference and applying Euler's method, we get the FTCS equation

$$\psi(x, t+h) = \psi(x, t) + h\frac{i\hbar}{2ma^2}\left[\psi(x+a, t) + \psi(x-a, t) - 2\psi(x, t)\right],$$

where a is the spacing of the spatial grid points and h is the size of the time-step. (Be careful not to confuse the time-step h with Planck's constant $\hbar$.) Performing a similar step in reverse, we get the implicit equation

$$\psi(x, t+h) - h\frac{i\hbar}{2ma^2}\left[\psi(x+a, t+h) + \psi(x-a, t+h) - 2\psi(x, t+h)\right] = \psi(x, t).$$

And taking the average of these two, we get the Crank–Nicolson equation for the

Schrödinger equation:

$$\psi(x,t+h) - h\frac{i\hbar}{4ma^2}\left[\psi(x+a,t+h) + \psi(x-a,t+h) - 2\psi(x,t+h)\right]$$
$$= \psi(x,t) + h\frac{i\hbar}{4ma^2}\left[\psi(x+a,t) + \psi(x-a,t) - 2\psi(x,t)\right].$$

This gives us a set of simultaneous equations, one for each grid point.

The boundary conditions on our problem tell us that $\psi = 0$ at $x = 0$ and $x = L$ for all t. In between these points we have grid points at a, $2a$, $3a$, and so forth. Let us arrange the values of ψ at these interior points into a vector

$$\psi(t) = \begin{pmatrix} \psi(a,t) \\ \psi(2a,t) \\ \psi(3a,t) \\ \vdots \end{pmatrix}.$$

Then the Crank–Nicolson equations can be written in the form

$$\mathbf{A}\psi(t+h) = \mathbf{B}\psi(t),$$

where the matrices $\mathbf{A}$ and $\mathbf{B}$ are both symmetric and tridiagonal:

$$\mathbf{A} = \begin{pmatrix} a_1 & a_2 & & & \\ a_2 & a_1 & a_2 & & \\ & a_2 & a_1 & a_2 & \\ & & a_2 & a_1 & \\ & & & & \ddots \end{pmatrix}, \quad \mathbf{B} = \begin{pmatrix} b_1 & b_2 & & & \\ b_2 & b_1 & b_2 & & \\ & b_2 & b_1 & b_2 & \\ & & b_2 & b_1 & \\ & & & & \ddots \end{pmatrix},$$

with

$$a_1 = 1 + h\frac{i\hbar}{2ma^2}, \quad a_2 = -h\frac{i\hbar}{4ma^2}, \quad b_1 = 1 - h\frac{i\hbar}{2ma^2}, \quad b_2 = h\frac{i\hbar}{4ma^2}.$$

(Note the different signs and the factors of 2 and 4 in the denominators.)

The equation $\mathbf{A}\psi(t+h) = \mathbf{B}\psi(t)$ has precisely the form $\mathbf{Ax} = \mathbf{v}$ of the simultaneous equation problems we studied in Chapter 6 and can be solved using the same methods. Specifically, since the matrix $\mathbf{A}$ is tridiagonal in this case, we can use the fast tridiagonal version of Gaussian elimination that we looked at in Section 6.1.6.

Consider an electron (mass $M = 9.109 \times 10^{-31}$ kg) in a box of length $L = 10^{-8}$ m. Suppose that at time $t = 0$ the wavefunction of the electron has the form

$$\psi(x,0) = \exp\left[-\frac{(x-x_0)^2}{2\sigma^2}\right]e^{i\kappa x},$$

where

$$x_0 = \frac{L}{2}, \quad \sigma = 1 \times 10^{-10}\,\text{m}, \quad \kappa = 5 \times 10^{10}\,\text{m}^{-1},$$

and $\psi = 0$ on the walls at $x = 0$ and $x = L$. (This expression for $\psi(x,0)$ is not normalized—there should really be an overall multiplying coefficient to make sure that the probability density for the electron integrates to unity. It's safe to drop the constant, however, because the Schrödinger equation is linear, so the constant cancels out on both sides of the equation and plays no part in the solution.)

a) Write a program to perform a single step of the Crank–Nicolson method for this electron, calculating the vector $\psi(t)$ of values of the wavefunction, given the initial wavefunction above and using $N = 1000$ spatial slices with $a = L/N$. Your program will have to perform the following steps. First, given the vector $\psi(0)$ at $t = 0$, you will have to multiply by the matrix $\mathbf{B}$ to get a vector $\mathbf{v} = \mathbf{B}\psi$. Because of the tridiagonal form of $\mathbf{B}$, this is fairly simple. The ith component of $\mathbf{v}$ is given by

$$v_i = b_1\psi_i + b_2(\psi_{i+1} + \psi_{i-1}).$$

You will also have to choose a value for the time-step h. A reasonable choice is $h = 10^{-18}$ s.

Second, you will have to solve the linear system $\mathbf{Ax} = \mathbf{v}$ for $\mathbf{x}$, which gives you the new value of ψ. You could do this using a standard linear equation solver like the function $\mathtt{solve}$ in $\mathtt{numpy.linalg}$, but since the matrix $\mathbf{A}$ is tridiagonal a better approach would be to use the fast solver for banded matrices given in Appendix E, which can be imported from the file $\mathtt{banded.py}$ (which you can find in the on-line resources). This solver works fine with complex-valued arrays, which you'll need to use to represent the wavefunction ψ and the matrix $\mathbf{A}$.

Once you have the code in place to perform a single step of the calculation, extend your program to perform repeated steps and hence solve for ψ at a sequence of times a separation h apart. Note that the matrix $\mathbf{A}$ is independent of time, so it doesn't change from one step to another. You can set up the matrix just once and then keep on reusing it for every step.

b) Extend your program to make an animation of the solution by displaying the real part of the wavefunction at each time-step. You can use the function $\mathtt{rate}$ from the package $\mathtt{visual}$ to ensure a smooth frame-rate for your animation—see Section 3.5 on page 117.

There are various ways you could do the animation. A simple one would be to just place a small sphere at each grid point with vertical position representing the value of the real part of the wavefunction. A more sophisticated approach would be to use the $\mathtt{curve}$ object from the $\mathtt{visual}$ package—see the on-line documentation at $\mathtt{www.vpython.org}$ for details. Depending on what coordinates you use for measuring x, you may need to scale the values of the wavefunction by an additional constant to make them a reasonable size on the screen. (If you measure your x position in meters then a scale factor of about 10^{-9} works well for the wavefunction.)

c) Run your animation for a while and describe what you see. Write a few sentences explaining in physics terms what is going on in the system.

9.9 The Schrödinger equation and the spectral method: This exercise uses the spectral method to solve the time-dependent Schödinger equation

$$-\frac{\hbar^2}{2M}\frac{\partial^2 \psi}{\partial x^2} = i\hbar \frac{\partial \psi}{\partial t}$$

for the same system as in Exercise 9.8, a single particle in one dimension in a box of length L with impenetrable walls. The wavefunction in such a box necessarily goes to zero on the walls and hence one possible (unnormalized) solution of the equation is

$$\psi_k(x,t) = \sin\left(\frac{\pi k x}{L}\right) e^{-iEt/\hbar},$$

where the energy E can be found by substituting into the Schrödinger equation, giving

$$E = \frac{\pi^2 \hbar^2 k^2}{2ML^2}.$$

As with the vibrating string of Section 9.3.4, we can write a full solution as a linear combination of such individual solutions, which on the grid points $x_n = nL/N$ takes the value

$$\psi(x_n, t) = \frac{1}{N}\sum_{k=1}^{N-1} b_k \sin\left(\frac{\pi k n}{N}\right) \exp\left(-i\frac{\pi^2 \hbar k^2}{2ML^2}t\right),$$

where the b_k are some set of (possibly complex) coefficients that specify the exact shape of the wavefunction and the leading factor of $1/N$ is optional but convenient.

Since the Schrödinger equation (unlike the wave equation) is first order in time, we need only a single initial condition on the value of $\psi(x,t)$ to specify the coefficients b_k, although, since the coefficients are in general complex, we will need to calculate both real and imaginary parts of each coefficient.

As in Exercise 9.8 we consider an electron (mass $M = 9.109 \times 10^{-31}$ kg) in a box of length $L = 10^{-8}$ m. At time $t = 0$ the wavefunction of the electron has the form

$$\psi(x,0) = \exp\left[-\frac{(x-x_0)^2}{2\sigma^2}\right]e^{i\kappa x},$$

where

$$x_0 = \frac{L}{2}, \qquad \sigma = 1 \times 10^{-10}\text{ m}, \qquad \kappa = 5 \times 10^{10}\text{ m}^{-1},$$

and $\psi = 0$ on the walls at $x = 0$ and $x = L$.

a) Write a program to calculate the values of the coefficients b_k, which for convenience can be broken down into their real and imaginary parts as $b_k = \alpha_k + i\eta_k$. Divide the box into $N = 1000$ slices and create two arrays containing the real and imaginary parts of $\psi(x_n, 0)$ at each grid point. Perform discrete sine transforms on each array separately and hence calculate the values of the α_k and η_k for all $k = 1 \ldots N-1$.

To perform the discrete sine transforms, you can use the fast transform function dst from the package dcst, which you can find in the on-line resources in the file named dcst.py. A copy of the code for the package can also be found in Appendix E. The function takes an array of N real numbers and returns the discrete sine transform as another array of N numbers.[4]

b) Putting $b_k = \alpha_k + i\eta_k$ in the solution above and taking the real part we get

$$\operatorname{Re}\psi(x_n, t) = \frac{1}{N}\sum_{k=1}^{N-1}\left[\alpha_k \cos\left(\frac{\pi^2\hbar k^2}{2ML^2}t\right) + \eta_k \sin\left(\frac{\pi^2\hbar k^2}{2ML^2}t\right)\right]\sin\left(\frac{\pi kn}{N}\right)$$

for the real part of the wavefunction. This is an inverse sine transform with coefficients equal to the quantities in the square brackets. Extend your program to calculate the real part of the wavefunction $\psi(x, t)$ at an arbitrary time t using this formula and the inverse discrete sine transform function idst, also from the package dcst. Test your program by making a graph of the wavefunction at time $t = 10^{-16}$ s.

c) Extend your program further to make an animation of the wavefunction over time, similar to that described in part (b) of Exercise 9.8 above. A suitable time interval for each frame of the animation is about 10^{-18} s.

d) Run your animation for a while and describe what you see. Write a few sentences explaining in physics terms what is going on in the system.

---

[4]Note that the first element of the input array should in principle always be zero for a sine transform, but if it is not the dst function will simply pretend that it is. Similarly the first element of the returned array is always zero, since the $k = 0$ coefficient of a sine transform is always zero. So in effect, the sine transform really only takes $N - 1$ real numbers and transforms them into another $N - 1$ real numbers. In some implementations of the discrete sine transform, therefore, though not the one in the package dsct used here, the first element of each array is simply omitted, since it's always zero anyway, and the arrays are only $N - 1$ elements long.

Chapter 10

Random processes and Monte Carlo methods

SOME processes in physics are random, like radioactive decay for instance. It is believed to be fundamentally impossible to predict when a particular radioactive atom will decay. Quantum mechanics tells us the probability of decay per unit time, but the exact moment of decay is random and cannot be calculated.

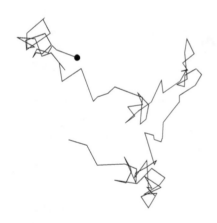

The seemingly random path followed by a particle undergoing Brownian motion.

There are also many other physical processes that are not truly random, but which we can treat as random for all intents and purposes. For instance, Brownian motion—the seemingly random movement of a particle in the air as it is buffeted by the gas molecules around it—is not actually random: if we knew the exact position and velocity of every gas molecule then, in principle at least, we could predict Brownian motion. If we don't know those things, however, then Brownian motion looks random to us, and we can model the physics well by assuming that it is.

In this chapter we will look at computational methods for studying random physical processes, both the truly random and the apparently random. To mimic randomness our computer programs themselves will need to have an element of randomness in them and for that we need random numbers.

10.1 RANDOM NUMBERS

To make programs behave randomly we use *random numbers*. Technically, in fact, we use *pseudorandom* numbers, which are not really random at all. They only look random, being generated by a deterministic formula referred to (inaccurately) as a *random number generator*.

10.1.1 RANDOM NUMBER GENERATORS

Consider the following equation

$$x' = (ax + c) \bmod m, \qquad (10.1)$$

where a, c, and m are integer constants and x is an integer variable. Given a value for x, this equation takes that value and turns it into a new integer value x'. Now suppose we take that new value and plug it back in on the right-hand side of the equation again and get another value, and so on, generating a stream of integers. Here's a program in Python to do exactly that, starting from the value $x = 1$:

```
from pylab import plot,show                              File: lcg.py

N = 100
a = 1664525
c = 1013904223
m = 4294967296
x = 1
results = []

for i in range(N):
    x = (a*x+c)%m
    results.append(x)
plot(results,"o")
show()
```

This program calculates the first 100 numbers in the sequence generated by Eq. (10.1) with $a = 1\,664\,525$, $c = 1\,013\,904\,223$, and $m = 4\,294\,967\,296$, and plots them as dots on a graph. If we run the program it produces the plot shown in Fig. 10.1. As we can see, the positions of the dots look pretty random.

This is a *linear congruential random number generator*. It generates a string of apparently random integers, one after another, simply by iterating the same equation over and over. It is probably the most famous of random number generators. There are a few things to notice about it:

1. First, it's obviously not actually random. The program is completely deterministic. If you know the equation it uses and the values of the constants a, c, and m, plus the starting value of x, then you know exactly the numbers the program will produce. If we run the program twice it will

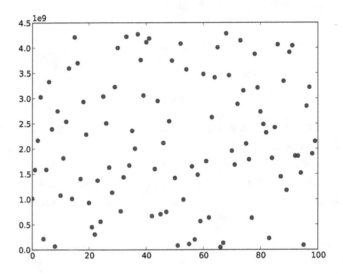

Figure 10.1: Output of the linear congruential random number generator. The vertical positions of the dots in this plot represent the first hundred values x generated by repeated iteration of the linear congruential generator equation, Eq. (10.1).

produce exactly the same graph both times. Nonetheless, the numbers generated may be "sufficiently random" to work just fine for physics calculations.

2. The numbers generated are always positive (or zero), and because they are calculated modulo m, they are always less than m. Thus all the numbers generated fall in the range from 0 to $m - 1$. Often we want numbers in some other range, in which case we must scale them appropriately. For instance, a common requirement is to generate real numbers r in the range $0 \le r < 1$. We could do this by taking the integers generated by Eq. (10.1) and dividing them by m. (Note that we don't divide by $m - 1$, because then we could get a value $r = 1$. Most often we want numbers that are strictly less than 1.)

3. It matters what values you choose for the constants a, c, and m. The values used above may look arbitrary but they were chosen with some care—these particular values have been widely tested and are known to work well. It's clear that some other choices would not be so good. For

instance, if c and m were both chosen to be even, then the process would generate only even numbers or only odd numbers, but not both, which is obviously not very random. Other choices may not have such obvious defects, but can give results that are poor in more subtle ways. It is wise to use only values of the constants that have been tested thoroughly, such as the ones in the program.

4. For a particular choice of the constants a, c, and m you can still get different sequences of random numbers by choosing different starting values for x. The initial value is called the *seed* for the random number generator; it specifies where the sequence will start.

Provided you are careful with these issues, the linear congruential generator returns pseudorandom numbers that are suitable for simple physics calculations, and it has been used very widely in computational physics over the years.

However, the linear congruential generator also has some serious flaws. The numbers it produces are, as these things go, quite bad random numbers. What does that mean? It means that there are correlations between the values of successive numbers, whereas true random numbers would be uncorrelated. These correlations can introduce errors into physics calculations that are of significant size yet hard to detect. For serious work in computational physics it is important that we use high-quality random numbers with little or no correlations, and for such work the linear congruential generator is not good enough.

Luckily, there are many other random number generators that give numbers of higher quality than the linear congruential generator. A large number of generators have been invented, studied, and used over the years, but the generator of choice for physics calculations these days seems to be the so-called *Mersenne twister*, which is a "generalized feedback shift-register generator." The Mersenne twister is quite complicated to program, but fortunately we don't have program it, because Python provides a version for us. It comes in the random package, which contains the following useful functions:[1]

---

[1]There are a variety of other functions in the random package that generate random numbers of other kinds, or perform random operations on given numbers. Most of them are somewhat specialized, but can be of use in certain circumstances. There is also a large selection of random number functions in the numpy package, including functions for choosing a random element from an array, generating a random permutation of integers, or randomly shuffling the contents of an array. See the on-line documentation for more details.

`random()`	Gives a random floating-point number uniformly distributed in the range from zero to one, including zero but not including one
`randrange(n)`	Gives a random integer from 0 to $n-1$
`randrange(m,n)`	Gives a random integer from m to $n-1$
`randrange(m,n,k)`	Gives a random integer in the range m to $n-1$ in steps of k

All of these functions generate their numbers using the Mersenne twister and they are considered good enough for serious physics calculations. Note that the function random has empty parentheses "()" after it—it has no arguments, but the parentheses are required nonetheless.

These functions are unusual among functions in Python in that they return a new random number each time you use them. Most functions we have seen, like sin or sqrt, return the same answer if you call them twice with the same argument. But if we do the following:

```
from random import randrange
print(randrange(10))
print(randrange(10))
print(randrange(10))
```

we might get something like

```
6
3
8
```

The function gives a different random number each time. If you want to use the *same* random number twice in your program for two different things then you will have to store it in a variable so you can come back to it:

```
from random import randrange
z = randrange(10)
print("The random number is",z)
print("That number again is",z)
```

Exercise 10.1: Rolling dice

a) Write a program that generates and prints out two random numbers between 1 and 6, to simulate the rolling of two dice.

b) Modify your program to simulate the rolling of two dice a million times and count the number of times you get a double six. Divide by a million to get the *fraction* of times you get a double six. You should get something close to, though probably not exactly equal to, $\frac{1}{36}$.

10.1.2 RANDOM NUMBER SEEDS

As we saw in the case of the linear congruential generator, a random number generator can have a "seed"—an input value that tells the generator where to start its sequence. Indeed, the seed specifies not only where the sequence starts but the entire sequence, since once the first number is fixed the rest of the sequence follows automatically. The seed in the case of the linear congruential generator is actually equal to the first number in the sequence, but this is not always the case. For many generators the seed specifies the starting point indirectly (via some mathematical formula) and is not actually equal to the first value. Yet still the seed fixes the whole of the sequence from the first value onward. The seed for the Mersenne twister generator works this way. The Python version of the twister is seeded with the function seed from the package random, which takes an integer seed as its single argument. Here, for example, is a program that seeds the random number generator with the integer 42, then generates some random numbers:

```
from random import randrange,seed
seed(42)
for i in range(4):
    print(randrange(10))
```

If you run this program you'll get something like:

```
2
5
3
6
```

And if you run the program again, you'll get the exact same sequence of numbers:

2
5
3
6

If you do not use a seed, if you just start generating random numbers right off the bat without using the seed function, then the random number generator will start at a different random point each time you run your program and generate a different sequence of numbers. By using the seed we tell the program where to start the random number sequence, so that if you use the same seed twice the program will produce the same sequence twice.

Why would you want to do this? Why would you ever want the same sequence of "random" numbers twice? The answer is that it is useful for getting your program working in the first place. When you are working with random programs one of the most frustrating things is that you run the program once and something goes wrong, but then you run it again, without changing anything in the program, and everything works fine. This happens because the program doesn't do the same thing every time you run it (which is the whole point of randomness). There might be some issue with the particular sequence of random numbers generated the first time around that caused your program to malfunction, but the second time around a different sequence was generated and the program worked fine. This kind of behavior can make it hard to track down problems.

Seeding the random number generator gets around this problem. It allows you to run the program in exactly the same way two or more times in a row, so that errors are reliably reproducible and you can work on making changes to the program until they are fixed. Once you get everything working properly you can get rid of the seed if you want to and then the random number generator will generate different random numbers on every run.

10.1.3 RANDOM NUMBERS AND SECRET CODES

This section is an aside. The material it contains is not needed for the rest of the chapter and you should feel free to skip it if you're not interested. It is, however, very cool.

In addition to their use in computational physics, random number generators play an important role in everyday life: they are used in cryptography, the creation of secret codes or ciphers. Ciphers, long associated in the public imagination with spies and skulduggery, have in the last couple of decades assumed a central role in business and commerce, for the transmission and safeguarding of financial information. For instance, when you make a purchase over the Internet and send your credit card number to a retailer it would be trivially easy for someone else to eavesdrop on the connection and steal that number. To prevent this, credit card numbers are routinely encrypted in transmission.

A simple example of a cipher, familiar to every schoolchild who ever had a secret club with their friends, is the *substitution cipher*. Take the letters of the alphabet and add a constant to them. If the constant is three, say, then you add three to A and you get D, B gives E, C gives F, and so forth. If you fall off the end of the alphabet, you wrap around to the beginning again, so X becomes A and Z becomes C. Under this cipher "Computational Physics" becomes "Frpsxwdwlrqdo Skbvlfv." Decoding the message again just involves subtracting the same constant from every letter to recover the original "clear-text" message.

This is fine for school kids, but the basic substitution cipher is an easy one to break. Indeed, in modern cryptography the rule is to assume the worst: assume that the thief trying to steal your credit card number knows the principles behind the cipher you are using. The cipher must be designed in such a way that *even so* they will not be able to read your message. In the simple cipher above, if you knew the basic principle, then even if you didn't know what the constant was that is added to each letter it would still be easy to crack the code. You could just program your computer to subtract every possible constant from 1 to 26 from the message until you found one that gave good English.

But here's a modification to the substitution cipher that makes it much harder to break. Instead of adding the same constant to each letter, let us change the constant. We will change it randomly for every letter. Then even if someone guesses the right constant for one letter it won't help them decode the rest of the message. This is the basic principle behind all modern ciphers and it reduces the entire problem of creating a cipher to one of creating good random numbers. If you can create a sequence of random numbers that is "very random," in the sense that it's very hard to guess what the next number in the sequence will be, then you have created a cipher that is hard to break.

The most secure form of cipher is the so-called *one-time pad*. In this cipher, two people who want to communicate with one another each have a copy of the same "pad" containing a list of random numbers. In practical situations the pad might actually be a computer file or a disk with the numbers stored on it. To make the cipher as secure as possible, the numbers should be truly random—generated, for instance, from the decays of radioactive atoms—so that it is impossible to guess them. Then the person sending the message takes the letters of the message, adds successive numbers in the random sequence onto successive letters, never using any number more than once, then sends the resulting coded message to the receiver. The receiver, referring to their copy of the pad, subtracts the same numbers again and recovers the clear-text message. If the random numbers are indeed truly random, then it is provably impossible to break this one-time pad code, so long as the pad is only ever used once (hence the name "one-time pad").[2]

Secure though it may be, this system has serious disadvantages. You have to create the pad. You have to get it to your friend without anyone else seeing it. Both parties have to guard it carefully against theft. And when it's used up you can't send any more messages. This would not be a good system for sending credit card numbers.

A much better system is to use pseudorandom numbers. Of course pseudorandom numbers are not truly random, but, as we have said, if you can make them random enough that it is very hard to predict the next number in the sequence, then that's usually good enough. Breaking the code is entirely equivalent to predicting the sequence of numbers, and if the latter is hard, so is the former. The nice thing about using pseudorandom numbers is that you don't need an entire pad to use them; you only need the seed for the random number generator, which in the cryptography world is called a *key*. Given the key and the random number generator itself, you can then encode your message and send it to your correspondent, safe in the knowledge that your communications will be impervious to the efforts of eavesdroppers, thieves, and other nefarious persons.[3]

---

[2]There are several caveats. First, you have to guard the pad itself. A trivial way to break the cipher is just to steal the pad from one of the people. Second, you don't put spaces in your message as we did above. That would be a big giveaway that would make it much easier to guess what the message was. In a real scenario, all characters, including digits, punctuation, and spaces, are given numbers and encoded using the one-time pad, so that there is no indication in the message of what's a letter and what's a space. Finally, the calculation is usually not done by addition as described here, although it could be in principle. Instead one usually uses an "exclusive OR" operation to combine the clear text with the random numbers from the pad.

[3]There remains the interesting problem of how you tell your correspondent what key you are using. You cannot use the same key (and hence the same sequence of random numbers) for every message you send. Just as with the one-time pad, it is important not to use the same sequence more than once, because that makes it easier to break the code. One can easily choose a new key for each message, but then you somehow need to tell the receiver what key you chose. Sending the key to the receiver in unencrypted form, such as in an email, would be a big mistake—a thief

Thus the problem of making a good secret code and the problem of making a good random number generator are basically the same problem. In fact, a good random number generator is sometimes said to be of "cryptographic quality," meaning it's so good you could use it to disguise your credit card number and still feel pretty safe. The Mersenne twister generator used in Python is *not* of cryptographic quality, although there are variants of it that are. This means that the twister is a reasonable random number generator but not the best. So why do we use it? The answer is that the really good generators are slower and use up too much computer time to be useful for fast physics calculations. Conversely, very simple random number generators like the linear congruential generator of Eq. (10.1) are very fast, but give poor random numbers. So, like so many other things in computational physics, the Mersenne twister is a compromise. It is reasonably fast and gives reasonably good results. It is for this reason that it has become the tool of choice for computational physicists.

10.1.4 PROBABILITIES AND BIASED COINS

It happens frequently when writing computer programs for physics calculations that you want some event to take place randomly with probability p. For instance, with probability 0.2 you want a particle to move, otherwise not, meaning it will move one fifth of the time and stay put the remainder. This kind of behavior (often called the "toss of a biased coin" by statisticians) is straightforward to create. You generate a random number between zero and one, using for instance the function random, and then you make the event happen if the resulting random number is *less than* p. Thus to make something happen with probability 0.2 (or not with probability 0.8) we might write

```
from random import random
if random()<0.2:
    print("Heads")
else:
    print("Tails")
```

could eavesdrop on that email, steal the key, and then break your code. So the key itself needs to be encrypted and it sounds like you are going to have a chicken-and-egg problem. Which comes first, the code or the key? Amazingly, it turns out that there is a solution to this problem and a very elegant one, which uses the ingenious technology of asymmetric ciphers and public key encryption. These subjects are outside the topic of this book, but if you're interested in learning more, you can find an entertaining and readable introduction in Singh, S., *The Code Book: The Secret History of Codes and Code Breaking*, Doubleday, New York (2000).

Since the function random produces a random number uniformly distributed between zero and one, its chance of being less than 0.2 is by definition 0.2. So this program will print "Heads" one fifth of the time (at random), and "Tails" the rest of the time. (This is a very biased coin.)

EXAMPLE 10.1: DECAY OF AN ISOTOPE

The radioisotope ^{208}Tl (thallium 208) decays to stable ^{208}Pb (lead 208) with a half-life of 3.053 minutes. Suppose we start with a sample of 1000 thallium atoms. Let us simulate the decay of these atoms over time, mimicking the randomness of that decay using random numbers.

On average we know that the number N of atoms in our sample will fall off exponentially over time according to the standard equation of radioactive decay:

$$N(t) = N(0)\, 2^{-t/\tau}, \tag{10.2}$$

where τ is the half-life. Then the fraction of atoms remaining after time t is $N(t)/N(0) = 2^{-t/\tau}$ and the fraction that have decayed, which is also equal to the probability $p(t)$ that any particular single atom has decayed, is one minus this number, or

$$p(t) = 1 - 2^{-t/\tau}. \tag{10.3}$$

Thus this number represents the probability that a single atom decays in a time interval of length t.

We will simulate the decay of our sample of 1000 atoms by dividing the atoms into two sets, one of thallium and one of lead. Initially all the atoms are in the thallium set. We will divide time into time-steps of 1 second each and in each time-step we will consider in turn each thallium atom and with the probability given by Eq. (10.3) decide whether or not it decays. In this way we work out the total number of thallium atoms that decay in each second, then we subtract this number from the total in the thallium set and add it to the total in the lead set.

Here's a program to perform the calculation and make a plot of the number of atoms of each type as a function of time for 1000 seconds:

File: decay.py

```
from random import random
from numpy import arange
from pylab import plot,xlabel,ylabel,show
```

```
# Constants
NTl = 1000              # Number of thallium atoms
NPb = 0                 # Number of lead atoms
tau = 3.053*60          # Half life of thallium in seconds
h = 1.0                 # Size of time-step in seconds
p = 1 - 2**(-h/tau)     # Probability of decay in one step
tmax = 1000             # Total time

# Lists of plot points
tpoints = arange(0.0,tmax,h)
Tlpoints = []
Pbpoints = []

# Main loop
for t in tpoints:
    Tlpoints.append(NTl)
    Pbpoints.append(NPb)

    # Calculate the number of atoms that decay
    decay = 0
    for i in range(NTl):
        if random()<p:
            decay += 1
    NTl -= decay
    NPb += decay

# Make the graph
plot(tpoints,Tlpoints)
plot(tpoints,Pbpoints)
xlabel("Time")
ylabel("Number of atoms")
show()
```

Figure 10.2 shows the plot produced by the program. The overall exponential decay of the sample is shown clearly, but we can also see a certain amount of "noise," randomly wiggles in the lines, which arise because of the inherently random nature of the decay process. This randomness is not captured by the standard exponential model for radioactive decay, but can be captured, as we see here, by a computer simulation of the process.

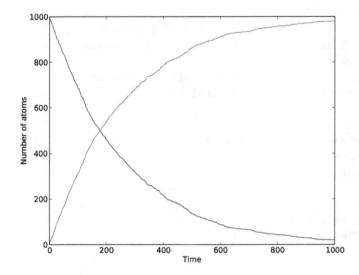

Figure 10.2: Decay of a sample of radioactive atoms. The two curves in this plot show the number of atoms of thallium and lead in the radioactive decay simulation of Example 10.1 as a function of time. The number of thallium atoms decays roughly exponentially, but some random variation around the exponential curve is also visible.

Exercise 10.2: Radioactive decay chain

This exercise looks at a more advanced version of the simple radioactive decay simulation in Example 10.1.

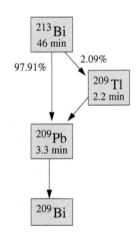

The isotope ^{213}Bi decays to stable ^{209}Bi via one of two different routes, with probabilities and half-lives as shown in the figure. (Technically, ^{209}Bi isn't really stable, but it has a half-life of more than 10^{19} years, a billion times the age of the universe, so it might as well be.)

Starting with a sample consisting of 10 000 atoms of ^{213}Bi, simulate the decay of the atoms as in Example 10.1 by dividing time into slices of length $\delta t = 1\,$s each and on each step doing the following:

a) For each atom of ^{209}Pb in turn, decide at random, with the appropriate probability, whether it decays or not. (The probability can be calculated from Eq. (10.3).) Count the total number that decay, subtract it from the number of ^{209}Pb atoms, and add it to the number of ^{209}Bi atoms.

b) Now do the same for ^{209}Tl, except that decaying atoms are subtracted from the total for ^{209}Tl and added to the total for ^{209}Pb.

c) For ^{213}Bi the situation is more complicated: when a ^{213}Bi atom decays you have to decide at random with the appropriate probability the route by which it decays. Count the numbers that decay by each route and add and subtract accordingly.

Note that you have to work up the chain from the bottom like this, not down from the top, to avoid inadvertently making the same atom decay twice on a single step.

Keep track of the number of atoms of each of the four isotopes at all times for 20 000 seconds and make a single graph showing the four numbers as a function of time on the same axes.

Exercise 10.3: Brownian motion

Brownian motion is the motion of a particle, such as a smoke or dust particle, in a gas, as it is buffeted by random collisions with gas molecules. Make a simple computer simulation of such a particle in two dimensions as follows. The particle is confined to a square grid or lattice $L \times L$ squares on a side, so that its position can be represented by two integers $i, j = 0 \ldots L - 1$. It starts in the middle of the grid. On each step of the simulation, choose a random direction—up, down, left, or right—and move the particle one step in that direction. This process is called a random walk. The particle is not allowed to move outside the limits of the lattice—if it tries to do so, choose a new random direction to move in.

Write a program to perform a million steps of this process on a lattice with $L = 101$ and make an animation on the screen of the position of the particle. (We choose an odd length for the side of the square so that there is one lattice site exactly in the center.)

Note: The visual package doesn't always work well with the random package, but if you import functions from visual first, then from random, you should avoid problems.

10.1.5 NONUNIFORM RANDOM NUMBERS

All the Python random number functions we have described generate uniformly distributed random numbers—all values that can be returned by these functions occur with equal probability. In physics, however, there are many processes that happen with nonuniform probability. Consider again, for instance, the situation we studied in Example 10.1, where we have N atoms of a radioisotope with half-life τ. As we showed, the probability that a particular atom decays in time t is $1 - 2^{-t/\tau}$ and hence the probability that it decays in a small time interval dt is

$$1 - 2^{-dt/\tau} = 1 - \exp\left(-\frac{dt}{\tau} \ln 2\right) = \frac{\ln 2}{\tau} \, dt, \qquad (10.4)$$

where we have dropped terms in $(dt)^2$ or smaller. Now we can ask what the probability is that a particular atom decays between times t and $t + dt$. In order to undergo such a decay, an atom must survive without decay until time t, which happens with probability $2^{-t/\tau}$, then decay in the following interval dt, which happens with the probability in Eq. (10.4) above. Thus the total probability $P(t) \, dt$ of decay between times t and $t + dt$ is

$$P(t) \, dt = 2^{-t/\tau} \frac{\ln 2}{\tau} \, dt. \tag{10.5}$$

This is an example of a nonuniform probability distribution. The decay times t are distributed in proportion to $2^{-t/\tau}$, so that earlier decay times are more probable than late ones, although all possible times can occur with some probability.

One way to calculate the decay of a sample of N atoms, which is significantly more efficient than the method we used in Example 10.1, would be to generate N random numbers drawn from this nonuniform probability distribution, to represent the time at which each of the atoms decays. Then creating curves like those in Fig. 10.2, to represent the number of atoms that have decayed as a function of time, is simply a matter of counting how many of the atoms decay before any given time.

The catch is that in order to do this we have to be able to generated nonuniform random numbers, drawn from the distribution in Eq. (10.5), which is really an exponential distribution, since $2^{-t/\tau} = e^{-t \ln 2/\tau}$. It turns out, however, that this is not too hard to do. It is possible to generate nonuniform random numbers from a wide range of different distributions, including the exponential distribution, using methods that build on the methods we have already seen for generating uniform numbers. If you have a source of uniform random numbers then you can turn them into nonuniform ones using any of several different techniques, the most common of which is the *transformation method*, which works as follows.

Suppose you have a source of random floating-point numbers z drawn from a distribution with probability density $q(z)$, meaning that the probability of generating a number in the interval z to $z + dz$ is $q(z) \, dz$. And suppose you have a function $x = x(z)$. Then when z is one of our random numbers, $x(z)$ is also a random number, but in general it will have some other distribution $p(x)$, different from $q(z)$. Our goal is to choose the function $x(z)$ so that x has the distribution we want.

The probability of generating a value of x between x and $x + dx$ is by definition equal to the probability of generating a value of z in the corresponding z interval. That is,

$$p(x) \, dx = q(z) \, dz \tag{10.6}$$

where the value of x is related to z by $x = x(z)$. The common situation is that we have a source of random numbers that are uniform on the interval from zero to one, as with the Python function random. In that case $q(z) = 1$ in the interval from zero to one, and 0 everywhere else. Then, integrating Eq. (10.6) on both sides, we find that

$$\int_{-\infty}^{x(z)} p(x') \, dx' = \int_0^z dz' = z. \tag{10.7}$$

If we can do the integral on the left, we end up with an equation that must be satisfied by the value $x(z)$. If we can solve this equation, then we have our function $x(z)$. There are two catches: we can't always do the integral, and even if we can we can't always solve the resulting equation. But if we can do both of these things then this method will give us what we want.

As an example, suppose we want to generate random real numbers x in the interval from zero to infinity with the exponential probability distribution

$$p(x) = \mu e^{-\mu x}. \tag{10.8}$$

(The leading factor of μ is necessary to make the distribution properly normalized.) This is the same distribution that arose in the radioactive decay problem above, where $\mu = \ln 2 / \tau$.

For this distribution Eq. (10.7) gives us

$$\mu \int_0^{x(z)} e^{-\mu x'} \, dx' = 1 - e^{-\mu x} = z. \tag{10.9}$$

Solving for x we then find that

$$x = -\frac{1}{\mu} \ln(1 - z). \tag{10.10}$$

So all we need to do is generate uniform random numbers z in the interval from zero to one and feed them into this equation to get exponentially distributed x values.[4]

---

[4]You might think that, since z is a random number distributed uniformly between zero and one, then so also is $1 - z$, so one could simply calculate $\ln z$ instead of $\ln(1 - z)$ in Eq. (10.10). This, however, would be a mistake. Recall that when we generate random numbers between zero and one using the random function, they can be equal to zero but they are never equal to one. Thus if we calculated $\ln z$ we would occasionally end up calculating the logarithm of zero, which is undefined and would cause the computer to give an error and the program to stop. It wouldn't happen very often, but if we generate enough random numbers our program will end up crashing occasionally, at random and unpredictable times. If we calculate $\ln(1 - z)$ then this problem is avoided.

Exercise 10.4: Radioactive decay again

Redo the calculation from Example 10.1, but this time using the faster method described in the preceding section. Using the transformation method, generate 1000 random numbers from the nonuniform distribution of Eq. (10.5) to represent the times of decay of 1000 atoms of ^{208}Tl (which has half-life 3.053 minutes). Then make a plot showing the number of atoms that have not decayed as a function of time, i.e., a plot as a function of t showing the number of atoms whose chosen decay times are greater than t.

Hint: You may find it useful to know that the package numpy contains a function sort that will rearrange the elements of an array in increasing order. That is, "b = sort(a)" returns a new array b containing the same numbers as a, but rearranged in order from smallest to largest.

10.1.6 GAUSSIAN RANDOM NUMBERS

A common problem in physics calculations is the generation of random numbers drawn from a Gaussian (or normal) distribution:

$$p(x) = \frac{1}{\sqrt{2\pi\sigma^2}} \exp\left(-\frac{x^2}{2\sigma^2}\right). \tag{10.11}$$

Here σ is the width, or standard deviation, of the distribution and the factor in front of the exponential is again for normalization. Applying Eq. (10.7) to this distribution we get

$$\frac{1}{\sqrt{2\pi\sigma^2}} \int_{-\infty}^{x} \exp\left(-\frac{x^2}{2\sigma^2}\right) dx = z. \tag{10.12}$$

Unfortunately it is not known how to perform this integral, so the transformation method fails in this case. However there is another trick we can use as follows.

Imagine we have two independent random numbers x and y, both drawn from a Gaussian distribution with the same standard deviation σ. The probability that the point with position vector (x, y) falls in some small element $dx\, dy$ of the xy plane is then

$$p(x)\, dx \times p(y)\, dy = \frac{1}{\sqrt{2\pi\sigma^2}} \exp\left(-\frac{x^2}{2\sigma^2}\right) dx \times \frac{1}{\sqrt{2\pi\sigma^2}} \exp\left(-\frac{y^2}{2\sigma^2}\right) dy$$

$$= \frac{1}{2\pi\sigma^2} \exp\left(-\frac{x^2 + y^2}{2\sigma^2}\right) dx\, dy. \tag{10.13}$$

We could alternatively express this in polar coordinates as the probability that the point falls in the elemental area $r \, dr \, d\theta$ with radial coordinate between r and $r + dr$ and angular coordinate between θ and $\theta + d\theta$. Making the appropriate substitutions $x^2 + y^2 \to r^2$ and $dx \, dy \to r \, dr \, d\theta$ we get

$$p(r, \theta) \, dr \, d\theta = \frac{1}{2\pi\sigma^2} \exp\left(-\frac{r^2}{2\sigma^2}\right) r \, dr \, d\theta = \frac{r}{\sigma^2} \exp\left(-\frac{r^2}{2\sigma^2}\right) dr \times \frac{d\theta}{2\pi}$$

$$= p(r) \, dr \times p(\theta) \, d\theta. \tag{10.14}$$

You can, if you wish, check that the distributions $p(r)$ and $p(\theta)$ of the two variables are correctly normalized to unity. If we generate random values of r and θ according to these two distributions and regard them as the polar coordinates of a point in two-dimensional space, then that point will have the same probability distribution as our original point (x, y). To put that another way, if we take the point and transform it back into Cartesian coordinates x and y, we will have two random numbers which are both independently distributed according to a Gaussian distribution with standard deviation σ, which solves our original problem—we will have generated a Gaussian random number, or actually two of them. If we don't need two, we can just throw one of them away, or save it for later.

Generating the θ variable is trivial—the distribution $p(\theta) = 1/2\pi$ is just a uniform distribution, so all we need to do is produce a uniformly distributed real number between zero and 2π, which we do by generating a number between zero and one and multiplying it by 2π. The radial coordinate r can be generated using the transformation method. Reading from the equation above, the correctly normalized distribution function for r is

$$p(r) = \frac{r}{\sigma^2} \exp\left(-\frac{r^2}{2\sigma^2}\right). \tag{10.15}$$

Putting this distribution in place of $p(x)$ in Eq. (10.7), we then have

$$\frac{1}{\sigma^2} \int_0^r \exp\left(-\frac{r^2}{2\sigma^2}\right) r \, dr = 1 - \exp\left(-\frac{r^2}{2\sigma^2}\right) = z. \tag{10.16}$$

Rearranging for r, we get

$$r = \sqrt{-2\sigma^2 \ln(1 - z)}. \tag{10.17}$$

With this value for r and our random value for θ, our two Gaussian random numbers are given by converting back to Cartesian coordinates thus:

$$x = r \cos\theta, \qquad y = r \sin\theta. \tag{10.18}$$

461

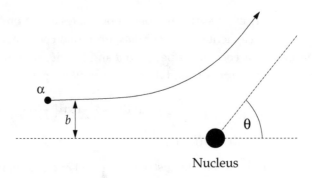

Figure 10.3: Rutherford scattering. When a positively charged α particle passes close to the nucleus of an atom (which is also positively charged) it will be repelled electrically, getting deflected through an angle θ given by Eq. (10.19).

EXAMPLE 10.2: RUTHERFORD SCATTERING

At the beginning of the 20th century, Ernest Rutherford and his collaborators showed that when an α particle, i.e., a helium nucleus of two protons and two neutrons, passes close to an atom as shown in Fig. 10.3, it is scattered, primarily by electrical repulsion from the positively charged nucleus of the atom, and that the angle θ through which it is scattered obeys

$$\tan \tfrac{1}{2}\theta = \frac{Ze^2}{2\pi\epsilon_0 Eb} \tag{10.19}$$

where Z is the atomic number of the nucleus, e is the electron charge, ϵ_0 is the permittivity of free space, E is the kinetic energy of the incident α particle, and b is the impact parameter, i.e., the perpendicular distance between the particle's initial trajectory and the axis running through the nucleus.

Consider a beam of α particles with energy 7.7 MeV that has a Gaussian profile in both its x and y axes with standard deviation $\sigma = a_0/100$, where a_0 is the Bohr radius. The beam is fired directly at a gold atom. Let us simulate the scattering process for one million α particles and calculate the fraction of particles that "bounce back" on scattering, i.e., that scatter through angles greater than 90°. The threshold value of the impact parameter for which this happens can be calculated from Eq. (10.19) with $\theta = 90°$, which gives

$$b = \frac{Ze^2}{2\pi\epsilon_0 E}. \tag{10.20}$$

If b is less than this value then the particle bounces back.

Here is a program to do the calculation:

```
from math import sqrt,log,cos,sin,pi
from random import random

# Constants
Z = 79
e = 1.602e-19
E = 7.7e6*e
epsilon0 = 8.854e-12
a0 = 5.292e-11
sigma = a0/100
N = 1000000

# Function to generate two Gaussian random numbers
def gaussian():
    r = sqrt(-2*sigma*sigma*log(1-random()))
    theta = 2*pi*random()
    x = r*cos(theta)
    y = r*sin(theta)
    return x,y

# Main program
count = 0
for i in range(N):
    x,y = gaussian()
    b = sqrt(x*x+y*y)
    if b<Z*e*e/(2*pi*epsilon0*E):
        count += 1

print(count,"particles were reflected out of",N)
```

File: rutherford.py

In this program we first define our constants, then define a function that generates two Gaussian random numbers with the appropriate standard deviation using the method described in the previous section. In the main program we use these numbers to calculate a value for the impact parameter of an α particle and compare it to the threshold value given in Eq. (10.20). We then repeat these operations a million times and count the total number of particles that bounce back.

If we run the program it prints the following:

```
1549 particles were reflected out of 1000000
```

463

The number produced varies a little from run to run because of the randomness in the calculation, but this is a typical figure. In other words, about 0.15% of particles bounce back.[5]

Rutherford, who didn't yet know about the atomic nucleus when he did his experiments in 1909, was amazed to observe such back-scattering and famously wrote that having an α particle bounce back "was about as credible as if you had fired a 15-inch shell at a piece of tissue paper and it had come back and hit you."

10.2 MONTE CARLO INTEGRATION

In fact we could have done the calculation for the α particles in Example 10.2 analytically. For a particle to be scattered through more than 90° the impact parameter must, as we said, be less than the value given in Eq. (10.20), but the impact parameter is distributed according to the radial distribution function of Eq. (10.15). Thus the probability of scattering through more than 90° is

$$\frac{1}{\sigma^2} \int_0^b \exp\left(-\frac{r^2}{2\sigma^2}\right) r \, dr = 1 - \exp\left(-\frac{b^2}{2\sigma^2}\right)$$

$$= 1 - \exp\left(-\frac{Z^2 e^4}{8\pi^2 \epsilon_0^2 \sigma^2 E^2}\right). \tag{10.21}$$

If we take the values of the parameters from Example 10.2 and substitute them into this expression we get a fraction of reflected particles equal to 0.156%, in good agreement with our numerical result.

But now look at this another way: if Eq. (10.21) and our computer program both calculate the same thing, it implies that we can, if we want, calculate the value of the integral in (10.21)—an integral that has a known exact value—approximately by simulating a random process using random numbers. This is actually quite a deep result: we can calculate the answers to exact calculations by doing random calculations. Normally, we think of the process the other way around: we are interested in some physical process that has a random element, but instead we write down an exact, non-random calculation

---

[5]This program could be simplified a little. In calculating the two Gaussian random numbers, we first calculate the polar coordinates r and θ, then convert them to x and y. In the main program we use x and y to calculate the impact parameter $b = \sqrt{x^2 + y^2}$, but this value of b is nothing other than the polar coordinate r again. So we could have skipped a step in this case and simply set b directly equal to the value of r. For the sake of clarity, however, the program given here goes about things the long way.

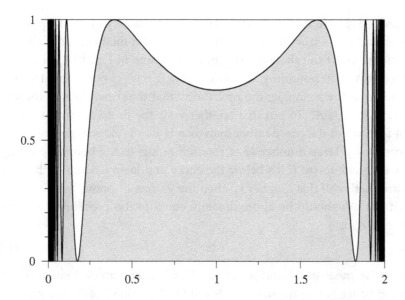

Figure 10.4: A pathological function. The integrand from Eq. (10.22), which varies arbitrarily fast at its edges, but nonetheless has a well defined integral, equal to the shaded area.

that gives an answer for its average behavior. But there's no reason in principle why we cannot make the argument in reverse: start with an exact problem, such as the calculation of an integral, and find an approximate solution to it by running a suitable random process on the computer. This leads us to a novel way of performing integrals, which works as follows.

Suppose that we want to evaluate the integral

$$I = \int_0^2 \sin^2\left[\frac{1}{x(2-x)}\right] dx. \tag{10.22}$$

The shape of the integrand is shown in Fig. 10.4. It is perfectly well behaved in the middle of its range, but it varies infinitely fast at the edges, which makes the integral challenging. On the other hand, since the entire function fits in a 2×1 rectangle, the value of the integral—the shaded area under the curve—is perfectly finite and must be less than 2, so in principle at least there is a well-defined answer for this integral. How can we find what it is? Neither basic methods like the trapezoidal rule nor more advanced ones like Gaussian quadrature are likely to work well, because they won't capture the infinitely fast variation of the function at the edges.

Here's a simple way of tackling the problem. It only gives an approximate answer, but that's true of all numerical integration methods, so it should not discourage us. If the shaded area under the curve in Fig. 10.4 is I and given that the area of the bounding rectangle is $A = 2$, if we choose a point uniformly at random in the rectangle, the probability that that point falls in the shaded region is $p = I/A$. To put that another way, the probability that a random point falls under the curve rather than over is $p = I/A$. So here's our scheme: we generate a large number N of random points in the bounding rectangle, check each one to see if it's below the curve and keep a count of the number that are. Let's call this number k. Then the *fraction* of points below the curve, which is k/N, should be approximately equal to the probability p. That is, $k/N \simeq I/A$, or

$$I \simeq \frac{kA}{N}. \tag{10.23}$$

Since we've measured k and we know A and N, we can evaluate this expression and hence get an approximate result for the value I of the integral.

This technique goes by the fanciful name of *Monte Carlo integration*, after the famous casino town in Monaco. It uses a random process to calculate the answer to an exact, nonrandom question. Monte Carlo integration is particularly useful for problems like this one where the integrand is pathological and also for multiple integrals in high dimensions—we'll see examples of the latter shortly.

Here's a program to perform the calculation described above with $N = 10000$ points:

File: mcint.py

```python
from math import sin
from random import random

def f(x):
    return (sin(1/(x*(2-x))))**2

N = 10000
count = 0
for i in range(N):
    x = 2*random()
    y = random()
    if y<f(x):
        count += 1
I = 2*count/N
print(I)
```

When run, the program prints a value for the integral of 1.4522, which is correct to two decimal places. (If you run it yourself, it may print a similar but slightly different value, because of the random elements in the calculation.)

The main disadvantage of the Monte Carlo method is that it doesn't give very accurate answers. For simple integrals where we can use the trapezoidal rule or Simpson's rule it normally gives worse results than those methods. How much worse? The probability that a single random point falls below the curve is, as we have said $p = I/A$, and the probability that it falls above is $1 - p$. The probability that a particular k of our points fall below and the remaining $N - k$ fall above is thus $p^k(1 - p)^{N-k}$. But there are $\binom{N}{k}$ ways to choose the k points out of N total, so the total probability $P(k)$ that we get exactly k points below the curve is

$$P(k) = \binom{N}{k} p^k (1 - p)^{N-k}, \tag{10.24}$$

which is called the *binomial distribution*. The variance of this distribution is[6]

$$\operatorname{var} k = Np(1 - p) = N\frac{I}{A}\left(1 - \frac{I}{A}\right), \tag{10.25}$$

and the standard deviation is the square root of the variance $\sqrt{\operatorname{var} k}$. This gives us an estimate of the expected variation or error in the value of k. Then, using Eq. (10.23), the expected error on the integral I itself is

$$\sigma = \sqrt{\operatorname{var} k}\,\frac{A}{N} = \frac{\sqrt{I(A - I)}}{\sqrt{N}}. \tag{10.26}$$

In other words the error varies with N as $N^{-1/2}$, which means the accuracy improves the more random samples we take. This is obviously a good thing in general, but $N^{-1/2}$ is not a very impressive rate of improvement. If we increase the number of samples by a factor of 100, the size of the error will only go down by a factor of 10. By contrast, recall that when we looked at the trapezoidal rule in Section 5.2 the total error on our integral was of order h^2, where h was the width of an integration slice (see Eq. (5.20) on page 151). For an integral from $x = a$ to $x = b$, the value of h is related to the number of

---

[6]The proof is straightforward. We calculate

$$\langle k \rangle = \sum_{k=0}^{N} kP(k) = Np, \qquad \langle k^2 \rangle = \sum_{k=0}^{N} k^2 P(k) = N(N-1)p^2 + Np,$$

and then $\operatorname{var} k = \langle k^2 \rangle - \langle k \rangle^2 = Np(1 - p)$.

samples N by $h = (b - a)/N$ and hence the error goes as N^{-2}, which is much better than the Monte Carlo method. When we increase the number of samples by a factor of 100 with the trapezoidal rule our error goes down by a factor of 10 000. Simpson's rule is better still. It has an error that goes as h^4, which is equivalent to N^{-4}.

To put it another way, if we do a Monte Carlo integral with a hundred samples total, we can expect our result to have an accuracy on the order of 10% because $1/\sqrt{100} = 0.1$. The exact figure for the error depends on the values of A and I via Eq. (10.26), but 10% is a reasonable rough guide. But for the same number of samples the accuracy of the trapezoidal rule will be around 0.01% and the accuracy of Simpson's rule will be around 0.000001%. Clearly, therefore, if we *can* use regular (non-Monte-Carlo) integration methods, we should. Monte Carlo integration should be used only for cases where other methods break down, which typically means for pathological integrands such as the one in this example or, as we will see, for integrals in high dimensions.

10.2.1   THE MEAN VALUE METHOD

Even among Monte Carlo methods, the method above is not a very good one for doing integrals. There are significantly better ways to perform the calculation, the most common of which is the *mean value method*. Let's take a general integration problem: we want to evaluate the integral

$$I = \int_a^b f(x)\,dx. \tag{10.27}$$

The average value of $f(x)$ in the range from a to b is by definition

$$\langle f \rangle = \frac{1}{b-a} \int_a^b f(x)\,dx = \frac{I}{b-a}. \tag{10.28}$$

Thus

$$I = (b-a)\langle f \rangle. \tag{10.29}$$

If we can estimate $\langle f \rangle$ then we can estimate I. A simple way to estimate $\langle f \rangle$ is just to measure $f(x)$ at N points $x_1 \ldots x_N$ chosen uniformly at random between a and b and calculate $\langle f \rangle \simeq N^{-1} \sum_{i=1}^N f(x_i)$. Then

$$I \simeq \frac{b-a}{N} \sum_{i=1}^N f(x_i). \tag{10.30}$$

This is the fundamental formula for the mean value method. As with the previous method it gives only an approximate estimate of the integral. How accurate is it? We can estimate the error using standard results for the behavior of random variables which tell us that the variance on the sum of N independent random numbers—no matter what their distribution—is equal to the sum of the variances of the individual numbers. The random numbers in this case are the values $f(x_i)$, and we can estimate the variance on a single one of them using the usual formula var $f = \langle f^2 \rangle - \langle f \rangle^2$, with

$$\langle f \rangle = \frac{1}{N} \sum_{i=1}^{N} f(x_i), \qquad \langle f^2 \rangle = \frac{1}{N} \sum_{i=1}^{N} [f(x_i)]^2. \tag{10.31}$$

Then the variance on the sum in Eq. (10.30) is N times the variance on a single term, or N var f, and the standard deviation on the sum is the square root of that, or $\sqrt{N \text{ var } f}$. Finally, we get the standard deviation on the integral I from Eq. (10.30), which gives

$$\sigma = \frac{b-a}{N} \sqrt{N \text{ var } f} = (b-a) \frac{\sqrt{\text{var } f}}{\sqrt{N}}. \tag{10.32}$$

Thus, once again, the error goes as $1/\sqrt{N}$. However, the leading constant is smaller in this case, which means the mean value method is always more accurate than our previous method. (See Exercise 10.6 below for a proof of this result.)

Exercise 10.5:

a) Write a program to evaluate the integral in Eq. (10.22) using the "hit-or-miss" Monte Carlo method of Section 10.2 with 10 000 points. Also evaluate the error on your estimate.

b) Now estimate the integral again using the mean value method with 10 000 points. Also evaluate the error.

You should find that the error is somewhat smaller using the mean value method.

Exercise 10.6: Construct a general proof that the mean value method always does better, or at least no worse, than the "hit-or-miss" method of Section 10.2, as follows.

a) For an integral of the form (10.27) with $f(x) \geq 0$ everywhere in the domain of integration, show that Eq. (10.23) can be rewritten as

$$I \simeq (b-a)H\langle s \rangle,$$

where H is the vertical height of the box enclosing the function to be integrated (so that the box's area is $A = (b-a)H$ and $\langle s \rangle$ is the average of variables s_i

defined such that $s_i = 1$ if the ith point in the Monte Carlo procedure was a "hit" (it fell below the curve of $f(x)$) and $s_i = 0$ if it was a "miss." Hence argue that the hit-or-miss method will never be more accurate than the mean value method if the variance of f in Eq. (10.32) satisfies $\text{var}(f) \leq H^2 \text{var}(s)$.

b) Show that the variance of a single variable s_i is $\text{var}(s) = p(1 - p)$, where $p = I/A$ as in Section 10.2. Show further that $p = \langle f \rangle / H$ and $H^2 \text{var}(s) = \langle f \rangle (H - \langle f \rangle)$ and thus that the hit-or-miss method will never be the more accurate method if $\langle f(f - H) \rangle \leq 0$. Given that the value of $f(x)$ never falls outside the interval from 0 to H, prove that this last condition is always true.

The hit-or-miss method can be extended to the case where $f(x)$ is not always positive by adding a constant onto $f(x)$ large enough to make it always positive, calculating the integral of the resulting function, then subtracting the constant again at the end. The proof above can be extended to this case by noting that the variance of f is unaffected by additive constants, and hence the mean value method is always the more accurate of the two integration methods for any function, positive or not.

10.2.2   INTEGRALS IN MANY DIMENSIONS

In addition to the integration of pathological functions, Monte Carlo integration is used for performing high-dimensional integrals. As we saw in Section 5.9, performing an integral over two variables by standard methods such as the trapezoidal rule requires us to take samples on a two-dimensional grid of points. Three-dimensional integrals require a three-dimensional grid, and so forth. If we have an integral over four or more variables, then the number of points on the grid can become very large and the standard integration methods can be very slow. If we use 100 points along each axis, for example, we'd need 100^4 integration points to do a four-dimensional integral, or 100 million points. This might just about be possible, though slow, but for more points or more than four dimensions things rapidly become impractical. So instead we turn to Monte Carlo integration, which gives reasonably good results for much smaller numbers of points—a few thousand is often adequate, as we've seen. Monte Carlo integration in high dimensions is an obvious extension of the technique in one dimension. The mean value method of Section 10.2.1, for instance, generalizes straightforwardly. The integral of a function $f(\mathbf{r})$ over a volume V in a high-dimensional space is given by the generalization of Eq. (10.30):

$$I \simeq \frac{V}{N} \sum_{i=1}^{N} f(\mathbf{r}_i),\tag{10.33}$$

where the points $\mathbf{r}_i$ are picked uniformly at random from the volume V.

An important application of this type of integral is in financial mathematics. The mathematical techniques used for predicting the values of portfolios of stocks and bonds require the evaluation of integrals over many variables, and those integrals need to be done rapidly in order to make quick trading decisions. Monte Carlo integration is a standard technique for doing this. Connections like this between physics and finance are one of the reasons why Wall Street and other trading centers often employ physicists.

Exercise 10.7: Volume of a hypersphere

This exercise asks you to estimate the volume of a sphere of unit radius in ten dimensions using a Monte Carlo method. Consider the equivalent problem in two dimensions, the area of a circle of unit radius:

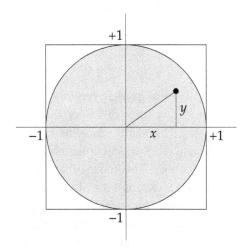

The area of the circle, the shaded area above, is given by the integral

$$I = \iint_{-1}^{+1} f(x,y)\, dx\, dy,$$

where $f(x,y) = 1$ everywhere inside the circle and zero everywhere outside. In other words,

$$f(x,y) = \begin{cases} 1 & \text{if } x^2 + y^2 \leq 1, \\ 0 & \text{otherwise.} \end{cases}$$

So if we didn't already know the area of the circle, we could calculate it by Monte Carlo integration. We would generate a set of N random points (x,y), where both x and y

are in the range from -1 to 1. Then the two-dimensional version of Eq. (10.33) for this calculation would be

$$I \simeq \frac{4}{N} \sum_{i=1}^{N} f(x_i, y_i).$$

Generalize this method to the ten-dimensional case and write a program to perform a Monte Carlo calculation of the volume of a sphere of unit radius in ten dimensions.

If we had to do a ten-dimensional integral the traditional way, it would take a very long time. Even with only 100 points along each axis (which wouldn't give a very accurate result) we'd still have $100^{10} = 10^{20}$ points to sample, which is impossible on any computer. But using the Monte Carlo method we can get a pretty good result with a million points or so.

10.2.3 IMPORTANCE SAMPLING

Monte Carlo integration is useful for integrating pathological functions, but there are still some such functions for which it doesn't work well. In particular, it gives problems if the function to be integrated contains a divergence. Consider the integral

$$I = \int_0^1 \frac{x^{-1/2}}{e^x + 1} \, dx, \tag{10.34}$$

which arises in the theory of Fermi gases. Even though the integrand diverges at $x = 0$, the integral is perfectly finite in value. But if you try to do it using the mean value method you'll run into problems because the value of $f(x_i)$ in Eq. (10.30) diverges when $x_i \to 0$, so occasionally you'll get a very large contribution to the sum in (10.30). This means that the estimated value of the integral can vary widely from one run of the algorithm to another. Sometimes you'll happen to get a large value (or several) in the sum, and sometimes you won't. Variations in the results are a normal part of the Monte Carlo method, which after all depends on random numbers that will be different from one run to another. But when the variations become large you have a problem. Another way to look at this is to say that the error σ on the estimate of the integral becomes large in such cases, and indeed you can show that it can even be formally infinite in some calculations.

We can get around these problems by drawing our points x_i non-uniformly from the integration interval. This technique is called *importance sampling* and it works as follows. For any general function $g(x)$, we can define a weighted average over the interval from a to b thus:

$$\langle g \rangle_w = \frac{\int_a^b w(x)g(x) \, dx}{\int_a^b w(x) \, dx}, \tag{10.35}$$

where $w(x)$ is any function we choose.

Now consider again the general one-dimensional integral

$$I = \int_a^b f(x)\,dx. \tag{10.36}$$

Setting $g(x) = f(x)/w(x)$ in Eq. (10.35), we have

$$\left\langle \frac{f(x)}{w(x)} \right\rangle_w = \frac{\int_a^b w(x)f(x)/w(x)\,dx}{\int_a^b w(x)\,dx} = \frac{\int_a^b f(x)\,dx}{\int_a^b w(x)\,dx} = \frac{I}{\int_a^b w(x)\,dx}. \tag{10.37}$$

So

$$I = \left\langle \frac{f(x)}{w(x)} \right\rangle_w \int_a^b w(x)\,dx. \tag{10.38}$$

This formula is analogous to Eq. (10.29) for the mean value method, but it allows us to calculate the value of the integral from a weighted average, rather than a standard uniform average.

And how do we calculate the weighted average? To see this, let us define a probability density function

$$p(x) = \frac{w(x)}{\int_a^b w(x)\,dx}, \tag{10.39}$$

which is like the weight function $w(x)$, but normalized so that its integral is 1. Let us sample N random points x_i nonuniformly with this density. That is, the probability of generating a value in the interval between x and $x + dx$ will be $p(x)\,dx$. Then the average number of samples that fall in this interval is $Np(x)\,dx$, and so for any function $g(x)$

$$\sum_{i=1}^N g(x_i) \simeq \int_a^b Np(x)g(x)\,dx, \tag{10.40}$$

with the approximation getting better as the number of samples becomes larger. Using this result, we can now write the general weighted average of the function $g(x)$ as

$$\langle g \rangle_w = \frac{\int_a^b w(x)g(x)\,dx}{\int_a^b w(x)\,dx} = \int_a^b p(x)g(x)\,dx \simeq \frac{1}{N}\sum_{i=1}^N g(x_i), \tag{10.41}$$

where the points x_i are chosen from the distribution (10.39). Putting this result together with Eq. (10.38), we then get

$$I \simeq \frac{1}{N}\sum_{i=1}^N \frac{f(x_i)}{w(x_i)} \int_a^b w(x)\,dx. \tag{10.42}$$

This is the fundamental formula of importance sampling. It is a generalization of the mean value method of Eq. (10.30)—if we choose $w(x) = 1$ we recover (10.30) again. The formula allows us to calculate an estimate of the integral I by calculating not the sum $\sum_i f(x_i)$, but instead the modified sum $\sum_i f(x_i)/w(x_i)$, where $w(x)$ is any function we choose. This is useful because it allows us to choose a $w(x)$ that gets rid of pathologies in the integrand $f(x)$. For instance, if $f(x)$ has a divergence, we can factor that divergence out and hence get a sum that is well behaved. The price we pay for this flexibility is that we have to draw our samples x_i from a nonuniform distribution of random numbers instead of a uniform one, which makes the programming more complex—typically we have to use the transformation method to generate the samples.

We can also calculate the error σ on Eq. (10.42), which again uses the properties of variances of random variables. We won't go through the derivation, but the result is that

$$\sigma = \frac{\sqrt{\mathrm{var}_w(f/w)}}{\sqrt{N}} \int_a^b w(x)\, dx, \tag{10.43}$$

where $\mathrm{var}_w g = \langle g^2 \rangle_w - \langle g \rangle_w^2$.

We can use the importance sampling method to evaluate the integral in Eq. (10.34), while avoiding problems with divergences. If we choose $w(x) = x^{-1/2}$ then $f(x)/w(x) = (e^x + 1)^{-1}$ in this case, which is finite and well-behaved over the domain of integration. Using the transformation method, we would write a program that draws random x values in the range from zero to one from the distribution

$$p(x) = \frac{x^{-1/2}}{\int_0^1 x^{-1/2}\, dx} = \frac{1}{2\sqrt{x}}, \tag{10.44}$$

then plug them into Eq. (10.42) to get a value for the integral, noting that

$$\int_0^1 w(x)\, dx = \int_0^1 x^{-1/2}\, dx = 2. \tag{10.45}$$

Exercise 10.8 asks you to write a program to perform this calculation and find a value for the integral.

Importance sampling has other uses as well. For instance, it can be used to evaluate integrals over infinite domains. An integral from 0 to ∞ can't be done using the ordinary mean value method, because we would have to generate random points x_i uniformly between 0 and ∞, which is impossible because their distribution would not be integrable. But importance sampling allows us to generate points from some other, nonuniform distribution, one that extends out to infinity but is integrable, such as an exponential distribution. Thus if, for example, we choose our weight function $w(x) = e^{-x}$, then Eq. (10.42) becomes

$$I \simeq \frac{1}{N} \sum_{i=1}^{N} e^{x_i} f(x_i) \int_0^\infty e^{-x} \, dx = \frac{1}{N} \sum_{i=1}^{N} e^{x_i} f(x_i). \qquad (10.46)$$

In effect, this equation says that we can sample points nonuniformly provided we compensate by weighting more heavily those points that fall in regions where there are fewer of them.

To be fair, as we saw in Section 5.8, a one-dimensional integral out to infinity can also be performed—and probably with greater accuracy—using conventional methods such as the trapezoidal rule in combination with a suitable change of variables. Once again, however, if the integrand took a pathological form, or if the integral were in a high-dimensional space, then Monte Carlo integration might become an attractive method.

In this situation the nonuniform distribution of points is no longer a downside of the importance sampling method but has become a positive advantage that we exploit to perform an integral that otherwise would not be possible. We will see another example of this type of approach in the next section, when we look at the technique of Monte Carlo simulation, particularly as applied to thermal physics.

Exercise 10.8: Calculate a value for the integral

$$I = \int_0^1 \frac{x^{-1/2}}{e^x + 1} \, dx,$$

using the importance sampling formula, Eq. (10.42), with $w(x) = x^{-1/2}$, as follows.

a) Show that the probability distribution $p(x)$ from which the sample points should be drawn is given by

$$p(x) = \frac{1}{2\sqrt{x}}$$

and derive a transformation formula for generating random numbers between zero and one from this distribution.

b) Using your formula, sample $N = 1\,000\,000$ random points and hence evaluate the integral. You should get a value around 0.84.

10.3 MONTE CARLO SIMULATION

Monte Carlo simulation is the name given to any computer simulation that uses random numbers to simulate a random physical process in order to estimate something about the outcome of that process. An example is the calculation we did of the decay of radioactive isotopes in Example 10.1. Another is the Rutherford scattering calculation in Exercise 10.2.

Although Monte Carlo simulation finds uses in every branch of physics, there is one area where it is used more than any other, and that is statistical mechanics. Because statistical mechanics is fundamentally about random (or apparently random) processes, Monte Carlo simulation assumes a particular importance in the field. In this section we look at Monte Carlo simulation, focusing particularly on statistical mechanics.

10.3.1 IMPORTANCE SAMPLING AND STATISTICAL MECHANICS

Recall that the fundamental problem of statistical mechanics is to calculate the average (or expectation) value of a quantity of interest in a physical system in thermal equilibrium at temperature T. We don't know the exact state of the physical system. Instead all we know is that a system at temperature T will pass through a succession of states such that at any particular moment the probability of its occupying state i with energy E_i is given by the Boltzmann formula

$$P(E_i) = \frac{e^{-\beta E_i}}{Z}, \qquad \text{with} \qquad Z = \sum_i e^{-\beta E_i}, \tag{10.47}$$

where $\beta = 1/k_B T$ and k_B is Boltzmann's constant. Then the average value of a quantity X that takes the value X_i in the ith state is

$$\langle X \rangle = \sum_i X_i P(E_i). \tag{10.48}$$

In a few cases we can calculate this sum exactly (for example in the quantum simple harmonic oscillator), but in most cases we can't, in which case numerical calculation is an attractive alternative. Normally, however, we cannot simply evaluate the sum over states directly because the number of states is far too large.[7] For instance, there are about 10^{23} molecules in a single mole of a gas, and even if each had only two possible quantum states (in fact they have many

---

[7]We did see one case where a direct attack is possible in Example 4.2 on page 135, but this is rare.

more than this) then the total number of states of the whole system would be $2^{10^{23}}$, which is a stupendously large number. Even the fastest supercomputer cannot evaluate a sum with anything like this many terms.

So instead we take an approach analogous to the Monte Carlo integration of Section 10.2. In that case we were evaluating an integral whereas now we're evaluating a sum, but the difference is not that big and the same basic approach works. In the standard mean value method for Monte Carlo integration, represented by Eq. (10.30), we evaluate an integral by choosing a set of points at random in the integration window and adding up the values of the integrand at those points. The equivalent for a sum is to choose a set of terms from the sum at random and add up those. Thus, instead of summing over all the states of the system in Eq. (10.48), we would choose N states at random, denoted by $k = 1 \ldots N$, and then calculate

$$\langle X \rangle \simeq \frac{\sum_{k=1}^{N} X_k P(E_k)}{\sum_{k=1}^{N} P(E_k)}. \qquad (10.49)$$

Notice the denominator, which is needed to normalize the weighted average correctly. There is no equivalent denominator in Eq. (10.48) because the expression is already normalized—the sum over all states $\sum_i P(E_i)$ is 1 by definition.

Unfortunately, Eq. (10.49) doesn't work. The problem is that the Boltzmann probability (10.47) is exponentially small for any state with energy $E_i \gg k_B T$, which is most of the states in most cases. That means that almost all of the states we choose for Eq. (10.49) will be ones that make very little contribution to the original sum in Eq. (10.48). There is typically just a small fraction of states that actually contribute significantly to the value of the sum, so few that we are unlikely to pick them in our random sample of terms. This means that the value of the sum we estimate from Eq. (10.49) will not be a good approximation to the true value, since the most important terms are missing. In the worst case there are so many states that make small contributions and so few that make large ones that we end up picking not one single one of the large ones, and our approximation to the sum will be very bad indeed.

But we have already seen a method that can solve this problem for us: importance sampling. Recall that importance sampling allowed us, for example, to get around divergences in integrals and other pathologies, but it did so at the expense of forcing us to choose our random samples from a nonuniform distribution. In Section 10.2.3 these nonuniform samples were mostly an annoyance—something we had to put up with to get the other benefits of the method. Now, however, this very nonuniformity becomes an advantage.

Importance sampling lets us calculate a correct value of an integral or a sum using a set of samples drawn nonuniformly. We have said that in our statistical mechanics calculations only a small fraction of the terms in our sum contribute significantly. Let us choose our nonuniform distribution of samples so that it focuses precisely on this small set, while ignoring the many terms that contribute little. It is, in fact, this application of importance sampling that gives the method its name. Importance sampling chooses the samples that are most important in the sum and neglects the ones that matter less.

Here's how it works in detail. For any quantity g_i that depends on the states i, we can define a weighted average over states

$$\langle g \rangle_w = \frac{\sum_i w_i g_i}{\sum_i w_i},\tag{10.50}$$

where w_i is any set of weights we choose. Making the particular choice $g_i = X_i P(E_i)/w_i$ this gives

$$\left\langle \frac{X_i P(E_i)}{w_i} \right\rangle_w = \frac{\sum_i w_i X_i P(E_i)/w_i}{\sum_i w_i} = \frac{\sum_i X_i P(E_i)}{\sum_i w_i} = \frac{\langle X \rangle}{\sum_i w_i},\tag{10.51}$$

where we have used Eq. (10.48). Rearranging this expression we get

$$\langle X \rangle = \left\langle \frac{X_i P(E_i)}{w_i} \right\rangle_w \sum_i w_i.\tag{10.52}$$

We can evaluate this expression approximately by selecting a set of N sample states randomly but nonuniformly, such that the probability of choosing state i is

$$p_i = \frac{w_i}{\sum_j w_j},\tag{10.53}$$

in which case the equivalent of Eq. (10.41) is

$$\langle g \rangle_w \simeq \frac{1}{N} \sum_{k=1}^{N} g_k,\tag{10.54}$$

where g_k is the value of the quantity g in the kth state we select. Combining this result with Eq. (10.52), we then get

$$\langle X \rangle \simeq \frac{1}{N} \sum_{k=1}^{N} \frac{X_k P(E_k)}{w_k} \sum_i w_i.\tag{10.55}$$

Note that the first sum here is over only those states k that we sample, but the second is over all states i. This second sum is usually evaluated analytically.

Our goal is to choose the weights w_i so that most of the samples are in the region where $P(E_i)$ is big and very few are in the region where it is small, so that we pick up the big terms in the sum and ignore the small ones. We should also choose w_i so that the sum $\sum_i w_i$ in Eq. (10.55) can be performed analytically—if we can't do this sum, then the equation is useless.

But both of these requirements are easy to satisfy. Let us simply choose $w_i = P(E_i)$. Then we have $\sum_i w_i = \sum_i P(E_i) = 1$ by definition, and Eq. (10.55) simplifies to

$$\langle X \rangle \simeq \frac{1}{N} \sum_{k=1}^{N} X_k. \tag{10.56}$$

In other words, we just choose N states in proportion to their Boltzmann probabilities and take the average of the quantity X over all of them, and that's it. The more states we use the more accurate the answer will be, but even with only a few states we already get a much more accurate answer than the purely random samples we considered to begin with.

There is another way of looking at this process, which is more physical and less mathematical (and therefore probably appeals more to physicists). In the *real* system we are studying—in the real world—the system passes, as we have said, through a succession of states with probabilities $P(E_i)$, and the average value of X is, by definition, the average of the values in the states that the system passes through. If we do our Monte Carlo calculation according to the recipe above then we are, in effect, just simulating this process: we choose a set of states k of the system in proportion to $P(E_i)$. You can think of these as the states the system passes through and we calculate the average value of X in those states. So our calculation really is just an imitation of nature, and hence it should not be surprising that it gives a correct value for the quantity we are interested in. Just as our simulation of radioactive decay in Example 10.1 was an imitation on the computer of the physics of real radioactive decay, so our simulation of a thermal system is an imitation of the physics of a real thermal system. This is why we call this technique Monte Carlo "simulation."

10.3.2 THE MARKOV CHAIN METHOD

Unfortunately, we're not done yet. There is a catch. The catch is that it's not easy to pick states with probability $P(E_i)$. We have a formula for $P(E_i)$ in Eq. (10.47), but it includes the factor Z, the so-called *partition function*, which is defined as a sum over states. We don't know the value of this sum in general and it's not easy to calculate. After all, the whole point of Monte Carlo simulation is to calculate sums over states and if such sums were easy we wouldn't be doing the calculation in the first place. So if we don't know the value of the partition function and we can't easily calculate it, how do we choose states with probability $P(E_i)$? Remarkably, it turns out that we can do it without knowing the partition function, using a device called a *Markov chain*.

We want to choose a set of states for the sum (10.56). We will do so by generating a string of states one after another—the Markov chain. Consider a single step in this process and suppose that the previous state in the chain, the state for the step before this one, was state i. For the new state, instead of choosing randomly from all possible states, let us make some change, usually small, to state i so as to create a new state. If the system we are studying is a gas, for instance, we might move one of the molecules in the gas to a new quantum level, but leave the other molecules where they are. The choice of the new state is determined probabilistically by a set of *transition probabilities* T_{ij} that give the probability of changing from state i to state j. It turns out that, if we choose T_{ij} right for all i and j, we can arrange that the probability of visiting any particular state on any step of the Markov chain is precisely the Boltzmann probability $P(E_i)$, so that when we take many steps and generate the entire chain the complete set of states that we move through is a correct sample of our Boltzmann distribution. Then all we have to do to evaluate Eq. (10.56) is measure the quantity X of interest in each state we move through and take the average. The trick lies in how we choose T_{ij} to achieve this.

Given that we must end up in some state on every step of the Markov chain it must be the case that

$$\sum_j T_{ij} = 1. \tag{10.57}$$

The secret of the Markov chain method is to choose the T_{ij} so that they satisfy this condition and also so that

$$\frac{T_{ij}}{T_{ji}} = \frac{P(E_j)}{P(E_i)} = \frac{e^{-\beta E_j}/Z}{e^{-\beta E_i}/Z} = e^{-\beta(E_j - E_i)}. \tag{10.58}$$

In other words, we are choosing a particular value for the ratio of the probability to go from i to j and the probability to go back again from j to i. Note that the value of the partition function Z cancels out of this equation, so that we don't need to know it in order to satisfy the equation—this gets around the problem we had before of not knowing the value of Z.

Suppose we can find a set of probabilities T_{ij} that satisfy Eq. (10.58) and suppose that the probability we are in state i on one particular step of the Markov chain is equal to the Boltzmann probability $P(E_i)$, for all i. In that case, making use of Eq. (10.58), the probability of being in state j on the next step is the sum of the probabilities that we got there from every other possible state i:

$$\sum_i T_{ij} P(E_i) = \sum_i T_{ji} P(E_j) = P(E_j) \sum_i T_{ji} = P(E_j), \qquad (10.59)$$

where we have used Eqs. (10.57) and (10.58). In other words, if we have the correct probability of being in every state on one step, then we have the correct probability on the next step too, and hence we will have it for ever afterwards.

This proves that if we can once get a Boltzmann distribution over states then we will keep a Boltzmann distribution. We say that the Boltzmann distribution is a *fixed point* of the Markov chain. We have not, however, proved that if we start with some other distribution over states we will *converge* to the Boltzmann distribution, but in fact we can prove this: we can prove that if we wait long enough the distribution of states in the Markov chain will always converge to the Boltzmann distribution. The proof is not complicated, but it's quite long, so it is relegated to Appendix D. The important point is that if you start off the system in any random state and run the Markov chain for long enough, the distribution over states will converge to the Boltzmann distribution. This is exactly what we need to make our calculation work.

We still need to work out what the values of the T_{ij} should be. We have assumed that it's possible to find a set of values that satisfy Eq. (10.58), but we haven't actually done it. In fact, it's not that hard: Equation (10.58) leaves us quite a lot of latitude about how we choose the probabilities, and there are many different choices that will get the job done. The most common choice by far, however, is the choice that leads to the so-called *Metropolis algorithm*.[8] To understand how this algorithm works, note first that we are allowed to visit the same state more than once in our Markov chain, and indeed we can even visit the same state twice on two consecutive steps. Another way to say this is that the probability T_{ii} of moving from a state *to itself* is allowed to be nonzero, in which case the "move" is no move at all—we just stay in the state we're already in. Bearing this in mind, the Metropolis algorithm then works as follows.

Suppose once again that the previous state in the chain is state i and we generate a new state j by making some change to state i. We will choose the

---

[8]Named after its coinventor, the early computational physicist Nicholas Metropolis, and also, in a more general formulation due to Keith Hastings, sometimes called the Metropolis–Hastings algorithm.

particular change we make uniformly at random from a specified set of possible changes, sometimes called a *move set*. To take our previous example of the gas, for instance, we might choose to change the state of one molecule of the gas, with the molecule chosen uniformly at random and the state we are moving to chosen randomly to be up or down one state in energy from the present one. A uniform choice of this kind does not, generally, satisfy Eq. (10.58). But then we either *accept* or *reject* the new state with *acceptance probability* P_a given by

$$P_a = \begin{cases} 1 & \text{if } E_j \le E_i, \\ e^{-\beta(E_j - E_i)} & \text{if } E_j > E_i. \end{cases} \qquad (10.60)$$

If the move is rejected then we do nothing—the system remains in its old state i for one more step, which is equivalent to moving from state i to the same exact state i, something that, as we have said, is allowed. If the move is accepted, on the other hand, then we change the system to the new state.

Another way of expressing Eq. (10.60) is to say that if the proposed move will decrease the energy of the system or keep it the same (i.e., $E_j \le E_i$), then we definitely accept it—we change the molecule to its new state, or whatever the move requires. On the other hand, if the proposed move will increase the energy, then we *may* still accept it—with the probability given in Eq. (10.60)—otherwise we reject it.

Under this scheme the total probability T_{ij} of making a move from state i to state j is the probability that we choose that move out of all possibilities, which is just $1/M$ if there are M possibilities, times the probability that we accept the move. So for instance if $E_j > E_i$ then

$$T_{ij} = \frac{1}{M} \times e^{-\beta(E_j - E_i)} \qquad \text{and} \qquad T_{ji} = \frac{1}{M} \times 1, \qquad (10.61)$$

and the ratio of the two is

$$\frac{T_{ij}}{T_{ji}} = \frac{e^{-\beta(E_j - E_i)}/M}{1/M} = e^{-\beta(E_j - E_i)}, \qquad (10.62)$$

which agrees with Eq. (10.58). Conversely, if $E_i \ge E_j$ then

$$T_{ij} = \frac{1}{M} \times 1, \qquad T_{ji} = \frac{1}{M} \times e^{-\beta(E_i - E_j)}, \qquad (10.63)$$

and the ratio of the two is

$$\frac{T_{ij}}{T_{ji}} = \frac{1/M}{e^{-\beta(E_i - E_j)}/M} = e^{-\beta(E_j - E_i)}, \qquad (10.64)$$

which again agrees with Eq. (10.58). Either way, we have found a set of transition probabilities T_{ij} that satisfies Eq. (10.58).

Our complete Markov chain Monte Carlo simulation now involves the following steps:

1. Choose a random starting state.
2. Choose a move uniformly at random from an allowed set of moves, such as changing a single molecule to a new state.
3. Calculate the value of the acceptance probability P_a given in Eq. (10.60).
4. With probability P_a accept the move, meaning the state of the system changes to the new state; otherwise reject it, meaning the system stays in its current state for one more step of the calculation.
5. Measure the value of the quantity of interest X in the current state and add it to a running sum of such measurements.
6. Repeat from step 2.

When we have done many such steps, we take our running sum and divide it by the total number of steps we took to get an estimate of the average value $\langle X \rangle$, Eq. (10.56).

A simple way to implement Eq. (10.60) in a program is to calculate the change in energy $E_j - E_i$ and then generate a random number z uniformly between zero and one and accept the move if $z < e^{-\beta(E_j - E_i)}$. Note that if $E_j \leq E_i$ then the exponential will be greater than or equal to one, which means the move will always be accepted—in agreement with Eq. (10.60)—while if $E_j > E_i$ the move will be accepted with probability $e^{-\beta(E_j - E_i)}$—again in agreement with Eq. (10.60). Thus, for instance, if we have a variable deltaE in our program that stores the change in energy $E_j - E_i$, then a single if statement of the form

```
if random()<exp(-beta*deltaE):
```

will accept or reject moves with the correct probabilities.

Although the Metropolis method is straightforward in practice, it does have some subtleties:

1. The steps where you reject a move and don't end up changing to a new state still count as steps. When you calculate your running sum of X values, you need to add to the sum even on the steps where you stay in the state you're already in. This means that when the system stays in the same state for two steps you will add the same number X to the sum two times. This may seem odd, but it's the correct thing to do.

2. In order for Eq. (10.62) to be correct it's crucial that the number of possible

moves M that you are choosing between be the same when going from i to j as it is when going from j to i. Care must be taken when choosing the move set to make sure that this is the case.

3. You also need to choose the move set so that you can actually get to every possible state. It is possible to choose sets such that Eq. (10.58) is satisfied but nonetheless some states are inaccessible—there is no path of moves that will take you there from your starting state. A move set for which all states are accessible is said to be *ergodic*, and ergodicity is a requirement for the Metropolis algorithm to work. See Appendix D for some additional discussion of this point.

4. Although we have said that the Markov chain always converges to the correct Boltzmann probability distribution, we haven't said how long it takes to do it—how long the simulation takes to *equilibrate*. There is no universal rule of thumb telling you how long equilibration takes but, as we'll see in the following example, it's often obvious just by looking at the results.

EXAMPLE 10.3: MONTE CARLO SIMULATION OF AN IDEAL GAS

The quantum states of a particle or atom of mass m in a cubic box of length L on each side have three integer quantum numbers $n_x, n_y, n_z = 1 \ldots \infty$ and energies given by

$$E(n_x, n_y, n_z) = \frac{\pi^2\hbar^2}{2mL^2}(n_x^2 + n_y^2 + n_z^2). \tag{10.65}$$

An ideal gas is a gas of N such atoms that do not interact with one another in any way, so that their total energy is just the sum of the energies of the individual particles:

$$E = \sum_{i=1}^{N} E(n_x^{(i)}, n_y^{(i)}, n_z^{(i)}), \tag{10.66}$$

where $n_x^{(i)}$ is the value of the quantum number n_x for the ith atom, and similarly for $n_y^{(i)}$ and $n_z^{(i)}$. In this example we will perform a Monte Carlo simulation of such an ideal gas and use it to calculate the internal energy of the gas.

First we need to choose a move set. In this case we will let the move set be the set of all moves of a single atom to one of the six neighboring states where n_x, n_y, or n_z differs by ± 1. In other words, on each step of the Monte Carlo simulation we will choose a random particle, choose one of the three quantum numbers n_x, n_y, or n_z, and choose a random change, either $+1$ or -1, for the quantum number. When we make such a move, just one term in the total energy, Eq. (10.66), will change, the term for the atom i that changes state. If, for example, n_x for atom i increases by 1, then the change will be

$$
\begin{aligned}
\Delta E &= \frac{\pi^2\hbar^2}{2mL^2}\left[(n_x+1)^2 + n_y^2 + n_z^2\right] - \frac{\pi^2\hbar^2}{2mL^2}\left(n_x^2 + n_y^2 + n_z^2\right) \\
&= \frac{\pi^2\hbar^2}{2mL^2}\left[(n_x+1)^2 - n_x^2\right] = \frac{\pi^2\hbar^2}{2mL^2}(2n_x+1).
\end{aligned}
\tag{10.67}
$$

Similarly if n_x decreases by 1 we have

$$
\Delta E = \frac{\pi^2\hbar^2}{2mL^2}(-2n_x+1).
\tag{10.68}
$$

We will write a program to perform the simulation for $N = 1000$ particles when $k_B T = 10$. For simplicity, we work with units where $m = \hbar = 1$ and set the size of box to be $L = 1$. Since, as we have seen, it doesn't matter what state we start our system in, let us just start with all particles in the ground state $n_x = n_y = n_z = 1$. Then we select at random the particle, axis, and sign for the first move, calculate the energy change $\Delta E = E_j - E_i$, and then either accept or reject the move according to the Metropolis probability, Eq. (10.60). Then we simply repeat the whole process for many moves. (Notice that if a particle is in an $n = 1$ state and we happen to choose to decrease the value of the quantum number n, then the move is always rejected, because there is no lower value of n.)

Here is a program to perform the entire calculation, and make a plot of the total energy of the gas as a function of time for 250 000 Monte Carlo moves:

```
from random import random,randrange
from math import exp,pi
from numpy import ones
from pylab import plot,ylabel,show


T = 10.0
N = 1000
steps = 250000

# Create a 2D array to store the quantum numbers
n = ones([N,3],int)

# Main loop
eplot = []
E = 3*N*pi*pi/2
```

File: mcsim.py

485

```
for k in range(steps):

    # Choose the particle and the move
    i = randrange(N)
    j = randrange(3)
    if random()<0.5:
        dn = 1
        dE = (2*n[i,j]+1)*pi*pi/2
    else:
        dn = -1
        dE = (-2*n[i,j]+1)*pi*pi/2

    # Decide whether to accept the move
    if n[i,j]>1 or dn==1:
        if random()<exp(-dE/T):
            n[i,j] += dn
            E += dE

    eplot.append(E)

# Make the graph
plot(eplot)
ylabel("Energy")
show()
```

Note how on each step of the calculation we choose one of the N particles using randrange(N), one of the three quantum numbers with randrange(3), and a direction, up or down, with 50:50 probability, by testing whether random()<0.5.

The program runs quickly—it takes just a few seconds to run through all 250 000 steps. Figure 10.5 shows the plot it produces. As we can see, the energy starts low at the beginning of the calculation (because we start it off in the ground state), but quickly rises and then levels off. Eyeballing the plot roughly, we might say that the calculation has equilibrated after about 50 000 steps, or perhaps 100 000 to be on the safe side. After this point, we could, for instance, start accumulating values of the energy to calculate an estimate of the average energy of the gas. If we did this, it looks like we'd get a value around $E = 25\,000$ (in the dimensionless units used in the calculation). In principle, we could then go back and run the calculation again for another value of the temperature, and another, perhaps making a graph of the calculated value of energy as a function of temperature.

Because the program is a random one it will not give the same answer every

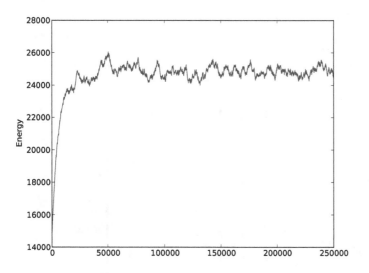

Figure 10.5: Internal energy of an ideal gas. The energy of an ideal gas of 1000 particles calculated by Monte Carlo simulation, as described in the text. From the shape of the curve it appears that the simulation equilibrates after about 50 000 steps.

time we run it, even if we use the same values of the parameters. But the longer we run it for, the more accurate our answers will be, and the more similar they will be from one run to another. If we want highly accurate answers, therefore, we should run for many Monte Carlo steps. If we are content with less accurate answers, then we can do a shorter run and get an answer sooner. Like many calculations in computational physics, Monte Carlo simulations are a compromise between speed and accuracy.

The Monte Carlo method can be extended to many other systems studied in statistical physics, and to the measurement of a wide variety of quantities. One can estimate heat capacity, entropy, free energy, pressure, magnetization, and many other things using straightforward extensions of the techniques described here.

Exercise 10.9: The Ising model

The Ising model is a theoretical model of a magnet. The magnetization of a magnetic material is made up of the combination of many small magnetic dipoles spread throughout the material. If these dipoles point in random directions then the overall magnetization of the system will be close to zero, but if they line up so that all or most of them point in the same direction then the system can acquire a macroscopic magnetic moment—it becomes magnetized. The Ising model is a model of this process in which the individual moments are represented by dipoles or "spins" arranged on a grid or lattice:

In this case we are using a square lattice in two dimensions, although the model can be defined in principle for any lattice in any number of dimensions.

The spins themselves, in this simple model, are restricted to point in only two directions, up and down. Mathematically the spins are represented by variables $s_i = \pm 1$ on the points of the lattice, $+1$ for up-pointing spins and -1 for down-pointing ones. Dipoles in real magnets can typically point in any spatial direction, not just up or down, but the Ising model, with its restriction to just the two directions, captures a lot of the important physics while being significantly simpler to understand.

Another important feature of many magnetic materials is that the individual dipoles in the material may interact magnetically in such a way that it is energetically favorable for them to line up in the same direction. The magnetic potential energy due to the interaction of two dipoles is proportional to their dot product, but in the Ising model this simplifies to just the product $s_i s_j$ for spins on sites i and j of the lattice, since the spins are one-dimensional scalars, not vectors. Then the actual energy of interaction is $-J s_i s_j$, where J is a positive interaction constant. The minus sign ensures that the interactions are *ferromagnetic*, meaning the energy is lower when dipoles are lined up. A ferromagnetic interaction implies that the material will magnetize if given the chance. (In some materials the interaction has the opposite sign so that the dipoles prefer to be antialigned. Such a material is said to be *antiferromagnetic*, but we will not look at the antiferromagnetic case here.)

Normally it is assumed that spins interact only with those that are immediately adjacent to them on the lattice, which gives a total energy for the entire system equal to

$$E = -J \sum_{\langle ij \rangle} s_i s_j ,$$

where the notation $\langle ij \rangle$ indicates a sum over pairs i, j that are adjacent on the lattice. On the square lattice we use in this exercise each spin has four adjacent neighbors with which it interacts, except for the spins at the edges of the lattice, which have either two or three neighbors.

Write a program to perform a Markov chain Monte Carlo simulation of the Ising model on the square lattice for a system of 20×20 spins. You will need to set up variables to hold the value ± 1 of the spin on each lattice site, probably using a two-dimensional integer array, and then take the following steps.

a) First write a function to calculate the total energy of the system, as given by the equation above. That is, for a given array of values of the spins, go through every pair of adjacent spins and add up the contributions $s_i s_j$ from all of them, then multiply by $-J$. Hint 1: Each unique pair of adjacent spins crops up only once in the sum. Thus there is a term $-J s_1 s_2$ if spins 1 and 2 are adjacent to one another, but you do not also need a term $-J s_2 s_1$. Hint 2: To make your final program to run in a reasonable amount of time, you will find it helpful if you can work out a way to calculate the energy using Python's ability to do arithmetic with entire arrays at once. If you do the calculation step by step, your program will be significantly slower.

b) Now use your function as the basis for a Metropolis-style simulation of the Ising model with $J = 1$ and temperature $T = 1$ in units where the Boltzmann constant k_B is also 1. Initially set the spin variables randomly to ± 1, so that on average about a half of them are up and a half down, giving a total magnetization of roughly zero. Then choose a spin at random, flip it, and calculate the new energy after it is flipped, and hence also the change in energy as a result of the flip. Then decide whether to accept the flip using the Metropolis acceptance formula, Eq. (10.60). If the move is rejected you will have to flip the spin back to where it was. Otherwise you keep the flipped spin. Now repeat this process for many moves.

c) Make a plot of the total magnetization $M = \sum_i s_i$ of the system as a function of time for a million Monte Carlo steps. You should see that the system develops a "spontaneous magnetization," a nonzero value of the overall magnetization. Hint: While you are working on your program, do shorter runs, of maybe ten thousand steps at a time. Once you have it working properly, do a longer run of a million steps to get the final results.

d) Run your program several times and observe the sign of the magnetization that develops, positive or negative. Describe what you find and give a brief explanation of what is happening.

e) Make a second version of your program that produces an animation of the system using the visual package, with spheres or squares of two colors, on a regular grid, to represent the up and down spins. Run it with temperature $T = 1$ and observe the behavior of the system. Then run it two further times at temperatures $T = 2$ and $T = 3$. Explain briefly what you see in your three runs. How and why does the behavior of the system change as temperature is increased?

10.4 SIMULATED ANNEALING

One of the most interesting uses of the Monte Carlo method in recent years has been in numerical optimization. In Section 6.4 we looked at methods for finding maxima or minima of functions, methods such as golden ratio search and gradient descent. As we pointed out, however, those methods are of limited use if one wants to find the *global* maximum or minimum—they can find a local minimum, a point in a function that is lower than the immediately surrounding points, but they do not guarantee that it is the lowest point anywhere. There are many problems, both in physics and elsewhere, that require us to find the highest or lowest overall value of a function—the global maximum or minimum—or at least an approximation to it. If, for example, we are looking for the ground state energy of a quantum system then it is important to find the overall minimum of the energy. An engineer trying to find the design parameters of an automobile that maximize its fuel efficiency wants the global maximum, not merely a local maximum. For such optimization problems the methods of Chapter 6 will not work and another approach is needed. Global optimization problems are among the hardest of computational tasks and it can demand both substantial computing resources and considerable ingenuity to solve them. One of the most promising approaches, proposed in 1985 by the physicist Scott Kirkpatrick, borrows ideas from statistical physics and goes under the name of *simulated annealing*.

The point indicated by the arrow is a local minimum, but not it is not the overall lowest point on the curve.

For a physical system in equilibrium at temperature T, the probability that at any moment the system is in a state i is given by the Boltzmann probability of Eq. (10.47), which for convenience we reproduce here:

$$P(E_i) = \frac{e^{-\beta E_i}}{Z}, \qquad \text{with} \qquad Z = \sum_i e^{-\beta E_i}, \qquad (10.69)$$

where $\beta = 1/k_B T$. Let us assume our system has a single unique ground state and let us choose our energy scale so that $E_i = 0$ in the ground state and $E_i > 0$ for all other states. Now suppose we cool down the system to absolute zero. In the limit $T \to 0$ we have $\beta \to \infty$ so $e^{-\beta E_i} \to 0$ except for the ground state, where $e^{-\beta E_i} = e^0 = 1$. Thus in this limit $Z = 1$ and

$$P(E_i) = \begin{cases} 1 & \text{for } E_i = 0, \\ 0 & \text{for } E_i > 0. \end{cases} \qquad (10.70)$$

In other words, the system will definitely be in the ground state. Thus one way to find the ground state of the system is to cool it down to $T = 0$ (or as close to it as we can get) and see what state it lands in. This in turn suggests a computational strategy for finding the ground state: let us simulate the system at temperature T, using the Markov chain Monte Carlo method of Section 10.3, then lower the temperature to zero and the system should find its way to the ground state. Kirkpatrick's insight was to realize that this same approach could be used to find the minimum of any function, not just the energy of a physical system. We can take any mathematical function $f(x, y, z, ...)$ and treat the independent variables x, y, z as defining a "state" of the "system" and f as being the energy of that system, then perform a Monte Carlo simulation. Taking the temperature to zero will again cause the system to fall into its ground state, i.e., the state with the lowest value of f, and hence we find the minimum of the function. This is the basic idea behind simulated annealing. The method can also be used to maximize functions—we treat $-f$ as the energy of the system (instead of $+f$), so that the minimum of energy corresponds to the maximum of f.

There is a catch, however. Recall that on each step of a thermal Monte Carlo simulation we choose a move from a given set of possibilities, changing the state of the system to some other nearby state, then we accept or reject that move with probability given by Eq. (10.60). Now suppose that the system finds itself in a local minimum of the energy, a state in which all other nearby states have higher energy. In that case, all proposed Monte Carlo moves will be to states with higher energy, and if we then set $T = 0$ the acceptance probability of Eq. (10.60) becomes zero for every move, because $\beta \to \infty$ and $E_j > E_i$. This means that every move will be rejected and the system will never be able escape from its local minimum. In principle there is still a global minimum that has a lower energy, but our computer program will never find it because it gets trapped in the local minimum. Luckily there is a trick that can get around this problem, and once again it is inspired by our knowledge of the physical world.

It has been well known for centuries, among glassworkers, metalworkers, ceramicists, and others, that when you are working with hot materials, like molten glass or metal, one must cool the materials slowly to produce a hardy, solid final product. Glass cooled too quickly will develop defects and weaknesses, or even shatter. Metal cooled too quickly will be soft or brittle. If one cools slowly, on the other hand—if one *anneals* the material—the end result will be rigid and sturdy. Why is this? The answer is that when materials are hot their atoms or molecules, which are in a rapid state of motion, are disordered; the motion randomizes their positions and prevents them coming to rest in an orderly arrangement. If one cools such a material rapidly, that disorder gets frozen into the material, and this creates flaws in the structure, cracks and defects that weaken the material. In the language of thermal physics, the system has become trapped at a local minimum of the energy, in a manner exactly analogous to what happens in our Monte Carlo simulation. If, on the other hand, one cools the material slowly then the disorder no longer gets frozen in and the atoms have time to rearrange and reach an ordered state, such as the regular lattice seen in the crystal structure of metals, which is the ground state of the system and contains no flaws, and is correspondingly stronger.

Thus while a rapidly cooled system can get stuck in a local energy minimum, an annealed system, one that is cooled sufficiently slowly, can find its way to the ground state. Simulated annealing applies the same idea in a computational setting. It mimics the slow cooling of a material on the computer by using a Monte Carlo simulation with a temperature parameter that is gradually lowered from an initially high value towards zero. It can be shown rigorously that this method is guaranteed to find the ground state if we cool slowly enough. Unfortunately, the rate of cooling required can often be so slow as to be impractical, so instead we typically just cool as slowly as we reasonably can, in the knowledge that we may not find the exact ground state of the system, but we will probably find a state close to it.

Some of the most interesting applications of simulated annealing are not to physics problems, but to problems in other areas of science and engineering. Like the problem of optimizing an automobile's fuel consumption, many technical challenges can be phrased in terms of finding the global minimum or maximum of a function of one or more parameters. Simulated annealing is one of the most widely used methods for solving such problems and has been applied in engineering, computer science, biology, economics, mathematics, statistics, and chemistry, as well as, of course, physics.

In practice, implementation of the method is quite straightforward—we perform a Monte Carlo simulation of the system of interest just as in Section 10.3.2 and slowly lower the temperature until the state of the system stops changing. The final state that the system comes to rest in is our estimate of the ground state. The only questions we need to answer are what the initial temperature should be and how fast we should cool.

The initial temperature should be chosen so that the system equilibrates quickly. To achieve this we should choose the thermal energy $k_B T$ to be significantly greater than the typical energy change accompanying a single Monte Carlo move. Then we will have $\beta(E_j - E_i) \ll 1$ in Eq. (10.60) and hence $P_a \simeq 1$ for all Monte Carlo moves, meaning that most moves will be accepted and the state of the system will be rapidly randomized, no matter what the starting state. As for the rate of cooling, one typically specifies a *cooling schedule*, i.e., a trajectory for the temperature as a function of time, and the most common choice by far is the exponential cooling schedule

$$T = T_0 e^{-t/\tau}, \tag{10.71}$$

where T_0 is the initial temperature and τ is a time constant. Some trial and error may be necessary to find a good value for the time constant. Typically one gets better results the slower the cooling, i.e., the larger the value of τ, but slower cooling also means that it takes the system longer to reach a temperature low enough that its state no longer changes (at which point the simulation can stop, since nothing more is happening). As with most computer calculations, therefore, the choice of τ is often a compromise between having a program that gives good answers and having one that runs in reasonable time.

EXAMPLE 10.4: THE TRAVELING SALESMAN

As an example of the use of simulated annealing, we will solve one of the most famous of optimization problems, the *traveling salesman problem*, which involves finding the shortest route that visits a given set of locations on a map. The traveling salesman problem is of interest not because we actually care about traveling salesmen—most of us don't. It's interesting because it belongs to the class of *NP-hard* problems, problems that are unusually difficult to solve computationally. Thus the problem provides a special challenge for us as computational scientists and a method that can find a good solution to this problem will probably work well for other problems too.

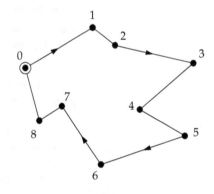

Figure 10.6: The traveling salesman problem. Starting from a given city (circled) a salesman wants to find the shortest tour that visits each city once and then returns to the starting point.

The definition of the problem is as follows. A traveling salesman, hawking his employer's products, wishes to visit N given cities. We'll assume that he can travel (let's say by air) in a straight line between any pair of cities. Given the coordinates of the cities, the problem is to devise the shortest tour, in terms of total distance traveled, that starts and ends at the same city and visits all N cities—see Fig. 10.6.

To make the problem simple, we will consider it in a flat two-dimensional space (as opposed to the surface of the Earth, which is curved, making the problem harder), and we'll solve it for the case where the N city locations are chosen at random within a square of unit length on each side.

Let us number the cities in the order in which the salesman visits them, in Python fashion starting from zero—see Fig. 10.6 again—and let us denote the position of city i by the two-dimensional vector $\mathbf{r}_i = (x_i, y_i)$, with $\mathbf{r}_N = \mathbf{r}_0$, since the tour ends where it begins. Then the total distance D traveled by the salesman over the whole tour is

$$D = \sum_{i=0}^{N-1} \left| \mathbf{r}_{i+1} - \mathbf{r}_i \right|. \tag{10.72}$$

We want to minimize this quantity over the set of all possible tours, i.e., over all possible choices of the order of the cities. To do this using simulated annealing we need to choose a set of moves for the Markov chain. In this case a suitable set of moves is swaps of pairs of cities in the tour, so the calculation will involve setting up an initial tour and repeatedly trying out a swap of two cities. If a swap shortens the tour, then we always accept it and proceed to the next move. If it lengthens the tour then we accept it with the Metropolis probability given in Eq. (10.60), with the energy E replaced by the distance D, otherwise we reject it. If the move is rejected we need to swap the two cities back to where they were previously before we try another move. Also on each step we need to calculate the new value of the temperature according to the cooling schedule, Eq. (10.71).

Here is a full program to perform the calculation and display an animated picture of the tour on the screen as it runs:

File: salesman.py

```
from math import sqrt,exp
from numpy import empty
from random import random,randrange
from visual import sphere,curve,display,rate

N = 25
R = 0.02

Tmax = 10.0
Tmin = 1e-3
tau = 1e4

# Function to calculate the magnitude of a vector
def mag(x):
    return sqrt(x[0]**2+x[1]**2)
```

```
# Function to calculate the total length of the tour
def distance():
    s = 0.0
    for i in range(N):
        s += mag(r[i+1]-r[i])
    return s

# Choose N city locations and calculate the initial distance
r = empty([N+1,2],float)
for i in range(N):
    r[i,0] = random()
    r[i,1] = random()
r[N] = r[0]
D = distance()

# Set up the graphics
display(center=[0.5,0.5])
for i in range(N):
    sphere(pos=r[i],radius=R)
l = curve(pos=r,radius=R/2)

# Main loop
t = 0
T = Tmax
while T>Tmin:

    t += 1
    T = Tmax*exp(-t/tau)     # Cooling

    if t%100==0:             # Update the visualization every 100 moves
        l.pos = r
        rate(25)

    # Choose two cities to swap and make sure they are distinct
    i,j = randrange(1,N),randrange(1,N)
    while i==j:
        i,j = randrange(1,N),randrange(1,N)

    # Swap them and calculate the change in distance
    oldD = D
    r[i,0],r[j,0] = r[j,0],r[i,0]
    r[i,1],r[j,1] = r[j,1],r[i,1]
    D = distance()
    deltaD = D - oldD
```

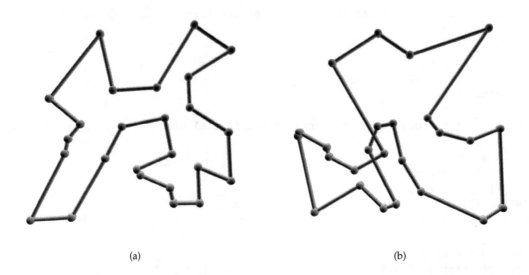

(a)   (b)

Figure 10.7: Solutions of the traveling salesman problem. (a) A solution to one of the random traveling salesman problems generated by the program given in the text, as found by the program using simulated annealing. (b) The solution found by the program for another instance of the problem. In this case the solution is clearly not perfect—it could be improved by swapping the two cities near the center, where the paths cross.

```
# If the move is rejected, swap them back again
if random()>=exp(-deltaD/T):
    r[i,0],r[j,0] = r[j,0],r[i,0]
    r[i,1],r[j,1] = r[j,1],r[i,1]
    D = oldD
```

For a tour of 25 cities the program takes about 30 seconds to run on the author's computer. Figure 10.7a shows the result from a typical run, where the computer appears to have found a reasonable tour around the given set of cities. As we have said, however, simulated annealing usually involves compromises between speed of execution and quality of the results and can occasionally return imperfect answers. Figure 10.7b shows an example: in this case the computer has found a pretty good solution to the problem but it's not perfect. The tour could clearly be improved by swapping the two cities in the middle, where the paths cross.

These results are typical of the simulated annealing method: it gives good but not necessarily perfect results for a wide range of problems, and is a good, general-purpose method for performing optimization on the computer.

Exercise 10.10: Global minimum of a function

Consider the function $f(x) = x^2 - \cos 4\pi x$, which looks like this:

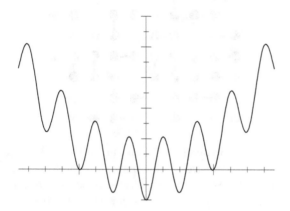

Clearly the global minimum of this function is at $x = 0$.

a) Write a program to confirm this fact using simulated annealing starting at, say, $x = 2$, with Monte Carlo moves of the form $x \to x + \delta$ where δ is a random number drawn from a Gaussian distribution with mean zero and standard deviation one. (See Section 10.1.6 for a reminder of how to generate Gaussian random numbers.) Use an exponential cooling schedule and adjust the start and end temperatures, as well as the exponential constant, until you find values that give good answers in reasonable time. Have your program make a plot of the values of x as a function of time during the run and have it print out the final value of x at the end. You will find the plot easier to interpret if you make it using dots rather than lines, with a statement of the form plot(x,".") or similar.

b) Now adapt your program to find the minimum of the more complicated function $f(x) = \cos x + \cos \sqrt{2}x + \cos \sqrt{3}x$ in the range $0 < x < 50$.

Hint: The correct answer for part (b) is around $x = 16$, but there are also competing minima around $x = 2$ and $x = 42$ that your program might find. In real-world situations, it is often good enough to find any reasonable solution to a problem, not necessarily the absolute best, so the fact that the program sometimes settles on these other solutions is not necessarily a bad thing.

Exercise 10.11: The dimer covering problem

A well studied problem in condensed matter physics is the *dimer covering problem* in which dimers, meaning polymers with only two atoms, land on the surface of a solid, falling in the spaces between the atoms on the surface and forming a grid like this:

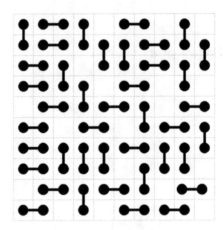

No two dimers are allowed to overlap. The question is how many dimers we can fit in the entire $L \times L$ square. The answer, in this simple case, is clearly $\frac{1}{2}L \times L$, but suppose we did not know this. (There are more complicated versions of the problem on different lattices, or with differently shaped elements, for which the best solution is far from obvious, or in some cases not known at all.)

a) Write a program to solve the problem using simulated annealing on a 50×50 lattice. The "energy" function for the system is *minus* the number of dimers, so that it is minimized when the dimers are a maximum. The moves for the Markov chain are as follows:

   i) Choose two adjacent sites on the lattice at random.

   ii) If those two sites are currently both empty, place a dimer on them.

   iii) If the two sites are currently occupied by a single dimer, remove the dimer from the lattice with the appropriate probability (which you will have to work out).

   iv) Otherwise, do nothing.

   Create an animation of the state of the system over time as the simulation runs.

b) Try exponential cooling schedules with different time constants. A reasonable first value to try is $\tau = 10\,000$ steps. For faster cooling schedules you should see that the solutions found are poorer—a smaller fraction of the lattice is filled with dimers and there are larger holes in between them—but for slower schedules the calculation can find quite good, but usually not perfect, coverings of the lattice.

FURTHER EXERCISES

10.12   A random point on the surface of the Earth: Suppose you wish to choose a random point on the surface of the Earth. That is, you want to choose a value of the latitude and longitude such that every point on the planet is equally likely to be chosen. In a physics context, this is equivalent to choosing a random vector direction in three-dimensional space (something that one has to do quite often in physics calculations).

Recall that in spherical coordinates θ, ϕ (where θ is the angle from the north pole and ϕ is the azimuthal or longitudinal angle) the element of solid angle is $\sin\theta \, d\theta \, d\phi$, and the total solid angle in a whole sphere is 4π. Hence the probability of our point falling in a particular element is

$$p(\theta, \phi) \, d\theta \, d\phi = \frac{\sin\theta \, d\theta \, d\phi}{4\pi}.$$

We can break this up into its θ part and its ϕ part thus:

$$p(\theta, \phi) \, d\theta \, d\phi = \frac{\sin\theta \, d\theta}{2} \times \frac{d\phi}{2\pi} = p(\theta) \, d\theta \times p(\phi) \, d\phi.$$

a) What are the ranges of the variables θ and ϕ? Verify that the two distributions $p(\theta)$ and $p(\phi)$ are correctly normalized—they integrate to 1 over the appropriate ranges.

b) Find formulas for generating angles θ and ϕ drawn from the distributions $p(\theta)$ and $p(\phi)$. (The ϕ one is trivial, but the θ one is not.)

c) Write a program that generates a random θ and ϕ using the formulas you worked out. (Hint: In Python the function `acos` in the `math` package returns the arc cosine in radians of a given number.)

d) Modify your program to generate 500 such random points, convert the angles to x, y, z coordinates assuming the radius of the globe is 1, and then visualize the points in three-dimensional space using the `visual` package with small spheres (of radius, say, 0.02). You should end up with a three-dimensional globe spelled out on the screen in random points.

10.13   Diffusion-limited aggregation: This exercise builds upon Exercise 10.3 on page 457. If you have not done that exercise you should do it before doing this one.

In this exercise you will develop a computer program to reproduce one of the most famous models in computational physics, *diffusion-limited aggregation*, or DLA for short. There are various versions of DLA, but the one we'll study is as follows. You take a square grid with a single particle in the middle. The particle performs a random walk from square to square on the grid until it reaches a point on the edge of the system, at which point it "sticks" to the edge, becoming anchored there and immovable:

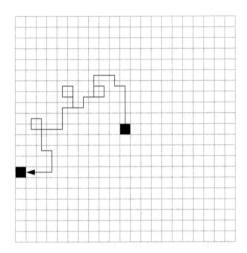

Then a second particle starts at the center and does a random walk until it sticks either to an edge or to the other particle. Then a third particle starts, and so on. Each particle starts at the center and walks until it sticks either to an edge or to any anchored particle.

a) Make a copy of the Brownian motion program that you wrote for Exercise 10.3. This will serve as a starting point for your DLA program. Modify your program to perform the DLA process on a 101×101 lattice—we choose an odd length for the side of the square so that there is one lattice site exactly in the center. Repeatedly introduce a new particle at the center and have it walk randomly until it sticks to an edge or an anchored particle.

You will need to decide some things. How are you going to store the positions of the anchored particles? On each step of the random walk you will have to check the particle's neighboring squares to see if they are outside the edge of the system or are occupied by an anchored particle. How are you going to do this? You should also modify your visualization code from the Brownian motion exercise to visualize the positions of both the randomly walking particles and the anchored particles. Run your program for a while and observe what it does.

b) In the interest of speed, change your program so that it shows only the anchored particles on the screen and not the randomly walking ones. That way you need update the pictures on the screen only when a new particle becomes anchored. Also remove any `rate` statements that you added to make the animation smooth.

Set up the program so that it stops running once there is an anchored particle in the center of the grid, at the point where each particle starts its random walk. Once there is a particle at this point, there's no point running any longer because any further particles added will be anchored the moment they start out.

Run your program and see what it produces. If you are feeling patient, try modifying it to use a 201×201 lattice and run it again—the pictures will be more impressive, but you'll have to wait longer to generate them.

A nice further twist is to modify the program so that the anchored particles are shown in different shades or colors depending on their age, with the shades or colors changing gradually from the first particle added to the last.

c) If you are feeling particularly ambitious, try the following. The original version of DLA was a bit different from the version above—and more difficult to do. In the original version you start off with a single *anchored* particle at the center of the grid and a new particle starts from a random point on the perimeter and walks until it sticks to the particle in the middle. Then the next particle starts from the perimeter and walks until it sticks to one of the other two, and so on. Particles no longer stick to the walls, but they are not allowed to walk off the edge of the grid.

Unfortunately, simulating this version of DLA directly takes forever—the single anchored particle in the middle of the grid is difficult for a random walker to find, so you have to wait a long time even for just one particle to finish its random walk. But you can speed it up using a clever trick: when the randomly walking particle does finally find its way to the center, it will cross any circle around the center at a random point—no point on the circle is special so the particle will just cross anywhere. But in that case we need not wait the long time required for the particle to make its way to the center and cross that circle. We can just cut to the chase and start the particle on the circle at a random point, rather than at the boundary of the grid. Thus the procedure for simulating this version of DLA is as follows:

   i) Start with a single anchored particle in the middle of the grid. Define a variable r to record the furthest distance of any anchored particle from the center of the grid. Initially $r = 0$.

   ii) For each additional particle, start the particle at a random point around a circle centered on the center of the grid and having radius $r + 1$. You may not be able to start exactly on the circle, if the chosen random point doesn't fall precisely on a grid point, in which case start on the nearest grid point outside the circle.

   iii) Perform a random walk until the particle sticks to another one, except that if the particle ever gets more than $2r$ away from the center, throw it away and start a new particle at a random point on the circle again.

   iv) Every time a particle sticks, calculate its distance from the center and if that distance is greater than the current value of r, update r to the new value.

   v) The program stops running once r surpasses a half of the distance from the center of the grid to the boundary, to prevent particles from ever walking outside the grid.

Try running your program with a 101×101 grid initially and see what you get.

Chapter 11

Using what you have learned

In the preceding chapters of this book we have studied the main pillars of computational physics, including numerical integration and differentiation, linear algebra, the solution of ordinary and partial differential equations, and Monte Carlo methods, and seen how to put the theoretical concepts into practice using the Python programming language. The material in this book forms a solid foundation for the solution of physics problems using computational methods.

That is not to say, however, that the book covers everything there is to know about computational physics. Certainly it does not. Computational physics is a huge field, heavily researched and developed over the decades, and it encompasses an enormous range of knowledge and approaches, some general, some specialized. If you continue with the study of computational physics, or if you do computational work in a professional situation in industry, business, or academia, you will almost certainly have to learn additional techniques not covered in this book. Here is just a taste of some of the advanced techniques currently in use in different areas of physics:

1. In Chapter 5 we studied several methods for performing integrals, including the trapezoidal rule and Gaussian quadrature, but there are many other methods that can prove useful under certain circumstances. Two important examples are *Clenshaw–Curtis quadrature* and *Gauss–Kronrod quadrature*, which are similar to Gaussian quadrature in using unevenly spaced integration points to improve on simple rules like the trapezoidal rule. They are in general less accurate than Gaussian quadrature, but they allow nesting of integration points in a manner similar to the adaptive integration method of Section 5.3, which allows us, for example, to make estimates of the errors on our integrals with relatively little computational effort. Gauss–Kronrod quadrature is discussed briefly in Appendix C. For multidimensional integrals there are also a wide range of

techniques available, some specialized to integration domains of particular shapes, others designed to get accurate answers in a minimum of computation time. The *low-discrepancy point sets* mentioned in Section 5.9 are an important example of the latter.

2. In Chapter 6 we studied the solution of linear systems of equations. An important class of linear equation problems that arises in physics are those involving *sparse equations*, whose matrix representation contains a large number of zeros and only a few nonzero elements. We saw one example in Section 6.1.6, where we looked at tridiagonal and banded matrices, which have zeros everywhere except close to the diagonal, but there are many more general examples also, which may have nonzero elements anywhere in the matrix but nonetheless have only a few nonzero elements overall. For such matrices the standard methods of equation solving, inversion, and calculation of eigenvalues are inefficient because they waste time performing computations on all the zeros that have no effect on the final result. Luckily, as with tridiagonal matrices, there are special techniques for sparse matrices that are more efficient, such as the *conjugate gradient method* and the *Lanczos algorithm*. These methods find use particularly in condensed matter physics, where they are employed to solve problems involving the lattices of atoms in a solid.

3. In Chapter 9 we studied the solution of partial differential equations using finite difference methods and spectral methods, but we omitted an important third class of methods, the *finite element methods*. These methods, though complex and challenging to use, are some of the most powerful available for the general solution of partial differential equations, and are particularly useful for problems such as the solution of Maxwell's equations or the Navier–Stokes equations of fluid dynamics.

4. In our discussion of Monte Carlo simulation in Chapter 10, we mentioned only the most common and important of sampling algorithms, the Metropolis algorithm, but there are others that may be appropriate for particular problems, including the *heat-bath algorithm*, rejection-free methods like the *Swendsen–Wang algorithm*, and continuous-time methods like the *Bortz–Kalos–Lebowitz algorithm*.

Entire books have been written on each of these topics and on many others as well, and while it's certainly not practical to read all of them, there are plenty of opportunities to learn more about a particular problem or class of problems when the need arises.

Another important way in which your future experiences with computational physics might differ from the material in this book is in the computer language employed. We have used the Python language exclusively, which is a good choice—it's easy to learn, powerful, and contains many features that are well suited to physics calculations. It does have its disadvantages, however, the primary one being that it is slower than some other languages, and significantly so for certain types of calculations. If you find yourself doing computational work in the future, it's quite likely that you will have to use a language other than Python, perhaps because you need greater speed, or because you are working as part of a team that routinely uses another language. The most likely alternative languages for physics work are C and Fortran, but you might also use C++, Java, IDL, Julia, or Matlab, among others.

Luckily, these computer languages do not differ greatly from one another. To be sure, there are differences in the details—the exact words and symbols you use to represent a statement—but the concepts of programming are similar in all of them and you will find, once you know how to program in Python, that switching to another language is not difficult. Armed with your knowledge of Python you can probably pick up the main points of a new language in just a day or so.

Consider, for example, the following short Python program, which calculates the integral $\int_0^1 e^{-x^2}\,dx$ using the trapezoidal rule with 1000 points:

```
from math import exp

def f(x):
    return exp(-x*x)

a = 0.0
b = 1.0
N = 1000
h = (b-a)/N

s = 0.5*(f(a)+f(b))
for k in range(1,N):
    s += f(a+k*h)
print(h*s)
```

Now here is the same program written in the C programming language:

```
#include <math.h>
#include <stdio.h>

double f(double x)
{
  return exp(-x*x);
}

main()
{
  double a,b,h,s;
  int N,k;

  a = 0.0;
  b = 1.0;
  N = 1000;
  h = (b-a)/N;

  s = 0.5*(f(a)+f(b));
  for (k=1; k<N; k++) s += f(a+k*h);
  printf("%g\n",h*s);
}
```

There are some obvious differences between the two programs. Note how almost every line in the C program ends with a semicolon. In C you use the semicolon to show when a statement is finished, and a statement that does not end with a semicolon is considered to continue on the following line. In Python, rather than using a special symbol to show when a line ends we use one to show when it *doesn't* end—recall that when we want a statement to continue on the next line we use a backslash symbol "\" (see Section 3.4). Either approach works fine. Both allow you to use both single-line and multi-line statements as needed. But the two languages make different choices about the details. These are the kinds of things you will need to learn if you have to work in a different computer language.

Another important difference between Python and C, visible in the example programs above, is that in C all variables have to be *declared* before they can be used. For example, the statement "double a,b,h,s;" in the C program above tells the computer that we are going to be using floating-point variables called a, b, h, and s. (In C the floating-point variable type is called "double" for historical reasons.[1]) If we omit this statement, the program will not run.

However, there are strong similarities between the two programs as well. In the C program the statements at the beginning that start with "#include" are the equivalent of import statements in Python. They tell the computer that we are going to be using functions imported from other modules. The definition of the function f(x) in the C program is similar to the Python definition, except for the occurrence of the word "double" again, which tells the computer that both the variable x and the result returned by the function are of floating-point type. In the main program we go through the same steps in the C and Python versions, first assigning values to various variables, then using a for loop to calculate the trapezoidal rule sum, and finally printing out the result. The details of the individual statements vary, but the overall logical structure is closely similar between the two languages.

Apart from differences in the programs themselves, another difference between Python and C (and some other languages, such as Fortran) is that the latter makes use of a *compiler*. At their lowest level, computers speak to themselves in a native language called a *machine language* or *machine code*, which is well suited for computers but very unintuitive for human beings. You can write programs in machine code, but it's challenging. Instead, therefore, we write our programs in more human-compatible languages like Python or C and then have the computer translate them into machine code for us. There are two general schemes for doing this. The first is to use an *interpreter*, which is a program that translates the lines of code one by one as the program is executed. This is the scheme Python uses. The second approach is to use a compiler, which translates the entire program into machine code in one go, then we run the machine-code version. This is the scheme that C uses. The advantage of using a compiler is that the final program runs significantly faster, because no translation needs to be done while the program is running. This is the main reason why C programs are faster than Python programs. However, using a compiler has disadvantages too. The compiler has to be used and the program translated every time we make even the smallest change to our code, even just changing the value of a single parameter, and this can be time-consuming and

---

[1]The floating-point type in C was originally called "float", which is more logical, but there was also another type, called "double", short for "double-precision," which was a more accurate floating-point type that could store larger numbers to greater precision, at the expense of slower calculations. On modern computer hardware, however, calculations are equally fast with either type, and hence the float type has become obsolete and is no longer used, since the double type is more accurate and just as fast.

inconvenient. Large programs can take minutes or even hours to translate. Languages like C that use a compiler also tend to be more cumbersome and difficult to work with than languages that use an interpreter. The requirement that we declare all our variables before we use them, for instance, is a result of the way the compiler works. In Python there is no such requirement—one can use any variable one likes at any time, and this makes programming significantly simpler and easier. So both approaches have their pluses and minuses. Naturally this means that each may be better in some situations than the other, and this is precisely why we have more than one computer language in the first place.

Nonetheless, the similarities between computer languages are much greater than their differences, and, as we have said, you should have little difficulty applying what you have learned about Python to another language if and when the need arises. Armed with the principles of computational physics, the practical experience you've acquired here, and a good fast computer, you should be able to go out and start doing physics.

Appendix A

Installing Python

THIS appendix explains how to install on your computer the software you will need for programming in the Python programming language. All of the software is distributed by its makers for free and is available for download on the Internet. To make best use of the book there are four software packages you should install: the Python language itself, and the packages "numpy", "matplotlib", and "visual" (also called "VPython" in some places).

There are currently two different versions of Python in circulation, version 2 and version 3. This book uses version 3, which is the most recent and up-to-date version. This appendix describes how to install all the necessary software to use version 3 of Python on your computer.

You can also use version 2 with this book if you wish. Some computers come with version 2 already installed, or you may have installed it yourself in the past and want to go on using it. If so, you can do that, but you need to take one extra precaution. There are a few small differences between versions 2 and 3 that affect some of the programs in the book, but you can get around this by the following simple trick. If you are using version 2 of Python, add the following line at the very beginning of every program you run:

```
from __future__ import division,print_function
```

(Note the two underscore characters "__" on either side of the word "future".) If you include this line in your programs, it makes version 2 of Python behave essentially the same as version 3. Add it to the start of any of the example programs in the book and they will work with version 2. Add it to the start of the programs you yourself write, and they will work as they would in version 3. You can find further discussion of the differences between Python versions 2 and 3 in Appendix B.

Bearing this in mind, the simplest way to install Python and the additional packages needed for this book is as follows.

1. Open your web browser and go to www.python.org. Click on "download", and on the downloads page find, download, and then install the version of Python you want. As discussed above, you should install version 3, unless you know for certain that you want to use version 2. At the time of writing the most recent sub-version was version 3.2, which will work fine. If you are installing on a Windows computer then (at least at the time of writing) you must use the 32-bit version of Python, not the 64-bit version, which is incompatible with some of the other packages you will need. Even if you have a 64-bit computer, you should still install the 32-bit version. (Python is also available for users of the Linux operating system. The installation procedure is different for Linux, and moreover varies between Linux varieties, but the most widely used varieties, including Ubuntu and Fedora, come with Python packages that can be installed using the standard software installer.)

2. Having installed the Python language you should next install the visual package, also sometimes called VPython. Go to www.vpython.org, click on the download link for your operating system (Windows or Mac), then download and install the appropriate file. There are different versions of VPython corresponding to the different versions of Python. If, for instance, you installed Python 3.2, you should choose the corresponding version of VPython. If you are using Windows, installing VPython will automatically install the numpy package for you as well. If you are using a Mac you will need to install numpy separately—see step 4 below.

3. Next you need to install the package matplotlib, which you can find at matplotlib.org/downloads.html. You should click on the link for the latest version of matplotlib, which at the time of writing is version 1.2.1, and you will be presented with a list of packages for different computers. Select and install the one that corresponds to your computer and the version of Python that you installed. For instance, you would click on matplotlib-1.2.1.win32-py3.2.exe for a Windows computer with Python version 3.2 installed.

4. If you use a Mac, you will also need to install the package numpy. (If you use Windows, numpy will already have been installed for you when you installed VPython.) You can download the latest version of numpy from sourceforge.net/projects/numpy/files/NumPy and install.

APPENDIX **B**

DIFFERENCES BETWEEN PYTHON VERSIONS

T HE Python programming language is continually being updated and improved by its creators. The most recent version is version 3, though version 2 is still available and finds wide use. (The much earlier version 1, which dates back to the 1980s, is now seen very rarely.)

The programs that appear in this book are written using Python version 3, although, as noted in Appendix A, they can also be used with version 2 if you prefer. If you use version 2 you should include at the beginning of all your programs the statement

```
from __future__ import division,print_function
```

(Note the two underscore characters "__" on either side of the word "future".) This statement tells version 2 of Python to behave in the way version 3 does with respect to the two most important differences between versions, the differences in the division of integers and the structure of the print command.

The remainder of this appendix describes the main technical differences between Python versions 2 and 3, for those readers who are interested.

Division returns a floating-point value: In version 2 of Python the division of two integers, one by another, returns another integer, rounding down if necessary. Thus 3/2 gives 1, not 1.5. In version 3 the same operation gives the floating-point value 1.5. Furthermore, in version 3 even if the result of a division is in fact an integer, the operation will still give a floating-point value. Thus 4/2 gives a floating-point 2.0, not an integer 2.

If you are using version 2 of Python, you can duplicate the behavior of version 3 with respect to division by including the statement

```
from __future__ import division
```

at the start of your program. If you are using version 3 and you want the rounding-down behavior of version 2, you can get it by using the integer division operation "//"—see page 23.

Print is a function: In version 3 of Python the print command is a *function*, where in version 2 it is a *statement*. The main practical difference between the two is that in version 3 you must enclose the argument(s) of a print command within parentheses, while in version 2 you should not. Thus in version 2 you might say

```
print "The energy is",E
```

while in version 3 you would say

```
print("The energy is",E)
```

In most other respects the two commands behave in the same way.

If you are using version 2 of Python, you can duplicate the behavior of the version 3 print function by including the statement

```
from __future__ import print_function
```

at the start of your program.

If you wish to duplicate the behavior of version 3 with respect to both division and the print function in the same program, you can use the single statement

```
from __future__ import division,print_function
```

at the start of your program (as mentioned previously).

Input returns a string: In version 3 of Python the input function always returns a *string*, no matter what you type in, even if you type in a number. In version 2, by contrast, the input statement takes what you type and evaluates it as an algebraic expression, then returns the resulting value. Thus if you write a program that includes the statement

```
x = input()
```

and you enter "2.5", the result will be different in Python versions 2 and 3. In version 2, x will be a floating-point variable with numerical value 2.5, while in version 3 it will be a string with string value "2.5". In version 2 if you entered an actual string like "Hello" you would get an error message, while in version 3 this works just fine.

Version 2 of Python includes another function called `raw_input`, which behaves the same way that `input` does in version 3. Thus if you are using version 2 you can still duplicate the behavior of version 3 by using `raw_input` everywhere that version 3 programs would use `input`. (In version 3 the function `raw_input` no longer exists.)

There is only one integer type: In version 2 of Python there are two types of integer variables called int and long. Variables of type int are restricted to numbers in the range $\pm 2^{31}$, but arithmetic using them is very fast; variables of type long can store numbers of arbitrary size but arithmetic using them is slower. In version 3 of Python there is only one type of integer variable, called int, which subsumes both the earlier types. For smaller integer values version 3 will automatically use old-style ints with their fast arithmetic, while for larger values it will automatically use old-style longs but slower arithmetic. You do not need to worry about the distinction between the two—Python takes care of it for you.

In fact, this change appeared earlier than version 3 of Python, starting in version 2.4. If you are using version 2 of the language, it's most likely that you are using either version 2.6 or 2.7, in which case you don't need to worry about this point—you already have the improved behavior of Python 3 with respect to integers.

Iterators: An *iterator* is an object in Python that behaves something like a list. It is a collection of values, one after another, but it differs from a list in that the values are not stored in the memory of the computer waiting for you to look them up; instead they are calculated on the fly, which saves memory.

In version 2 of Python the function `range` generates an actual list of numbers, which occupies space in the computer memory. This can cause problems if the list is very large. For instance, in version 2 on most computers the statement

```
for n in range(10000000000):
```

will give an error message because there is not enough memory to store the huge list generated by the `range` function. To get around this problem version 2 provides another function called `xrange`, which acts like `range` but produces an iterator. Thus "`xrange(100)`" behaves in many respects like a list of 100 elements, but no actual list is created. Instead, each time you ask for the next element in the list the computer just works out what that element ought to be and hands the value to you. The value is never stored anywhere. Thus you could say

```
for n in xrange(10000000000):
```

and the program would run just fine without crashing (although it would take a long time to finish because the loop is so long).

In version 3 of Python range behaves the way xrange does in version 2, producing an iterator, not a true list. Since the most common use of range by far is in for loops, this is usually an improvement: it saves memory and often makes the program run faster. Sometimes, however, you may want to generate an actual list from a range. In that case you can use a statement of the form

```
x = list(range(100))
```

which will create an iterator then convert it into a list. In version 2 of Python you do not need to do this (although it will work fine if you do).

(The function arange in the package numpy, which is similar to range but works with arrays rather than lists, really does create an array, not an iterator. It calculates all the values of the array and stores them in memory, rather than calculating them on the fly. This means that using arange with large arguments can slow your program or cause it to run out of memory, even in Python version 3.)

Another situation in which iterators appear in version 3 is the map function, which we studied in Section 2.4.1. Recall that map applies a given function to each element of a list or array. Thus in version 2 of Python

```
from math import log
r = [ 1.0, 1.5, 2.2 ]
logr = map(log,r)
```

applies the natural logarithm function separately to each element of the list [1.0,1.5,2.2] and produces a new list logr with the three logarithms in it. In version 3, however, the map function produces an iterator. If you need a real list, you would have to convert the iterator like this:

```
logr = list(map(log,r))
```

There are a number of other differences between versions 2 and 3 of Python, but the ones above are the most important for our purposes. A full description of the differences between versions can be found on-line at the main Python web site www.python.org.

APPENDIX C

GAUSSIAN QUADRATURE

THIS appendix gives a derivation of the fundamental formulas for Gaussian quadrature, which were discussed but not derived in Section 5.6.2.

Gaussian quadrature, defined over the standard domain from -1 to 1, makes use of an integration rule of the form

$$\int_{-1}^{1} f(x)\,dx \simeq \sum_{k=1}^{N} w_k f(x_k). \tag{C.1}$$

The derivation of the positions x_k of the sample points and the weights w_k is based on the mathematics of Legendre polynomials. The Legendre polynomial $P_N(x)$ is an Nth degree polynomial in x that has the property

$$\int_{-1}^{1} x^k P_N(x)\,dx = 0 \qquad \text{for all integer } k \text{ in the range } 0 \leq k < N \tag{C.2}$$

and satisfies the normalization condition

$$\int_{-1}^{1} \left[P_N(x)\right]^2 dx = \frac{2}{2N+1}. \tag{C.3}$$

Thus, for instance, $P_0(x) = $ constant, and the constant is fixed by (C.3) to give $P_0(x) = 1$. Similarly, $P_1(x)$ is a polynomial of degree one $ax + b$ satisfying

$$\int_{-1}^{1} (ax + b)\,dx = 0. \tag{C.4}$$

Carrying out the integral, we find that $b = 0$ and a is fixed by (C.3) to be 1, giving $P_1(x) = x$. The next two polynomials are $P_2(x) = \frac{1}{2}(3x^2 - 1)$ and $P_3(x) = \frac{1}{2}(5x^3 - 3x)$, and you can find tables on-line or elsewhere that list them to higher order.

Now suppose that $q(x)$ is a polynomial of degree less than N, so that it can be written $q(x) = \sum_{k=0}^{N-1} c_k x^k$ for some set of coefficients c_k. Then

$$\int_{-1}^{1} q(x) P_N(x)\,dx = \sum_{k=0}^{N-1} c_k \int_{-1}^{1} x^k P_N(x)\,dx = 0, \tag{C.5}$$

514

by Eq. (C.2). Thus, for any N, $P_N(x)$ is orthogonal to every polynomial of lower degree. A further property of the Legendre polynomials, which we will use shortly, is that for all N the polynomial $P_N(x)$ has N real roots that all lie in the interval from -1 to 1. That is, there are N values of x in this interval for which $P_N(x) = 0$.

Returning now to our integral, Eq. (C.1), suppose that the integrand $f(x)$ is a polynomial in x of degree $2N - 1$ or less. If we divide $f(x)$ by the Legendre polynomial $P_N(x)$, then we get

$$f(x) = q(x)P_N(x) + r(x), \tag{C.6}$$

where $q(x)$ and $r(x)$ are both polynomials of degree $N - 1$ or less. Thus our integral can be written

$$\int_{-1}^{1} f(x)\,\mathrm{d}x = \int_{-1}^{1} q(x)P_N(x)\,\mathrm{d}x + \int_{-1}^{1} r(x)\,\mathrm{d}x = \int_{-1}^{1} r(x)\,\mathrm{d}x, \tag{C.7}$$

where we have used (C.5). This means that to find the integral of the polynomial $f(x)$ we have only to find the integral of the polynomial $r(x)$, which always has degree $N - 1$ or less.

But we already know how to solve this problem. As we saw in Section 5.6.1, for any choice of sample points x_k a polynomial of degree $N - 1$ or less can be fitted exactly using the interpolating polynomials $\phi_k(x)$, Eq. (5.53), and then the fit can be integrated to give a formula of the form

$$\int_{-1}^{1} f(x)\,\mathrm{d}x = \int_{-1}^{1} r(x)\,\mathrm{d}x = \sum_{k=1}^{N} w_k r(x_k), \tag{C.8}$$

where

$$w_k = \int_{-1}^{1} \phi_k(x)\,\mathrm{d}x. \tag{C.9}$$

(See Eq. (5.60) on page 167.) Note that, unlike Eq. (C.1), the equality in Eq. (C.8) is now an exact one (because the fit is exact).

Thus we have a method for integrating any polynomial of degree $2N - 1$ or less exactly over the interval from -1 to 1: we divide by the Legendre polynomial $P_N(x)$ and then integrate the remainder polynomial $r(x)$ using any set of N sample points we choose plus the corresponding weights.

This, however, is not a very satisfactory method. In particular the polynomial division is rather complicated to perform. However, we can simplify the procedure by noting that, so far, the positions of our sample points are unconstrained and we can pick them in any way we please. So consider again an integration rule of the form (C.1) and make the substitution (C.6), to get

$$\sum_{k=1}^{N} w_k f(x_k) = \sum_{k=1}^{N} w_k q(x_k) P_N(x_k) + \sum_{k=1}^{N} w_k r(x_k). \tag{C.10}$$

But we know that $P_N(x)$ has N zeros between -1 and 1, so let us choose our N sample points x_k to be exactly the positions of these zeros. That is, let x_k be the kth root of the Legendre polynomial $P_N(x)$. In that case, $P_N(x_k) = 0$ for all k and Eq. (C.10) becomes simply

$$\sum_{k=1}^{N} w_k f(x_k) = \sum_{k=1}^{N} w_k r(x_k). \tag{C.11}$$

Combining with Eq. (C.8), we then have

$$\int_{-1}^{1} f(x)\, dx = \sum_{k=1}^{N} w_k f(x_k), \tag{C.12}$$

where the equality is an exact one.

Thus we have a integration rule of the standard form that allows us to integrate any polynomial function $f(x)$ of degree $2N - 1$ or less from -1 to 1 and get an *exact* answer (except for rounding error). It will give the exact value for the integral, even though we only measure the function at N different points.

We have not derived the closed-form expression for the weights w_k given in Eq. (5.64). The derivation of this expression is lengthy and tedious, so we omit it here, but the enthusiastic reader can find it in Hildebrand, F. B., *Introduction to Numerical Analysis*, McGraw-Hill, New York (1956).

GAUSS–KRONROD QUADRATURE

A widely used variant of Gaussian quadrature is *Gauss–Kronrod quadrature*, which was mentioned briefly, but not defined, in Section 5.6.3. Gauss–Kronrod quadrature provides an additional set of sample points interlaced between those of ordinary Gaussian quadrature. By computing an estimate of an integral using just the ordinary Gaussian points, and then recomputing it using the two sets of points combined, one gets two values whose difference gives an estimate of the error on the result. Thus Gauss–Kronrod quadrature gives results of accuracy comparable with Gaussian quadrature plus an estimate of the error on the result (which Gaussian quadrature alone does not provide), but does so at the expense of some addition computational effort, since one must evaluate the integrand at all of the additional sample points.

The derivation of the Gauss–Kronrod formula is similar to that for ordinary Gaussian quadrature. Suppose we choose N sample points to be the roots of the Nth Legendre polynomial $P_N(x)$, as in standard Gaussian quadrature, and an additional $N + 1$ other points, which we are free to place anywhere we like, for a total of $2N + 1$ points. For any integrand $f(x)$ we can create a polynomial approximation of degree $2N$ that matches the integrand exactly at these $2N + 1$ points, for instance using the method of interpolating polynomials from Section 5.6.1, then integrate that approximation to get an approximation to the integral of $f(x)$. If $f(x)$ itself happens to be a polynomial of degree $2N$ or less, then the calculation will be exact, apart from rounding error.

But now we note that we have $N + 1$ degrees of freedom in the positions of our additional $N + 1$ sample points, which as we have said we can choose in any way we like, and this suggests that, if we choose those points correctly, we should be able to create an integration rule that is exact for polynomials of degree $N + 1$ higher, i.e., polynomials of degree $3N + 1$. Gauss–Kronrod quadrature tells us how to pick the additional $N + 1$ points to achieve this. The result is an integration rule with $2N + 1$ sample points that is accurate for polynomials up to degree $3N + 1$, which is not as good as Gaussian quadrature (which would be accurate up to degree $4N + 1$ on $2N + 1$ points), but it's the best we can do if we restrict our first N points to fall at the roots of $P_N(x)$.

To describe this another way, the N initial points at the roots of $P_N(x)$ are *nested* within (i.e., a subset of) the $2N + 1$ points for the Gauss–Kronrod quadrature, which is the crucial property that makes Gauss–Kronrod quadrature attractive. It means we can evaluate our integral using standard Gaussian quadrature on N points, and then again using Gauss–Kronrod quadrature on $2N + 1$, as described above, and the second calculation requires us to evaluate the integrand $f(x)$ only at the newly added sample points. For the rest of the points we can reuse the values from the first step. (Notice, however, that the weights w_k for the quadrature rule are different on the two steps, so one must recompute the sum, Eq. (C.1). One can reuse values of $f(x)$ on the second step, which can save a lot of time, but one cannot reuse the value of the complete sum.)

How then do we choose the additional $N + 1$ sample points for Gauss–Kronrod quadrature? Let us define a new polynomial $E_{N+1}(x)$ of degree $N + 1$ by

$$\int_{-1}^{1} x^k P_N(x) E_{N+1}(x) \, \mathrm{d}x = 0, \quad \text{for integer } k \text{ in the range } 0 \le k \le N. \quad \text{(C.13)}$$

This formula gives us $N + 1$ conditions on $E_{N+1}(x)$. If we also fix the normalization of $E_{N+1}(x)$ (using any method we like), then we have $N+2$ conditions, which is enough to fix all $N+2$ coefficients of the polynomial and hence uniquely specify it. The polynomials $E_N(x)$ are known as *Stieltjes polynomials*. In Gauss–Kronrod quadrature, we choose the additional $N+1$ sample points to be the roots of the polynomial $E_{N+1}(x)$. Thus the complete set of $2N+1$ sample points is the set of roots of $P_N(x)$ plus the roots of $E_{N+1}(x)$. The corresponding integration weights w_k can then be calculated using Eq. (5.60) on page 167.

Now suppose our integrand $f(x)$ is a polynomial of degree $3N+1$ or less. If we divide $f(x)$ by $P_N(x)E_{N+1}(x)$—which is a polynomial of degree $2N+1$—we get

$$f(x) = q(x)P_N(x)E_{N+1}(x) + r(x), \tag{C.14}$$

where $q(x)$ and $r(x)$ are polynomials of degree N or less. Then the integral of $f(x)$ over the standard interval from -1 to 1 is

$$\int_{-1}^{1} f(x)\,dx = \int_{-1}^{1} q(x)P_N(x)E_{N+1}(x)\,dx + \int_{-1}^{1} r(x)\,dx = \int_{-1}^{1} r(x)\,dx, \tag{C.15}$$

where we have used (C.13) to eliminate the first term. This integral can now be evaluated in the standard fashion

$$\int_{-1}^{1} r(x)\,dx = \sum_{k=1}^{2N+1} w_k r(x_k), \tag{C.16}$$

where, as we have said, the sample points x_k are the roots of the Legendre and Stieltjes polynomials. Since $r(x)$ is a polynomial of degree N or less and there are $2N+1$ sample points, Eq. (C.16) will always give an exact answer, to the limits set by rounding error.

But now, using Eq. (C.14), we can also write

$$\sum_{k=1}^{2N+1} w_k f(x_k) = \sum_{k=1}^{2N+1} w_k q(x_k)P_N(x_k)E_{N+1}(x_k) + \sum_{k=1}^{2N+1} w_k r(x_k)$$

$$= \sum_{k=1}^{2N+1} w_k r(x_k), \tag{C.17}$$

where the first sum has vanished because every sample point x_k falls at a zero of either $P_N(x)$ or $E_{N+1}(x)$, so every term in the sum is zero. Combining Eqs. (C.15) to (C.17), we have

$$\int_{-1}^{1} f(x)\,dx = \sum_{k=1}^{2N+1} w_k f(x_k), \tag{C.18}$$

where the equality is an exact one. Thus this particular choice of sample points does indeed give us an integration rule that is exact for all polynomial integrands of degree $3N + 1$ or less.

The Gauss–Kronrod integration method now involves the following steps:

1. We evaluate the integral of $f(x)$ first using standard Gaussian quadrature on N points.

2. We evaluate it again using Gauss–Kronrod quadrature on $2N + 1$ points. N of those points are the same as those for Gaussian quadrature—the roots of $P_N(x)$—so we do not have to recalculate $f(x)$ at these points. We can reuse the values from step 1. Only the values at the $N + 1$ new points have to be calculated, and this can save us a lot of time.

3. The second estimate of the integral (since it is the more accurate of the two) gives us our final result. And the difference between the two estimates gives us an estimate of the error, by analogy with Eq. (5.66). In fact, the difference only gives us an upper bound on the error. The actual error is probably significantly smaller, but unfortunately no precise expression for the error is known, so in practice one usually just uses the difference, bearing in mind that the true error may in fact be smaller than this.

APPENDIX D

CONVERGENCE OF MARKOV CHAIN MONTE CARLO CALCULATIONS

THIS appendix outlines the proof that the Markov chain Monte Carlo procedure of Section 10.3.2 always converges to the Boltzmann distribution. Recall that we start from any state of the system and repeatedly move to new states, one after another, with the probability T_{ij} of changing from state i to state j satisfying the conditions

$$\sum_j T_{ij} = 1 \tag{D.1}$$

and

$$\frac{T_{ij}}{T_{ji}} = \frac{P(E_j)}{P(E_i)}, \tag{D.2}$$

where $P(E_i)$ is the Boltzmann probability for a state with energy E_i. We wish to prove that if we continue the chain of states for long enough, the probability of being in state i will converge to $P(E_i)$.

The proof has several parts. First let us define $p_i(t)$ to be the probability that the Markov chain visits state i at step t. Then the probability that it visits state j at the next step is given by

$$p_j(t+1) = \sum_i T_{ij} p_i(t), \tag{D.3}$$

or in vector notation

$$\mathbf{p}(t+1) = \mathbf{T}\mathbf{p}(t), \tag{D.4}$$

where $\mathbf{p}$ is the vector with elements p_i and $\mathbf{T}$ is the matrix whose ijth element is equal to T_{ji}. (Notice that it is T_{ji}—the indices have to be backwards to make it work out.) Thus the vector of probabilities is simply multiplied by a constant matrix on each step, which means that

$$\mathbf{p}(t) = \mathbf{T}^t \mathbf{p}(0). \tag{D.5}$$

Now let us expand the vector $\mathbf{p}(0)$ as a linear combination of the right eigenvectors $\mathbf{v}_k$ of $\mathbf{T}$ thus:

$$\mathbf{p}(0) = \sum_k c_k \mathbf{v}_k. \tag{D.6}$$

Substituting this form into Eq. (D.5), we get

$$\mathbf{p}(t) = \mathbf{T}^t \sum_k c_k \mathbf{v}_k = \sum_k c_k \lambda_k^t \mathbf{v}_k = \lambda_1^t \sum_k c_k \left(\frac{\lambda_k}{\lambda_1}\right)^t \mathbf{v}_k, \tag{D.7}$$

where λ_k is the eigenvalue corresponding to $\mathbf{v}_k$ and λ_1 is the eigenvalue of largest magnitude. Note that our argument will work only if there is just a single such eigenvalue, i.e., if the leading eigenvalue is unique, which we will assume for the moment to be the case. This point is discussed further below.

Now we consider what happens in the limit of long time. Dividing Eq. (D.7) throughout by λ_1^t, taking the limit $t \to \infty$, and noting that $|\lambda_k/\lambda_1| < 1$ for all k except $k = 1$ if the leading eigenvalue is unique, then all the terms in the sum will tend to zero except for the $k = 1$ term, and we are left with

$$\lim_{t \to \infty} \frac{\mathbf{p}(t)}{\lambda_1^t} = c_1 \mathbf{v}_1. \tag{D.8}$$

In other words, in the limit of long time the vector of probabilities $\mathbf{p}$ is proportional to the leading eigenvector of the matrix $\mathbf{T}$. Thus the probabilities always converge to the same probability distribution in the end, given by the leading eigenvector. It remains to show that this leading eigenvector is, in fact, the Boltzmann distribution.

That the Boltzmann distribution is *an* eigenvector of $\mathbf{T}$ is straightforward to demonstrate. From Eq. (D.2) we have

$$\sum_j T_{ji} P(E_j) = \sum_j T_{ij} P(E_i) = P(E_i) \sum_j T_{ij} = P(E_i), \tag{D.9}$$

where we have used Eq. (D.1) in the last equality. Thus if we set $p_i = P(E_i)$ we have, in vector notation

$$\mathbf{T}\mathbf{p} = \mathbf{p}. \tag{D.10}$$

In other words, the vector of Boltzmann probabilities is indeed an eigenvector of $\mathbf{T}$, with eigenvalue 1.

We now need to show that this is the *leading* eigenvector, which we do as follows. First of all, we note that the eigenvector of Boltzmann probabilities in Eq. (D.10) has all elements strictly positive, since the Boltzmann probabilities are strictly positive. We also note that the *left* eigenvector corresponding to eigenvalue 1 is the vector $\mathbf{1} = (1, 1, 1, \ldots)$, the vector with all elements equal to 1. This is a result of Eq. (D.1), which in vector notation reads $\mathbf{1}^T \mathbf{T} = \mathbf{1}^T$, so that $\mathbf{1}^T$ is indeed the left eigenvector with eigenvalue 1. This result in turn implies that the corresponding right eigenvector, the vector of Boltzmann probabilities, must be the only eigenvector that has no negative elements, since all right eigenvectors with other eigenvalues must be orthogonal to the left eigenvector $(1, 1, 1, \ldots)$, and the only way to be orthogonal to a vector with all elements positive is to have some negative elements.

Second, we showed in Eq. (D.8) that, upon repeated multiplication by the matrix $\mathbf{T}$, the vector $\mathbf{p}$ converges to a value proportional to the leading eigenvector of the matrix. Suppose that the starting value of $\mathbf{p}$ has all elements nonnegative. For instance, we could choose one of them to be one and the others all to be zero, which corresponds to the system being in a single definite starting state. When we multiply such a vector by $\mathbf{T}$ we will get another vector with all elements nonnegative—we must because all the elements of $\mathbf{T}$ are also nonnegative. Thus if we multiply by $\mathbf{T}$ repeatedly we will always have a vector with nonnegative elements, and hence the leading eigenvector to which we converge in the limit of long time must also have all elements nonnegative. Since, as we have said, there is only one such vector, namely the vector of Boltzmann probabilities, the vector to which we converge must be precisely the Boltzmann probabilities.

This essentially completes the proof. The only remaining detail is that one might, in principle, have more than one left/right eigenvector pair associated with eigenvalue 1, in which case our argument above that only one eigenvector has all elements positive breaks down. As we said earlier, we are assuming this not to be the case; we assume that the leading eigenvalue is unique. We have not however proved this, and the reason why we haven't is that, in general, it's not true. It is entirely possible to create a reasonable matrix of transition probabilities $\mathbf{T}$ with more than one leading eigenvector. However, if the condition of ergodicity, discussed in Section 10.3.2, is enforced—the condition that every state of the system is reachable from every other by an appropriate sequence of moves—then the leading eigenvalue is unique, and the Markov chain Monte Carlo method converges reliably to the Boltzmann distribution.

Appendix E

Useful programs

THIS appendix contains some programs and functions that are used in the main text. All of these are also available in the on-line resources.

E.1 Gaussian quadrature

The function gaussxw below calculates sample points and weights for Gaussian quadrature. The points are defined on the standard interval $[-1, 1]$. To use the function you would write, for example "x,w = gaussxw(N)". The function returns two floating-point arrays of N elements each with x containing the positions of the sample points x_k and w containing the weights w_k such that $\sum_k w_k f(x_k)$ is the N-point Gaussian approximation to the integral $\int_{-1}^{1} f(x) \, dx$. To perform integrals over any other domain $[a, b]$ both the positions and the weights must be transformed according to

$$x'_k = \tfrac{1}{2}(b - a)x_k + \tfrac{1}{2}(b + a), \qquad w'_k = \tfrac{1}{2}(b - a)w_k. \qquad \text{(E.1)}$$

A second function gaussxwab is provided which calls the first to calculate the positions and weights then performs the transformation for you and returns arrays x and w for any interval $[a, b]$ that you specify. To use this function you would write "x,w = gaussxwab(N,a,b)". See Section 5.5 for further discussion and examples.

```
from numpy import ones,copy,cos,tan,pi,linspace
```

File: gaussxw.py

```
def gaussxw(N):

    # Initial approximation to roots of the Legendre polynomial
    a = linspace(3,4*N-1,N)/(4*N+2)
    x = cos(pi*a+1/(8*N*N*tan(a)))
```

523

```
        # Find roots using Newton's method
        epsilon = 1e-15
        delta = 1.0
        while delta>epsilon:
            p0 = ones(N,float)
            p1 = copy(x)
            for k in range(1,N):
                p0,p1 = p1,((2*k+1)*x*p1-k*p0)/(k+1)
            dp = (N+1)*(p0-x*p1)/(1-x*x)
            dx = p1/dp
            x -= dx
            delta = max(abs(dx))

        # Calculate the weights
        w = 2*(N+1)*(N+1)/(N*N*(1-x*x)*dp*dp)

        return x,w

    def gaussxwab(N,a,b):
        x,w = gaussxw(N)
        return 0.5*(b-a)*x+0.5*(b+a),0.5*(b-a)*w
```

E.2 SOLUTION OF TRIDIAGONAL OR BANDED SYSTEMS OF EQUATIONS

The function banded below calculates the solution for the vector **x** of a system of linear simultaneous equations of the form $\mathbf{Ax} = \mathbf{v}$ when the matrix **A** is tridiagonal, or more generally banded, as described in Section 6.1.6. To use it you say x = banded(A,v,up,down), where A is an array containing the banded matrix, v is the vector on the right-hand side of the equation, and the variables up and down specify how many nonzero elements there are above and below the diagonal, respectively, in each column of the matrix. More generally, v can be a two-dimensional array containing several independent right-hand sides to $\mathbf{Ax} = \mathbf{v}$, each appearing as a separate column of the array.

To save space storing the matrix **A**—given that most of its elements are zero—the array A is not in the usual form of a matrix, but instead contains the diagonals of the matrix along its rows. Suppose, for instance, that our banded matrix was like this:

$$\mathbf{A} = \begin{pmatrix} a_{00} & a_{01} & a_{02} & & \\ a_{10} & a_{11} & a_{12} & a_{13} & \\ & a_{21} & a_{22} & a_{23} & a_{24} \\ & & a_{32} & a_{33} & a_{34} \\ & & & a_{43} & a_{44} \end{pmatrix} \tag{E.2}$$

That is, it has two nonzero elements above the diagonal, one below, and four nonzero diagonals in all. We would represent this with a four-row array A having elements as follows:

$$A = \begin{pmatrix} - & - & a_{02} & a_{13} & a_{24} \\ - & a_{01} & a_{12} & a_{23} & a_{34} \\ a_{00} & a_{11} & a_{22} & a_{33} & a_{44} \\ a_{10} & a_{21} & a_{32} & a_{43} & - \end{pmatrix} \tag{E.3}$$

The values in the elements marked "$-$" do not matter—you can put anything in these elements and it will make no difference to the results. The vector $\mathbf{v}$ is stored in the array v in standard form—no special arrangement is used.

The function banded returns a single array with the same length as v containing the solution $\mathbf{x}$ to the equations, or, in the case of multiple right-hand sides, an array with the same shape as v with the solution for each of the right-hand sides in the corresponding column.

```
from numpy import copy
```
File: banded.py

```
def banded(Aa,va,up,down):

    A = copy(Aa)
    v = copy(va)
    N = len(v)

    # Gaussian elimination
    for m in range(N):

        # Normalization factor
        div = A[up,m]

        # Update the vector first
        v[m] /= div
        for k in range(1,down+1):
            if m+k<N:
                v[m+k] -= A[up+k,m]*v[m]
```

```
            # Now normalize and subtract the pivot row
            for i in range(up):
                j = m + up - i
                if j<N:
                    A[i,j] /= div
                    for k in range(1,down+1):
                        A[i+k,j] -= A[up+k,m]*A[i,j]

    # Backsubstitution
    for m in range(N-2,-1,-1):
        for i in range(up):
            j = m + up - i
            if j<N:
                v[m] -= A[i,j]*v[j]

    return v
```

E.3 DISCRETE COSINE AND SINE TRANSFORMS

Functions for performing (complex) discrete Fourier transforms (DFTs) are available in Python in the module numpy.fft, but not functions for performing discrete cosine and sine transforms. As discussed in Section 7.4.3, however, the cosine and sine transforms are simply DFTs of data that have a particular symmetry, either even or odd, about the midpoint of their interval. So one can calculate such transforms by first mirroring the data to create the required symmetry and then using the standard DFT functions in numpy.fft. The functions given below use this trick to perform both forward and inverse discrete cosine and sine transforms.

File: dcst.py

```
from numpy import empty,arange,exp,real,imag,pi
from numpy.fft import rfft,irfft

# 1D DCT Type-II
def dct(y):
    N = len(y)
    y2 = empty(2*N,float)
    y2[:N] = y[:]
    y2[N:] = y[::-1]
    c = rfft(y2)
    phi = exp(-1j*pi*arange(N)/(2*N))
    return real(phi*c[:N])
```

```
# 1D inverse DCT Type-II
def idct(a):
    N = len(a)
    c = empty(N+1,complex)
    phi = exp(1j*pi*arange(N)/(2*N))
    c[:N] = phi*a
    c[N] = 0.0
    return irfft(c)[:N]

# 1D DST Type-I
def dst(y):
    N = len(y)
    y2 = empty(2*N,float)
    y2[0] = y2[N] = 0.0
    y2[1:N] = y[1:]
    y2[:N:-1] = -y[1:]
    a = -imag(rfft(y2))[:N]
    a[0] = 0.0
    return a

# 1D inverse DST Type-I
def idst(a):
    N = len(a)
    c = empty(N+1,complex)
    c[0] = c[N] = 0.0
    c[1:N] = -1j*a[1:]
    y = irfft(c)[:N]
    y[0] = 0.0
    return y
```

One can build upon these functions to perform cosine and sine transforms in higher dimensions. Here are functions to perform forward and inverse two-dimensional cosine and sine transforms.

```
# 2D DCT                                              File: dcst.py
def dct2(y):
    M = y.shape[0]
    N = y.shape[1]
    a = empty([M,N],float)
    b = empty([M,N],float)
    for i in range(M):
        a[i,:] = dct(y[i,:])
    for j in range(N):
        b[:,j] = dct(a[:,j])
```

```
        return b

# 2D inverse DCT
def idct2(b):
    M = b.shape[0]
    N = b.shape[1]
    a = empty([M,N],float)
    y = empty([M,N],float)
    for i in range(M):
        a[i,:] = idct(b[i,:])
    for j in range(N):
        y[:,j] = idct(a[:,j])
    return y

# 2D DST
def dst2(y):
    M = y.shape[0]
    N = y.shape[1]
    a = empty([M,N],float)
    b = empty([M,N],float)
    for i in range(M):
        a[i,:] = dst(y[i,:])
    for j in range(N):
        b[:,j] = dst(a[:,j])
    return b

# 2D inverse DST
def idst2(b):
    M = b.shape[0]
    N = b.shape[1]
    a = empty([M,N],float)
    y = empty([M,N],float)
    for i in range(M):
        a[i,:] = idst(b[i,:])
    for j in range(N):
        y[:,j] = idst(a[:,j])
    return y
```

Copies of all these functions can be found in the on-line resources in the file dcst.py.

E.4 COLOR SCHEMES

As described in Section 3.3, one can change the color scheme—or *colormap* as it is known—used in density plots, to make individual plots clearer or more attractive. The pylab module defines a range of useful colormaps, some of which are listed in the table in Section 3.3. However, there are some others that are occasionally useful in physics applications that are not included in pylab. The following short package defines three additional color maps that I find useful. They are:

redblue	Goes from red to blue via black
redwhiteblue	Goes from red to blue via white
inversegray	Goes from white to black, the opposite of gray

Code defining these colormaps is given below and can be found in the on-line resources in the file colormaps.py. To use the redblue colormap, for example, you would say "from colormaps import redblue", then "redblue()" when you want to change the colormap. (For the technically minded, you can also import the colormap objects themselves. They are called cp_redblue, cp_redwhiteblue, and cp_inversegray.) The red/blue colormaps are useful for representing hot/cold distinctions and especially positive/negative distinctions in electric potentials. The inversegray colormap is sometimes useful when you are going to print out your density plot on paper.

```
from matplotlib.colors import LinearSegmentedColormap          File: colormaps.py
from matplotlib.cm import RdBu
from matplotlib.pyplot import set_cmap

cdict = {"red":   [(0.0,1.0,1.0),(0.5,0.0,0.0),(1.0,0.0,0.0)],
         "green": [(0.0,0.0,0.0),(1.0,0.0,0.0)],
         "blue":  [(0.0,0.0,0.0),(0.5,0.0,0.0),(1.0,1.0,1.0)]}
cp_redblue = LinearSegmentedColormap("redblue",cdict)

cp_redwhiteblue = RdBu

cdict = {"red":   [(0.0,1.0,1.0),(1.0,0.0,0.0)],
         "green": [(0.0,1.0,1.0),(1.0,0.0,0.0)],
         "blue":  [(0.0,1.0,1.0),(1.0,0.0,0.0)]}
cp_inversegray = LinearSegmentedColormap("inversegray",cdict)

def redblue():
    set_cmap(cp_redblue)
```

```
def redwhiteblue():
    set_cmap(cp_redwhiteblue)

def inversegray():
    set_cmap(cp_inversegray)
```

INDEX

Page numbers in **bold** denote definitions or principal references.

Printed in the USA
CPSIA information can be obtained
at www.ICGtesting.com
LVHW080256061223
765621LV00007B/311